Adaptation to a Changing Climate in the Arab Countries

MENA DEVELOPMENT REPORT

Adaptation to a Changing Climate in the Arab Countries

A Case for Adaptation Governance and Leadership in Building Climate Resilience

Dorte Verner, Editor

THE WORLD BANK
Washington, D.C.

1818 H Street NW, Washington DC 20433
Telephone: 202-473-1000; Internet: www.worldbank.org

1 2 3 4 15 14 13 12

ISBN (paper): 978-0-8213-9458-8
ISBN (electronic): 978-0-8213-9459-5
DOI: 10.1596/978-0-8213-9458-8

Cover photo: Dorte Verner.

Library of Congress Cataloging-in-Publication Data

Verner, Dorte.
Adaptation to a changing climate in the Arab countries : a case for adaptation governance and leadership in building climate resilience / Dorte Verner.
p. cm.—(MENA development report)
Includes bibliographical references.
ISBN 978-0-8213-9458-8 (alk. paper)—ISBN 978-0-8213-9459-5
1. Climatic changes—Arab countries. 2. Climatic changes—Social aspects—Arab countries. 3. Climatic changes—Environmental aspects—Arab countries. 4. Climatic changes—Economic aspects—Arab countries. 5. Climatic changes—Effect of human beings on—Arab countries. 6. Arab countries—Environmental conditions. I. Title.
QC903.2.A64V47 2012
363.738'7456109174927—dc23

2012020579

Contents

Boxes

Figures

Maps

Tables

Foreword

Adapting to climate change is not a new phenomenon for the Arab World. For thousands of years, the people in Arab countries have coped with the challenges of climate variability by adapting their survival strategies to changes in rainfall and temperature. Their experience has contributed significantly to the global knowledge on climate change and adaptation. But over the next century global climatic variability is predicted to increase, and Arab countries may well experience unprecedented extremes in climate. Temperatures may reach new highs, and in most places there may be a risk of less rainfall. Under these circumstances, Arab countries and their citizens will once again need to draw on their long experience of adapting to the environment to address the new challenges posed by climate change.

Arab countries have already experienced the effects of a changing climate; higher temperatures and extreme events such as drought and flash floods may well become the new norm. The year 2010 was already the warmest globally since records began in the late 1800s, with 19 countries setting new national temperature highs. Five of these were Arab countries, including Kuwait, which set a new record at 52.6 °C in 2010, only to be followed by 53.5 °C in 2011.

This report—prepared through a consultative process with Government and other stakeholders in the Arab World—assesses the potential effects of climate change on the Arab region and outlines possible approaches and measures to prepare for its consequences. It offers ideas and suggestions for Arab policy makers as to what mitigating actions may be needed in rural and urban settings to safeguard key areas such as health, water, agriculture, and tourism. The report also analyzes the differing impacts of climate change, with special attention paid to gender, as a means of tailoring strategies to address specific vulnerabilities.

The socioeconomic impact of climate change will likely vary from country to country, reflecting a country's coping capacity and its level of development. Countries that are wealthier and more economically diverse are generally expected to be more resilient. This finding is borne out by recent comparative research on the Syrian Arab Republic, Tunisia, and the Republic of Yemen, which focused on the impact of climate variability on income and poverty levels. Data show that over the next 30 to 40 years, if no compensating measures are taken, climate change may lead to a cumulative reduction in household incomes of about 7 percent in Syria and Tunisia, while the Republic of Yemen—mainly driven by a decline in agriculture—could potentially suffer a much larger reduction of 24 percent. While not addressed directly in this report, the impact of the ongoing conflict in Syria would likely add greater welfare losses and make the adaptation process even more difficult.

Climate change may also affect water availability and its impacts (potentially severe) in both rural and urban areas. By 2050, water runoff is expected to decrease by 10 percent. The gains in agricultural productivity over the past two decades may slow and even decline after about 2050. Recurrent droughts may spur increased rural-to-urban migration, adding additional stress to the already overcrowded cities. Flash floods that result from intense storms may particularly affect cities; 500,000 people have already been affected by flash floods in Arab countries.

The livelihoods of men and women are potentially at risk from climate change. Men and women have different vulnerabilities, largely based on their respective roles in society. Men in rural areas are likely to follow the current trend of moving to cities to seek paid employment, because their traditional livelihoods have become unsustainable. Rural women would then face the double challenge of having to devote more time to daily activities, such as fetching scarce water, while assuming the farming and community responsibilities of the absent men. Left unaddressed, these changes in the family structure have the potential of becoming significant sources of stress. Actions taken to help both men and women understand and adapt to these changes at the household level are important regarding the possible policy responses. Women, with their pivotal role in societies, can be a major influence on the attitudes and behavior needed to accommodate new forms of livelihood and social organization. Women's empowerment thus becomes a critical part of any adaptation strategy.

Fortunately, Arab countries can take steps to reduce the impacts of climate change. This report outlines measures that may help reduce the region's vulnerability and also contribute to more sustainable long-term development. The report offers an approach—an "Adaptation Pyramid Framework"—for strengthening public sector management in a changing climate, and assisting stakeholders in integrating climate risks and opportunities into all development activities. The report suggests that countries and households will need to diversify their production and income generation, integrate adaptation into all policy making and activities, and ensure a sustained national commitment to address the social, economic, and environmental consequences of climate variability. With these coordinated efforts, the Arab world can, as it has for centuries, successfully adapt and adjust to the challenges of a changing climate.

Inger Andersen
Regional Vice President
Middle East and North Africa Region
The World Bank

Acknowledgments

The report was developed and managed by Dorte Verner (World Bank). Mme. Fatma El-Mallah (League of Arab States) and Ian Noble (World Bank) served as senior advisers. The team is grateful to contributions from the following lead and contributing authors.

Chapter 1. Lead author: Dorte Verner (World Bank) and contributing authors Maximillian Ashwill (World Bank), Clemens Breisinger (International Food Policy Research Institute—IFPRI), Jakob Kronik (World Bank), Ian Noble (World Bank), and Banu Seltur (World Bank).

Chapter 2. Lead authors: Jens Hesselbjerg Christensen, Martin Stendel, Shuting Yang (Danish Meteorological Institute—DMI), contributing authors Ian Noble (World Bank) and Michael Westphal (World Bank).

Chapter 3. Lead author: Hamed Assaf (American University of Beirut, Lebanon), and contributing authors Wadid Erian (Arab Center for Study of Arid Zones and Dry Lands, Syria), Raoudha Gafrej (University of Tunis El Manar, Tunisia), Sophie Herrmann (World Bank), Rachael McDonnell (International Center for Biosaline Agriculture, United Arab Emirates, Oxford University, UK), and Awni Taimeh (University of Jordan, Jordan).

Chapter 4. Lead authors: Rachael McDonnell (International Center for Biosaline Agriculture, United Arab Emirates; Oxford University, UK) and Shoaib Ismail (International Center for Biosaline Agriculture, United Arab Emirates), and contributing authors Rami Abu Salman (IFAD), Julian Lampietti (World Bank), Maurice Saade (World Bank), William Sutton (World Bank), and Christopher Ward (IAIS, University of Exeter, UK).

Chapter 5. Lead authors: Amal Dababseh (Amman Institute for Urban Development/Greater Amman Municipality, Jordan) and Kristina Katich (World Bank) and contributing authors Maximillian Ashwill (World Bank), Hamed Assaf (American University of Beirut, Lebanon), Hesham Bassioni (Arab Academy of Science, Egypt), Tim Carrington (World Bank), Ibrahim Abdel Gelil (Arabian Gulf University, Bahrain), and Mohamed El Raey (Alexandria University, Egypt).

Chapter 6. Lead authors: Marina Djernaes (World Bank) and Mme. Fatma El-Mallah (League of Arab States).

Chapter 7. Lead author: Sarah Grey (International Center for Biosaline Agriculture and Dubai School of Government Gender and Public Policy Program, United Arab Emirates), and contributing authors Maximillian Ashwill (World Bank), Fidaa Haddad (International Union for Conservation of Nature, Jordan), Grace Menck Figueroa, and Sanne Tikjøb (World Bank).

Chapter 8. Lead author: Rima R. Habib (American University of Beirut, Lebanon) and contributing authors Enis Baris and Tamer Rabie (World Bank).

Chapter 9. Lead author: Dorte Verner (World Bank) and contributing authors Rami Abu Salman (IFAD), Kulsum Ahmed (World Bank), Hamed Assaf (American University of Beirut, Lebanon), Amal Dababseh (Amman Institute for Urban Development/Greater Amman Municipality, Jordan), Marina Djernaes (World Bank), Mme. Fatma El-Mallah (League of Arab States), Wadid Erian (Arab Center for Study of Arid Zones and Dry Lands, Syria), Raoudha Gafrej (University of Tunis El Manar, Tunisia), Habiba Gitay (World Bank), Sarah Grey (International Center for Biosaline Agriculture and Dubai School of Government Gender and Public Policy Program, United Arab Emirates), Rima R. Habib (American University of Beirut, Lebanon), Jens Hesselbjerg Christensen (Danish Meteorological Institute—DMI), Shoaib Ismail (International Center for Biosaline Agriculture, United Arab Emirates), Kristina Katich (World Bank), Tamara Levine (World Bank), Andrew Losos (World Bank), Rachael McDonnell (International Center for Biosaline Agriculture, United Arab Emirates), Grace Menck Figueroa (World Bank), Ian Noble (World Bank), Noemí Padrón Fumero (Universidad de Las Palmas de Gran Canaria, Spain), Tamer Rabie (World Bank), Martin Stendel (DMI), and Shuting Yang (DMI).

The lead authors of the Disaster Risk Management Spotlight (1) were Aditi Banerjee (World Bank), Mme. Fatma El-Mallah (League of Arab States, Egypt), Wadid Erian (Arab Center for Study of Arid Zones and Dry Lands, Syria), and contributing author Hesham Bassioni (Regional Center for Disaster Risk Reduction, Egypt).

The lead authors of Biodiversity and Ecosystem Services Spotlight (2) were Raoudha Gafrej (University of Tunis El Manar, Tunisia) and Habiba Gitay (World Bank); and the contributing authors were Rami Abu Salman (IFAD), Hamed Assaf (American University Beirut, Lebanon), and Balgis Osman-Elasha (AfDB, Sudan).

The team is grateful for the ideas, comments, and contributions provided by the advisers on this report: Ibrahim Abdel Gelil (Arabian Gulf University, Bahrain), Rami Abu Salman (IFAD), Emad Adly (Arab Network for Environment and Development—RAED, Egypt), Hesham Bassioni (Regional Center for Disaster Risk Reduction, Egypt), Wadid Erian (Arab Center for Study of Arid Zones and Drylands, Syria), Nadim Khouri (United Nations Economic and Social Commission for Western Asia—ESCWA, Lebanon), Rami Khouri (Issam Fares Institute, American University Beirut—IFI-AUB, Lebanon), Karim Makdisi (IFI-AUB, Lebanon), Mohammed Messouli (University of Marrakech, Morocco), Ziad Mimi (Birzeit University, West Bank and Gaza), Balgis Osman-Elasha (AfDB, Sudan), Najib Saab (Arab Forum for Environment and Development—AFED, Lebanon), and Shahira Wahbi (League of Arab States, Egypt).

The team is grateful to peer reviewers Marianne Fay (SDNVP), Erick C.M. Fernandes (LCSAR), and John Nash (LCSSD), as well as to the Chair of the Quality Enhancement Review and the Decision Meeting, Caroline Freund (MNACE).

The team appreciates the comments, ideas, and contributions received from the following: Yasser Mohammad Abd Al-Rahman (QHSE, Egypt), Perrihan Al-Riffai (IFPRI), Shardul Agrawala (OECD), Hala Al-Dosari (independent writer, Saudi Arabia), Haithem Ali (Amman Institute for Urban Development, Jordan), Salahadein Alzien (S.D.C., Libya), Nabeya Arafa (Egyptian Environmental Affairs Agency—EEAA, Egypt), Ruby Assad (GIZ Senior Advisor), Carina Bachofen (Red Cross/Red Crescent Climate Center), JoAnn Carmin (MIT, United States), Coastal Research Institute (CoRI, Egypt), Roshan Cooke (IFAD), Leila Dagher (American University Beirut, Lebanon), Suraje Dessai (University of Exeter, UK), Kamel Djemouai (Ministry of Environment, Algeria), Olivier Ecker (IFPRI), Jauad El Kharraz (EMWIS/Water Information Systems, France), Hacene Farouk (Ministry of Environment, Algeria), Jose Funes (IFPRI), Mohamed Konna El Karim (League of Arab States, Egypt), Bouchta El Moumni (Université A. ESSAÂDI, Morocco), Ahmed Farouk (Center for Development Services, Egypt), Bence Fülöp (Trinity Enviro, Hungary), Samia Galal Saad (High Institute of Public Health—Alexandria University, Egypt), Raja Gara (AfDB), Benjamin Garnaud (Institut du Développement Durable et des Relations Internationales—IDDRI, France), Brendan Gillespie (OECD), Hilary Gopnik (Emory University, United States), Flora Ijjas (Trinity Enviro, Hungary), Steen Lau Jorgensen (World Bank), Vahakn Kabakian (Ministry of Environment, Lebanon), Léa Kai Aboujaoudé (Ministry of Environment, Lebanon), Jack Kalpakian (Al Akhawayn University Ifrane, Morocco), Jakob Kronik (F7 Consult), Jeffrey Lecksell (World Bank), Hassan Machlab (ICARDA, Lebanon), Seeme Mallick (ONE UN Joint Programme on Environment, Pakistan), Carla Mellor (ICBA), George Mitri (University of Balamand, Lebanon), Amal Mosharrafieh (AFED, Lebanon), Adrian Muller (ETH

Zurich, Switzerland), Manal Nader (University of Balamand, Lebanon), Cristina Narbona (OECD), Frode Neergaard (OECD), Gerald Nelson (IFPRI), Asif Niazi (WFP, Egypt), Michele Nori (Independent Consultant on Agro-pastoral Livelihoods), Remy Paris (OECD), Noemí Padrón Fumero (Universidad de Las Palmas de Gran Canaria, Spain), Taoufik Rajhi (AfDB), Ricky Richardson (IFPRI), Richard Robertson (IFPRI), Saloua Rochdane (University of Marrakech, Morocco), Abdul-Rahman Saghir (Lebanon), Idllalène Samira (Université Cadi Ayyad, Morocco), Saeed Shami (IUCN), William Stebbins (World Bank), Nathalie Sulmont (OECD), Rianne C. ten Veen (Islamic Relief Worldwide, UK), Naoufel Telahigue (IFAD), Rainer Thiele (Kiel Institute for the World Economy—IfW, Germany), Manfred Wiebelt (IfW, Germany) Robert Wilby (Loughborough University, UK), Tingju Zhu (IFPRI), and Samira al-Zoughbi (National Agricultural Policy Center, Syria).

The team is grateful for comments and suggestions provided during presentations of initial ideas to the final consultations. Particularly we would like to thank the Ministries of Environment representatives from all the Arab countries and others present at the Joint Committee on Environment and Development in the Arab Region (JCEDAR), at the League of Arab States in October 2010 and 2011. We would also like to thank the Executive Directors of the World Bank from the Arab countries and their teams that guided us during the presentation at the World Bank in September 2011. Finally we would like to thank all invitees from all the Arab countries who participated in the consultations and the preparation of the report, including at the OECD, the Marseille Center for Mediterranean Integration (CMI), the United Nations Economic and Social Commission for Western Asia (ESWUA), and at the side event hosted by the government of Lebanon at the COP 17 in Durban.

The team gratefully acknowledges the helpful comments, ideas, and guidance from these individuals regarding global knowledge at the World Bank: Biesan Abu Kwaik, Hafed Al-Ghwell, Shamshad Akhtar, Sameer Akbar, Inger Andersen, Abdulhamid Azad, Anthony Bigio, Sidi Boubacar, Franck Bousquet, Marjory-Anne Bromhead, Ato Brown, Kevin Carey, Nadereh Chamlou, Diana Chung, Luis Constantino, Bekele Debele Negewo, Moira Enerva, Marcos Ghattas, Grace Hemmings-Gapihan, Santiago Herrera, Dan Hoornweg, Imane Ikkez, Gabriella Izzi, Willem Janssen, Junaid Kamal Ahmed, Mats Karlsson, Claire Kfouri, Hoonae Kim, Yoshiharu Kobayashi, Julian Lampietti, Hedi Larbi, Dale Lautenbach, Qun Li, Dahlia Lotayef, Laszlo Lovei, Ida Mori, Thoko Moyo, Dylan Murray, Lara Saade, Alaa Sarhan, Banu Setlur, William Sutton, Eileen Brainne Sullivan, Deepali Tewari, David Treguer, Jonathan Walters, and Andrea Zanon.

The team would also like to thank Marie-Francoise How Yew Kin, Josephine Onwuemene, Salenna Prince, Indra Raja, and Perry Radford for assisting the team very effectively throughout the process of the task, and Hilary Gopnik for editing and providing input. We would also like to thank Susan Graham, Daniel Nikolits, and Nora Ridolfi, all from the World Bank's Office of the Publisher, for efficiently managing the production of the publication. Additional thanks go to the country offices' staff for their support of our workshop and consultations: Faten Abdulfattah, Rola Assi, Zeina El Khalil, Lana Mourtada, Mohammed A. Sharief, Steve Tinegate, Natalie Abu-Ata, Claire Ciosi, Margarita Gaillochet, Olivier Lavinal, Mona Yafi, and Sabrina Zitouni (Center for Mediterranean Integration—CMI); and Eileen Murray. We extend final thanks to Chedly Rais and Narsreddine Jomma (OKIANOS, Tunisia).

The core World Bank team who worked on the task includes Johanne Holten, Kristina Katich, Tamara Levine, Grace Menck Figueroa, Dylan Murray, Ian Noble, Sanne Agnete Tikjoeb, and Dorte Verner.

Contributors

Hamed Assaf is an Arab water resource management specialist who focuses on climate change and sustainability issues. He participated in several international and regional water resources and climate change initiatives involving several international organizations including the Economic and Social Commission for Western Asia, the World Bank, the World Meteorological Organization, and the United Nations Development Programme (UNDP). He was a professor of water resources engineering at the American University of Beirut from 2003–11. Before this he was a senior water and risk analysis engineer at BC Hydro in Canada. He received his Ph.D. in civil engineering (water resources) from the University of British Columbia, Canada.

Aditi Banerjee is a Disaster Risk Management (DRM) specialist at the World Bank in the Middle East and North Africa region. She specializes in building institutional capacities and structures, particularly at local levels. Her experience in DRM includes work on the tsunami floods in India in 2005, the 2008 floods in the Republic of Yemen, and the L'Aquila earthquake in Italy in 2009. Before this Banerjee was Project Officer at the Environment Division of the UNDP in India. She holds a Master's degree in Public Policy and Econometrics from the George Washington University and a Bachelor's degree in Economics from Delhi University.

Hesham Bassioni is a professor of maritime transport and technology at the Arab Academy of Science in Alexandria, Arab Republic of Egypt. He is also the Executive Director of the Cairo-based Regional Center for Disaster Risk Reduction.

Jens Hesselbjerg Christensen is the scientific head at the Danish Climate Centre at the Danish Meteorological Institute and Director for the Centre for Regional Change in the Earth System. He has been a lead author or coordinating lead author on previous Intergovernmental Panel on Climate Change (IPCC) reports and is currently a coordinating lead author of the 5th IPCC assessment report due in 2013/14. He has a long scientific background in the field of regional climate change and has published more than 70 peer-reviewed articles in scientific journals. He holds a Ph.D. in Astrophysics from the Niels Bohr Institute at the University of Copenhagen.

Amal Aldababseh is a climate change and disaster risk reduction expert. She currently manages a consultancy firm that supports government in climate change and disaster risk reduction. She has worked as an international consultant for the World Bank and UNDP. When this book was being written, she was the Director of Sustainable Development at the Amman Institute for Urban Development and Greater Amman Municipality. Aldababseh has previous worked with the local nongovernmental organizations and UNDP in Jordan, becoming the Head of Environment and Energy Unit and the Disaster Risk Reduction Focal Point in 2007. She holds a B.S. with honors from the Hashemite University of Jordan and a M.S. with honors in Environmental Science and Management from the University of Jordan.

Marina Djernaes is an environmental consultant for the World Bank. She formerly held the positions of Finance Director at Greenpeace and Senior Advisor at the Danish Embassy in Washington, DC. Before that, she held several positions relating to strategic corporate management. Djernaes holds an MBA from American University and a Master's degree in Environmental Science and Policy from Johns Hopkins University.

Mme Fatma El Mallah has been an Advisor to the Secretary General of the League of Arab States on Climate Change. She joined the General Secretariat of the League of Arab States in 1974 and became a Senior Economist in 1990. She was Director of Environment, Housing and Sustainable Development as well as Director of the Technical Secretariat of the Council of Arab Ministries Responsible for the Environment from 1994–2009. She earned a Bachelor's degree in Economics from Cairo University in 1967 and a Master's degree in Economics from the American University in Cairo in 1971.

Wadid Erian is a professor of soil science and Acting Director of the Land and Water Use Division at the Arab Center for the Study of Arid Zones and Dry Lands in the Syrian Arab Republic. He is the team leader in several projects for soils, land use, land degradation, and drought studies. He is the Environment Coordinator for the League of Arab States' Summit of South American-Arab Countries. He was a lead author on the IPCC "Special Report on Managing the Risks of Extreme Events and Disasters to Advance Climate Change Adaptation" (SREX) report and has frequently provided expertise to the League of Arab States. He is also part of the Advisory Board for the Global Assessment Report on Disaster Risk Reduction.

Raoudha Gafrej is a water expert and assistant professor. Working as a hydraulic engineer in Tunisia, she has more than 15 years of experience as a national and international expert on water management, with a focus on economic and environmental aspects. She was involved in several strategic studies with the German Agency for International Cooperation (GIZ) and the Islamic Development Bank and has contributed to both the development of the Tunisian strategy for agriculture and ecosystems adaptation to climate change and to the Tunisian national climate change strategy. In 2002, Gafrej joined the faculty of the University of Tunis El Manar, where she is currently an assistant professor at the Higher Institute of Applied Biological Sciences of Tunis. She holds a Ph.D. in Earth Science from Pierre and Marie Curie University in Paris.

Habiba Gitay is a Senior Environmental Specialist in the Environment Department of the World Bank. Before joining the World Bank, she was a Senior Research Fellow and Lecturer at the Australian National University working on sustainable development and ecosystem management. Her experience includes being a convening lead author for chapters related to ecosystems, impacts of climate change, and adaptation options in five IPCC reports; lead author and capacity development lead on the Millennium Ecosystem Assessment; Vice-Chair of the Scientific and Technical Advisory Panel of the Global Environment Facility; and consultant to various environment-related conventions. She has a Ph.D. in Ecology from University of Wales.

Sarah Grey is a researcher on gender and development, specializing in the Arab world. She is currently a visiting scholar at the Centre for Middle Eastern Studies at Lund University, Sweden, and a non-resident research associate with the Gender and Public Policy Program at the Dubai School of Government. When this book was being written, she was also a researcher at the International Center for Biosaline Agriculture in Dubai, United Arab Emirates. She has undertaken research on gender-related topics in Arab countries, including gender and development, gender and globalization, gender and public policy, and Muslim women's historical roles. She holds an M.Phil. in modern Middle East studies from the University of Oxford and a B.A. in Arabic and Middle East studies from the University of Exeter.

Rima R. Habib is an Associate Professor of Environmental and Occupational Health in the Faculty of Health Sciences at the American University of Beirut. Recently, she authored a chapter in the book *Ecohealth Research in Practice: Innovative Applications of an Ecosystem Approach to Health* (2012). Besides peer-reviewed articles, he has also authored a number of policy briefs, one of which focused on addressing climate change and health research in the Eastern Mediterranean region. Habib has consulted for international organizations, including carrying out a study for the International Labour Organization on the occupational health and safety situation in 18 Arab countries. Habib currently assumes the position of Chair on the Technical Committee of Gender and Work at the International Ergonomics Association.

Shoaib Ismail is a research scientist in agronomy and university educator. He has more than 30 years of experience in integrated water resource management, with a focus on optimizing agricultural research and development under marginal conditions. Recently, his focus has been on the agriculture and water sector in relation to climate change. He currently works at the International Center for Biosaline Agriculture, and before this he was Associate Professor in the Department of Botany, University of Karachi, Pakistan. Ismail holds a Ph.D. in Plant Physiology from the University of Karachi.

Kristina Katich specializes in climate adaptation and disaster risk management for cities. She is trained as an architect and urban planner. Since 2009 she has consulted for the World Bank, and has worked on urban and environmental issues in the Middle East and North Africa, Latin America and the Caribbean, and East Asia and Pacific regions. Before joining the World Bank, Katich volunteered in the United States Peace Corps, where she spent two years building rural water systems in the Dominican Republic. She holds a graduate degree from the Massachusetts Institute of Technology's Department of Urban Studies and Planning.

Rachael A. McDonnell is a water policy and governance scientist at the International Center for Biosaline Agriculture (ICBA) and a senior research scientist with the University of Oxford's Water Futures Group. Before joining ICBA, she was the course director for the MSc in Water Science, Policy and Management at Oxford. She is currently the Monitoring Agriculture and Water Resources Development program leader, a USAID-funded initiative in partnership with NASA's Goddard Space

Flight Center, through which regional- and country-scale modeling of water resources and agricultural water use is being developed under current and future climate change conditions. She has also led projects on water governance, policy and regulation, and advised various Middle Eastern governments on future water policy directions. She holds a Ph.D. in geography from the University of Oxford.

Ian Noble is a climate change specialist and has recently been a member of the Climate Change and the World Development Report teams at the World Bank, with particular responsibility for the Bank's activities in adaptation to climate change. Before this he was Professor of Global Change Research at the Australian National University. He has played senior roles in the IPCC process, which prepares scientific assessments relevant to climate policy, and in international cooperative research on climate change as part of the International Geosphere-Biosphere Programme. In Australia he participated in the public and policy debate over responses to climate change and served as a Commissioner in an inquiry into the future of the Australian forests and forest industries.

Martin Stendel is a senior researcher at the Danish Climate Centre of the Danish Meteorological Institute. His expertise is on modeling climate and climate change with a focus on Africa and the Arctic, and he has authored or co-authored several publications in journals, as book chapters, and in scientific assessments. Further, he serves as editor for three scientific journals. He has been appointed as expert advisor in the Third World Academy of Sciences project on climate change in Sub-Saharan Africa and is currently serving as project leader, contributor, and supervisor for Ph.D. students in several projects in eastern and southern Africa. He holds a Ph.D. in Meteorology from the University of Cologne, Germany.

Dorte Verner is the Climate Coordinator and Senior Economist in the World Bank's Middle East and North Africa Region. Before taking on this position, she led the Social Implications of Climate Change program in the Latin America and the Caribbean Region. She has developed and led many World Bank projects and published extensively in the areas of poverty, rural issues, and climate change. She has written books and papers on poverty, labor markets, indigenous peoples, youth-at-risk issues, and climate change. Before she joined the World Bank, she worked in the Development Center of the Organisation for Economic Co-operation and Development and as a researcher at the European University Institute in Florence and the Sorbonne in Paris. She holds a Ph.D. in Macroeconomics and Econometrics from the European University Institute,

Italy, and a postgraduate degree in Economics from the University of Aarhus, Denmark.

Shuting Yang is a senior scientist at the Danish Climate Centre in Danish Meteorological Institute. She has research experience in the areas of climate modeling, climate variability, and climate change. Her research interests cover atmospheric circulation regimes, climate sensitivity and feedbacks, and climate prediction and projections with both the idealized/simplified circulation model and general circulation models. She has published several articles in academic journals, and is a contributing author to *Climate Impacts on Energy Systems: Key Issues for Energy Sector Adaptation*, published by the World Bank and the Energy Sector Management Assistance Program. Yang holds a B.A. in Meteorology from Beijing University, China, and a Ph.D. in Meteorology from Stockholm University, Sweden.

Abbreviations

AI	Aridity Index
AOGCM	atmosphere-ocean global climate model
AR4	IPCC Fourth Assessment Report
AR5	IPCC Fifth Assessment Report
ASDRR	Arab Strategy for Disaster Risk Reduction
BREEAM	Building Research Establishment Environmental Assessment Method
CCA	climate change adaptation
CGE	computable general equilibrium
CMIP5	Coupled Model Intercomparison Project
COMET	Community, Energy, and Technology (project) (West Bank and Gaza)
COP	Conference of Parties
CORDEX	Coordinated Regional Climate Downscaling Experiment
DALY	disability-adjusted life year
DELP	Desert Ecosystems and Livelihoods Project (Jordan)
DRM	disaster risk management
DRR	disaster risk reduction
EBA	ecosystem-based assessment
ECA&D	European Climate Assessment and Dataset
EMR	Eastern Mediterranean Region (WHO)
EMRO	Eastern Mediterranean Regional Office (WHO)
ESM	earth system model
GCC	Gulf Cooperation Council
GCM	global circulation model
GDP	gross domestic product
GFCS	Global Framework for Climate Services
GFDRR	Global Facility for Disaster Reduction and Recovery
GGCA	Global Gender and Climate Alliance
GIS	geographic information system
GIZ	German Agency for International Cooperation

GTZ	German Agency for Technical Cooperation
GWP	Global Water Partnership
HFA	Hyogo Framework for Action
ICPAC	IGAD Climate Prediction and Applications Centre
IDPs	internally displaced people
IGAD	Intergovernmental Authority on Development
INDH	National Initiative for Human Development (Morocco)
IPCC	Intergovernmental Panel on Climate Change
IRI	International Research Institute for Climate and Society
ITCZ	Inter-Tropical Convergence Zone
IUCN	International Union for Conservation of Nature
IWRM	Integrated Water Resources Management
LAS	League of Arab States
LDCs	least developed countries
LEED	Leadership in Energy and Environmental Design
MAWRED	Modeling and Monitoring Agriculture and Water Resources Development (program)
MMD	multimodel data set
NAO	North Atlantic Oscillation
NAPA	UN National Adaptation Programme of Action
NFC	Nile Forecast Center
NGO	nongovernmental organization
OECD	Organisation for Economic Co-operation and Development
OECD-DAC	Organisation for Economic Co-operation and Development's Development Assistance Committee
PDNA	postdisaster needs assessment
PES	payment for ecosystem services
PTOLEMY	Planning, Transport, and Land Use for the Middle East Economy
PV	photovoltaic
RAED	Arab Network for Environment and Development
RCM	regional climate model
RDM	robust decision making
RCP	representative concentration pathway
RREE	rural renewable energy electrification
SEA	strategic environmental assessment
SOI	Southern Oscillation Index
SRES	IPCC Special Report on Emission Scenarios
SWOT	strengths, weaknesses, opportunities, and threats
UfW	Unaccounted-for water

UN	United Nations
UNDP	United Nations Development Programme
UNFCCC	United Nations Framework Convention on Climate Change
UNISDR	United Nations International Strategy for Disaster Reduction
UNWTO	United Nations World Tourist Organization
VWC	village water committee
WG	weather generator
WHO	World Health Organization
WISP	Water Information Systems Platform
WMO	World Meteorological Organization

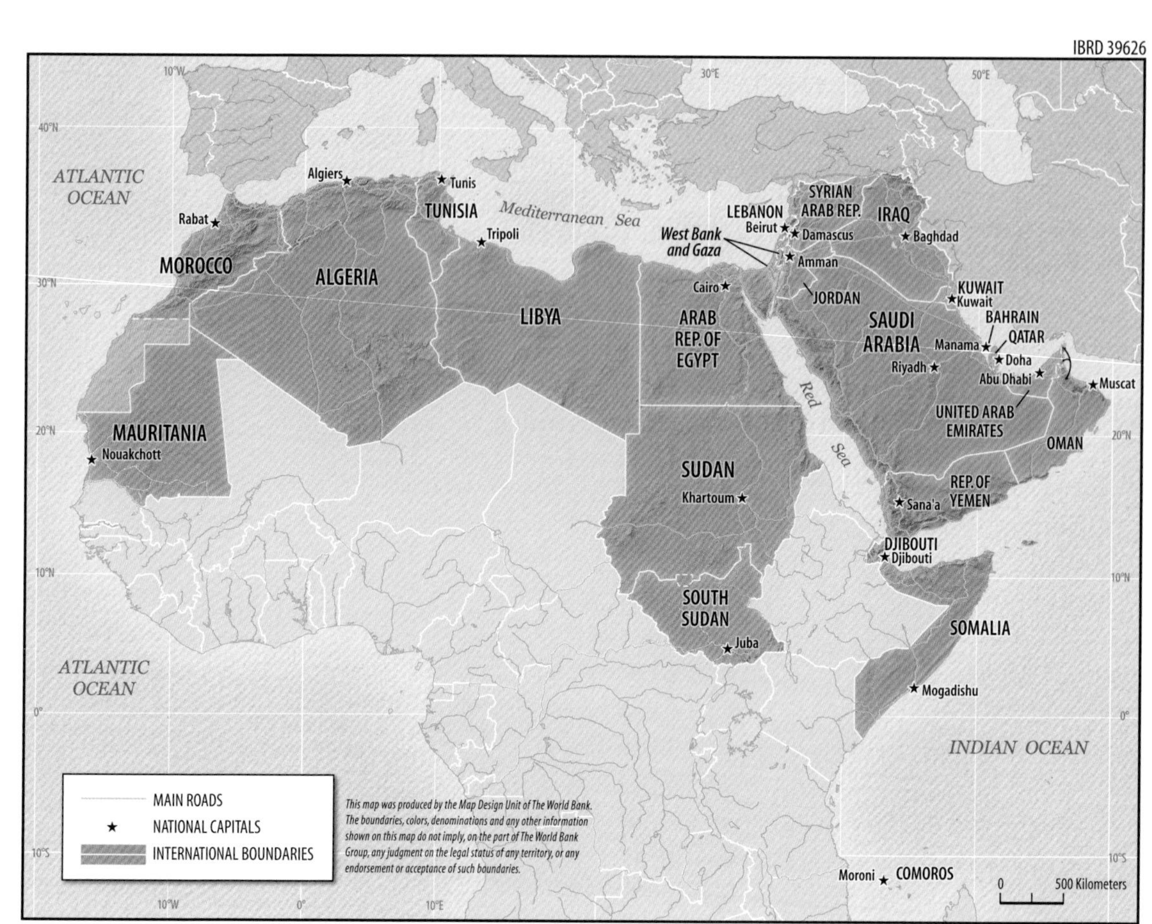
IBRD 39626
ATLANTIC OCEAN
Algiers
Tunis
Rabat
TUNISIA
Mediterranean Sea
Tripoli
MOROCCO
ALGERIA
LIBYA
LEBANON
Beirut
SYRIAN ARAB REP.
Damascus
IRAQ
Baghdad
West Bank and Gaza
Amman
JORDAN
Cairo
ARAB REP. OF EGYPT
KUWAIT
Kuwait
SAUDI ARABIA
BAHRAIN
Manama
QATAR
Doha
Riyadh
Abu Dhabi
Muscat
UNITED ARAB EMIRATES
OMAN
Red Sea
MAURITANIA
Nouakchott
SUDAN
Khartoum
REP. OF YEMEN
Sana'a
DJIBOUTI
Djibouti
SOUTH SUDAN
Juba
SOMALIA
Mogadishu
ATLANTIC OCEAN
INDIAN OCEAN
MAIN ROADS
NATIONAL CAPITALS
INTERNATIONAL BOUNDARIES
This map was produced by the Map Design Unit of The World Bank. The boundaries, colors, denominations and any other information shown on this map do not imply, on the part of The World Bank Group, any judgment on the legal status of any territory, or any endorsement or acceptance of such boundaries.
Moroni
COMOROS
0 500 Kilometers
NOVEMBER 2012

Overview

Climate Has Shaped the Cultures of Arab Countries

The first settlements in the world—farming communities and cities—began in this region, and all of them have changed in response to the variable climate. For thousands of years, the people of this region have coped with the challenges of climate variability by adapting their survival strategies to changes in rainfall and temperature. But the message is clear: over the next century this variability will increase and the climate of Arab countries will experience unprecedented extremes. Temperatures will reach new highs, and in most places there will be less rainfall. Water availability will be reduced, and with a growing population, the already water-scarce region may not have sufficient supplies to irrigate crops, support industry, or provide drinking water.

Fortunately, Arab countries can take steps to reduce climate change impacts and build resilience. In many cases, climate change is bringing attention to issues that were overlooked before. For example, low quality urban drainage systems have contributed to flooding in some Arab cities, and the threat of more flooding from climate change could be the impetus needed to finally improve this infrastructure. In rural areas, climate change is forcing communities to rethink long-standing gender roles that have perpetuated gender inequality. As a result, climate change presents many opportunities, not only to reduce vulnerability, but also to contribute to greater long-term development.

Climate change is happening now in the Arab countries. The year 2010 was the warmest since the late 1800s, when this data began to be collected, with 19 countries setting new national temperature highs. Five of these were Arab countries, including Kuwait, which set a record high of 52.6°C in 2010, only to be followed by 53.5°C in 2011. Extreme climate events are widely reported in local media, and a 2009 Arab region survey showed that over 90 percent of the people sampled agree that climate change is occurring and is largely due to human activities; 84 percent believe it is a serious challenge for their countries; and respondents were evenly split on whether their governments were acting appropriately to address climate change issues. The sample came mostly from the better-educated population, but it shows that there is a firm base and desire for action regarding climate change across the Arab region.

Water scarcity will increasingly be a challenge in the Arab countries. The Arab region has the lowest freshwater resource endowment in the world. All but four Arab countries (Arab Republic of Egypt, Iraq, Saudi Arabia, and Sudan) suffer from "chronic water scarcity," and over half of countries fall below the "absolute water scarcity" threshold. It is estimated that climate change will reduce water runoff by 10 percent by 2050. Currently, the region suffers a water deficit (demand is greater than supply), and with increasing populations and per capita water use, demand is projected to increase further, by 60 percent, by 2045.

Climate change will likely reduce agricultural production in Arab countries. Projections suggest that the rate of increase in agricultural production will slow over the next few decades, and it may start to decline after about 2050. Most of the Mediterranean region, which supports 80 percent of production, is projected to have less rainfall and hotter conditions. This will increase water use and likely limit the productivity of some crops. Other areas, such as the Nile Delta, will have to contend with saline intrusion from the sea. Farmers will face additional problems from higher temperatures. For example, the chilling requirements for some fruits may not be met; new pests will emerge; and soil fertility is likely to decline. This is alarming because almost half of the Arab region's population lives in rural areas, and 40 percent of employment is derived from agriculture. Compounding this vulnerability are troubling poverty rates: 34 percent of the rural population is poor, and unemployment is high, especially for women and youth.

Urban populations are rapidly growing. Currently 56 percent of Arab people live in urban centers, and by 2050 these populations will increase to 75 percent. Droughts have been shown to increase rural-to-urban migration in the region. A recent multi-year drought in the Syrian Arab Republic is estimated to have led to the migration of about one million people to informal settlements around the major cities. Many cities are already experiencing severe housing shortages because of this urban population growth.

Urban areas are vulnerable to climate change. Flash flooding is increasing in cities across the region as a result of more intense rainfall events, concrete surfaces that do not absorb water, inadequate and blocked drainage systems, and increased construction in low-lying areas and *wadis*. The number of people affected by flash floods has doubled over the last ten years to 500,000 people across the region. Climate change projections suggest that average temperatures in the Arab countries are likely to increase by up to 3°C by 2050. The urban heat island effect is projected to increase nighttime temperatures by an additional 3°C. In addition, providing water to urban areas is becoming increasingly difficult. Reasons for this include aging pipes, water loss from leakage of 40 percent or more in some major cities, and no water infrastructure in informal settlements.

Climate change threatens tourism, an important source of revenue and jobs. Tourism today contributes about US$50 billion per year to the Arab region, which is about 3 percent of its total gross domestic product (GDP), and tourism is projected to grow by about 3.3 percent per year for the next 20 years. It is also an important sector for jobs, because roughly 6 percent of the region's employment is tourism related. Higher temperatures are an obvious threat to tourism in a region that is already regarded as hot. Analyses of tourism patterns suggest that in the long-term, destinations on the north Mediterranean coast or within Europe will become more attractive than will the Arab region. Snowfall in Lebanon (for skiing), Red Sea coral reefs, and many ancient monuments across the region are threatened by climate change and severe weather. Ecotourism is an expanding sector, but the ecosystems (coral reefs, mountains, and oases) on which it depends will have to be managed carefully as they adjust to a changing climate. Extreme events, such as heavy rains, or more chronic pressures, such as increased salinity in groundwater, can threaten the region's historic buildings, paintings, and artifacts. Some destinations, such as Alexandria, will be further threatened by seawater inundation as sea levels rise. In most cases, there is already a need to better conserve and protect these cultural sites. Climate change increases its urgency.

Climate change threatens progress to achieve gender equity in the Arab region. Men and women possess unique vulnerabilities to climate change impacts, largely based on their respective roles in society. However, in the majority of cases, rural women tend to be vulnerable in more ways than are rural men. Climate change will further affect rural livelihoods, and more men will feel obligated to move to cities to seek paid employment, which is mostly unskilled and temporary, with little security, low wages, crowded living conditions, and poor health support. As a result, on top of their already heavy workload of domestic tasks and local natural resource management, rural women assume the departed male's community role, but with additional challenges. Women tend to have less education; they find travel difficult because of cultural norms, pregnancy, and child care; and women often lack the cultural and legal authority to assert their rights. For example, their access to credit might be limited, access to and control of water is usually ceded to the landowner—rarely a woman—and even access to rural organizations and support systems is often thwarted. Women's representation in Arab governments is only 9 percent, or half of the global average.

Women are active agents of adaptation. Because of their central role in family, household, and rural activities, women are in a position to change the attitudes, behaviors, and livelihoods that are needed for successful adaptation. A focus on gender is not an add-on to policy formulation but an essential part of any development strategy. Effective adaptation can only be achieved if the many barriers to gender equity are removed and, in particular, women are empowered to contribute. While women still have a literacy rate 15 percent lower than men and little voice in decision making, there is evidence that this is changing. For example, in some Gulf countries, more women than men graduate from universities.

The impacts of climate change on human health are varied and often indirect. Higher temperatures are known to lead directly to increased morbidity (deaths) through heat stress and indirectly to strokes and heart-related deaths. Warmer conditions also affect the geographic range of disease vectors, such as mosquitoes. A warmer climate will expose new human populations to diseases, such as malaria and dengue, for which they are unprepared. In the Arab region, disruptions to existing agricultural practices will lead to more widespread malnutrition, because of higher food prices and greater exposure to diseases and other health problems—especially if greater migration to unsanitary, informal settlements is triggered. The impacts of malnutrition on children are particularly troublesome because they lead not only to increased child mortality, but also to developmental and long-term physical and mental impediments.

While experts agree on climatic trends, it is less clear what the socioeconomic impacts of climate change will be. The diverse Arab region includes six least developed countries (LDCs) with predominantly rural populations and an annual per capita GDP as low as US$600 (Somalia). By comparison, the GDP in Kuwait, Qatar, and the United Arab Emirates is more than US$50,000 per capita annually, with 80–90 percent of the people living in cities. It is likely that all economies will be increasingly affected by climate change as time passes. This is illustrated by background case studies prepared for this report on income, livelihoods, well-being, and poverty in Syria, Tunisia, and the Republic of Yemen. Nevertheless, results show that over the next 30–40 years, climate change is likely to lead to a cumulative reduction in household incomes of about 7 percent in Syria and Tunisia, and 24 percent in the Republic of Yemen.

Arab countries can take action to reduce their vulnerability to climate change. For example, this report proposes an Adaptation Pyramid (figure 1) Framework that assists stakeholders in Arab countries in integrating climate risks and opportunities into development activities. It is based on an adaptive management approach, but it also highlights the

FIGURE 1

The Adaptation Pyramid: A Framework for Action on Climate Change Adaptation

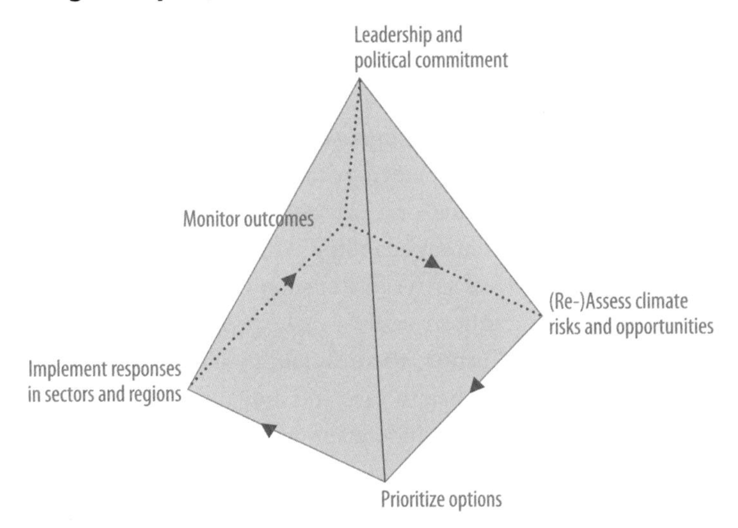

Source: Authors.

importance of leadership, without which adaptation efforts are unlikely to achieve the necessary commitment to be successful. The framework begins by assessing climate risks and opportunities and identifying options within the context of other development planning. The next step is to identify and prioritize adaptation options within the context of national, regional, and local priorities. Finally, adaptation responses will be implemented and outcomes monitored over time. It is important to take into account the long-term consequences of these decisions, because short-term responses may not be efficient or could lead to maladaptive outcomes. Other important measures for Arab region policy makers to implement are discussed below.

1. **Facilitate the development of publicly accessible and reliable information related to climate change.** Access to quality weather and climate data is essential for policy making. Without reliable data on temperature and precipitation levels, it is difficult to assess the current climate and make reliable weather forecasts and climate predictions. For example, information on river flows, groundwater levels, and water quality and salinity is critical for assessing current and future water availability. However, climate stations across most of the Arab region are very limited compared to most other parts of the world and what data exists is often not digitized or publicly available. Conflict in parts of the region disrupts both the collection and sharing of data. Information on food production and the main food supply chains (such as

changes in agricultural yields and production for important crops, forage, and livestock) needs to be linked with weather and water data to better monitor and understand the effects of a changing climate. In addition, socioeconomic data (including household and census data) and other economic data related to the labor market and production should be collected and made available.

2. **Build climate resilience through social protection and other measures.** Resilience is determined by factors such as an individual's age, gender, and health status, or a household's asset base and degree of integration with the market economy. Underinvestment in social safety nets—public services such as water supply and wastewater treatment, and housing and infrastructure—make people more vulnerable to a changing climate. Further, there should be measures in place to ensure equitable access to health care and a quality education. Such social protection measures include insurance schemes, pensions, access to credit, cash transfer programs, relocation programs, and other forms of social assistance. These investments and instruments facilitate economic and social inclusion, which creates co-benefits between adaptation and development goals.

3. **Develop a supportive policy and institutional framework for adaptation.** Basic conditions for effective development, such as the rule of law, transparency and accountability, participatory decision-making structures, and reliable public service delivery that meets international quality standards are conducive to effective development and adaptation action. In addition, climate change adaptation requires new or revised climate-smart policies and structures at all levels.

Sound adaptation planning, strong governmental/nongovernmental cooperation, and plentiful financial resources are all important for building resilience to climate change. Developing national adaptation strategies are important for prioritizing adaptation activities that respond to urgent and immediate needs, and for setting forth guiding principals in the effort to cope with climate change. National governments have a key role in developing these strategies and as a result play an important role in promoting collaboration and cooperation. This cooperation should include the government, civil society, the private sector, and international institutions. Within governments, inter-ministerial coordination is especially critical, because adaptation responses often require activities involving multiple ministries and sectors. Finally, to do any of the activities above it is important to secure the necessary financial resources. There are many sources for adaptation funding, but first the Arab countries will need to build their capacity to analyze their financial needs and generate and manage these resources.

By nature, adaptation to climate change is a dynamic process, and so is the governance of adaptation. Political change, including those changes originating from the Arab Spring, can provide an opportunity to increase civil society participation in adaptation governance and a move toward a more inclusive approach to addressing climate change issues and building climate resilience.

This report is about climate change, its impacts on people, the systems upon which we depend, and how we might adapt to climate change. It highlights a number of issues and areas that are being affected by climate change. One important message of this report is that climate change should be taken into account in all activities—however, this report cannot provide solutions or options for all issues. For example, the transboundary water issues are already being addressed by international task forces; this report can deal only with how climate change might affect their decisions. Anticipation of climate change can be the stimulus for improving interventions and accelerating action, which has been seen in countries such as Australia, where water laws and management were extensively changed in response to a prolonged drought and the anticipation of further climate change issues.

This report can be used as a road map. This report seeks to provide, for the first time, a coherent assessment of the implications of climate change to the Arab region and the resultant risks, opportunities, and actions needed. The information highlighted within explains the potential impacts of climate change and the adaptation responses needed in key sectors such as water, agriculture, tourism, gender, and health, as well as in urban and rural settings. This report attempts to advance the discussion by providing adaptation guidance to policy makers in Arab countries. It does this in three ways. First, it proposes the Adaptation Pyramid Framework on how to move forward on this agenda. Second, it presents a typology of policy approaches relevant to the region, which would facilitate effective policy responses by decision makers. Third, a matrix is provided, which outlines key policy recommendations from each of the chapters.

CHAPTER 1

Climate Change Is Happening Now, and People Are Affected in Arab Countries

A Harsh Environment Has Shaped the Cultures in the Region

Climate change is a defining element of today's development challenges. In Arab countries[1] and around the world, climate change is already damaging people's livelihoods and well-being. It is a threat to poverty reduction and economic growth and may unravel many of the development gains made in recent decades. Both now and over the long term, climate variability and change threaten development by restricting the fulfillment of human potential and disempowering people and communities, thereby constraining their ability to protect and enrich their livelihoods. This situation calls for action. As the *World Development Report 2010* put it: we need to "act now, act together, and act differently" (World Bank 2010).

Climate has shaped the cultures of Arab countries. The first settlements, farming communities, and cities in the world all began in this region, and all have changed in response to a variable climate. For thousands of years, people of the region have coped with the challenges of climate variability by adapting their survival strategies to changes in rainfall and temperature. But the message in chapter 2 is clear: over the next century this variability will increase, and the climate of Arab countries will experience unprecedented extremes. Temperatures will reach record highs in many places, and there will be less rainfall. As discussed in chapter 3, water availability will be reduced because of lower precipitation, increased temperatures, and a growing population. The already

Photograph by Dorte Verner

water-scarce region may not have enough water to irrigate crops, support industry, or provide drinking water. As climate variability increases, so does human vulnerability to it—especially for the poor and those heavily dependent on natural resources (such as the farmers and pastoralists who are the subject of chapter 4). People living in cities (chapter 5) and those working in tourism (chapter 6) must also cope with the degraded resources that sustain urban communities or tourist destinations. Gender dynamics and public health systems in Arab societies will also be challenged (chapters 7 and 8). But among these unprecedented challenges, there are new opportunities and approaches that the Arab people can take advantage of today. This report proposes applying an Adaptation Pyramid, which is based on the process of assessing the climate, reacting to the perceived challenges, implementing cross-sectoral responses, and monitoring progress—all with strong leadership (see figure 1.4, later in this chapter, and chapter 9).

Many climate adaptation strategies that people have used throughout history have become less viable. In about 2200 BCE, a temporary climate shift created 300 years of reduced rainfall and colder temperatures, which forced people to abandon their rainfed fields in what is now the northeast Syrian Arab Republic. As people migrated to the south or turned to pastoralism to survive, whole cities were deserted and covered in the dust of drought (Weiss and Bradley 2001). Today, despite technological gains, the ability of climate-affected people to migrate in the face of these challenges is limited, partly because of borders that are difficult to cross and property rights that are difficult to leave behind or attain in new locations. As discussed in chapter 5, often the only choice left for people faced with depleted assets and less productive livelihoods in drought-stricken areas is to move to cities or towns, where their rural skills are hard to deploy. The Tuaregs in southern Algeria during the prolonged drought of the 1970s were such a case: a large number of families moved, for example, to the town of In Guezzam. A similar example happened more recently among the Bedouin in Syria (discussed later in this chapter).

Climate change has already affected—or will soon affect—most of the 340 million people in the Arab region, but the roughly 100 million poor people are the least resilient to the negative impacts from these changes.[2,3] Although they have contributed the least to the causes of climate change—some do not have the electricity to power a single light bulb—the poor have the fewest resources at their disposal to adapt to these changes.[4] Climate change adaptations, including the development of innovative technology, diversification, and alternate livelihood and social protection schemes, can reduce the poor's vulnerability to negative impacts of climate change (see chapter 9).

Climate Change Is Happening Now

Arab countries are located in a hyperarid to arid region—less than 0.2 on the Aridity Index (AI)—with pockets of semiarid areas (between 0.2 and 0.5 AI). There are some temperate zones in coastal North Africa, the eastern Mediterranean, and equatorial areas in southern Somalia and the Comoros. There are snow-classified areas in the mountains of Algeria, Iraq, Lebanon, and Morocco. Environmental challenges in the Arab world include water scarcity, with the lowest freshwater resource endowment in the world;[5] very low and variable precipitation; and excessive exposure to extreme events, including drought and desertification. This demanding environment, combined with a high poverty rate, makes the Arab region among the world's most vulnerable regions to climate change. If no drastic measures are taken to reduce the impacts of climate change, the region will be exposed to reduced agricultural productivity and incomes, a higher likelihood of drought and heat waves, a long-term reduction in water supplies, and the loss of low-lying coastal areas through sea-level rise. This climate exposure will have considerable implications for human settlements and socioeconomic systems (IPCC 2007).

Climate change is already being felt in Arab countries. Globally, 2010 tied 2005 as the warmest year since climate data began to be collected in the late 1800s. Of the 19 countries that set new national temperature highs in 2010, 5 were Arab countries. Temperatures in Kuwait reached 52.6°C only to be followed by 53.5°C in 2011.[6] In addition to the warming climate, the frequency of extreme weather events is increasing. For example, in June 2010, the Arabian Sea experienced the second-strongest tropical cyclone on record—Cyclone Phet—which peaked at category 4 strength with winds at 145 miles per hour, killing 44 people and causing US$700 million in damages to Oman.[7] Also, 2010 was the second-worst year on record for coral reef dieback, which was caused by near-record highs in summer ocean water temperatures. A snapshot of the scientific and media reports on climate change in Arab countries shows its increasing profile (Allison et al. 2009; Chomiz 2011; Füssel 2009; Parry, Canziani, Palutikof, van der Linden, and Hanson 2007):

- Higher temperatures and more frequent and intense heat waves threaten lives, crops, terrestrial biodiversity, and ecosystems such as coral reefs and fisheries.
- Less but more intense rainfall causes both more droughts and more frequent flash flooding.
- Loss of winter precipitation storage in snow masses induces summer droughts.

- Increased frequency of prolonged droughts leads to losses in livelihoods, incomes, and human well-being.
- Sea-level rise threatens river deltas, coastal cities, wetlands, and small island nations such as Bahrain and the Comoros with storm surges, saltwater intrusion, flooding, and subsequent human impacts.
- More intense cyclones put human life and property at risk.
- Changing rainfall patterns and temperatures create new areas exposed to dengue, malaria, and other vector- and waterborne diseases affecting people's health and productivity.
- Inequality between males and females increases as females assume many of the new burdens associated with climate change.

There is increasing evidence that climate change will have severe negative impacts on the economic and social development of Arab countries. As discussed in chapters 7 and 8, climate change threatens to stall and reverse progress toward poverty reduction, better health, gender equality, and social inclusion (see, for example, Kronik and Verner 2010; Mearns and Norton 2009; Verner 2010). Yet research on the socioeconomic dimensions of climate change in the Arab region is in only the early stages (Tolba and Saab 2009). This report aims to assess the impacts of climate variability and change and to fill knowledge gaps so that practitioners can better respond to Arab government requests for assistance in understanding and identifying successful adaptation policies and programs. Such strategies will assist these countries and their people in building resilience to climate change, particularly for the poor and vulnerable (see box 1.1 on definitions of climate change adaptation).

This report also serves as a resource for researchers to begin to assess climate risks, opportunities, and actions. The information highlighted explains the potential impacts of climate change in key sectors such as water, agriculture, tourism, and health, as well as in urban and rural settings, and then goes on to discuss possible policy options to reduce climate risk and better adapt to climate variability and change. The chapters are mostly led by Arab region specialists. In the preparation of each chapter, an effort was made to draw upon the regional literature—whether in Arabic, French, or English.

This report is organized into nine chapters. Each chapter reviews the literature and core issues and presents relevant policy options.[8] This chapter sets the stage for the report by discussing some of the impacts and challenges of climate change. It also previews a framework for building the capacity to adapt to climate change and guide national- and local-level project planning, hence facilitating adaptation governance. Chapter 2 ex-

BOX 1.1

Intergovernmental Panel on Climate Change Definitions: Climate, Climate Change, and Climate Variability

Climate in a narrow sense is usually defined as the average weather, or more rigorously, as the statistical description in terms of the mean and variability of relevant quantities over a period of time ranging from months to thousands or millions of years. The classical period is 30 years, as defined by the World Meteorological Organization. These quantities are most often surface variables such as temperature, precipitation, and wind. Climate in a wider sense is the state, including a statistical description, of the climate system.

Climate change refers to a statistically significant variation in either the mean state of the climate or in its variability, persisting for an extended period (typically decades or longer). Climate change may be due to natural internal processes or external forcings, or to persistent anthropogenic changes in the composition of the atmosphere or in land use.

Climate variability refers to variations in the mean state and other statistics (such as standard deviations and the occurrence of extremes) of the climate on all temporal and spatial scales beyond those of individual weather events. Variability may be due to natural internal processes within the climate system (internal variability), or to variations in natural or anthropogenic external forcings (external variability).

Source: IPCC 2001, Glossary.

amines the projected climate variability and changes that will likely occur in the region. It provides a checklist for policy makers and project managers on how to take climate change into account when preparing any new project, regulation, or policy. The following six chapters address key areas that are affected by climate variability and change. Chapter 3 discusses water stress and options to overcome water-related challenges; chapter 4 examines the rural sector, particularly as it relates to livelihoods, agriculture, and food security; chapter 5 looks at urban areas; chapter 6 discusses tourism; chapter 7 examines gender; and chapter 8 discusses public health.

Finally, in chapter 9 this report attempts to move the discussion on adaptation governance one step further by providing guidance to policy makers in Arab countries. It does this in three ways. First, the chapter provides a Framework for Action on Climate Change Adaptation, represented by an Adaptation Pyramid, which illustrates the key elements of adaptation decision making. Second, it puts forward a typology of policy approaches that are relevant to such decision making in the Arab region. Third, it provides a policy matrix that outlines the key policy options covered in each of the areas of this report.

The Effects of Climate Change Are Socially Differentiated

Climate change affects all people and countries in the Arab region (box 1.2). Still, the effects of climate change are regionally and socially unequal, within and among countries, and across the Arab region as a whole. While wealthier people continue to enjoy the benefits of high-carbon-emitting lifestyles that include air-conditioned houses and cars, millions of the poor are disproportionately threatened by climate change, compared with their negligible contribution to its causes. Asset-poor communities, such as the Bedouin[9] from the arid areas of the Arabian Peninsula and North Africa, have few resources but have some capacity to adapt to the changing climate. Many manage to take action by diversifying their livelihood, moving, pursuing education, and so forth. Climate change is superimposed over the preexisting risks and vulnerabilities that poor and marginalized groups typically face.

Many studies have suggested that the poor are the most vulnerable to climate change because of the following conditions:

- Dependence on natural resources, which are exposed to the climate
- Lack of assets, which hinders effective adaptation
- Settlement in at-risk areas, which are less productive and are also vulnerable to floods or droughts or other severe events
- Migrant status, which can prevent the poor from accessing certain social services
- Low levels of education, which prevents them from developing more climate-resilient skills or livelihood strategies
- Minority status, which deters policy makers from making them the focus of adaptation policies

The Arab people increasingly do not know what to expect regarding the climate and, hence, what decisions to make. This lack of information is most visible among climate-dependent activities such as rainfed agriculture, given changes in the timing and intensity of rainfall. A key asset of farmers is their traditional knowledge of the environment, but this

BOX 1.2

Water-Related Impacts: Selected Examples from the Arab Region

Alexandria, Arab Republic of Egypt. A half-meter increase in sea levels is predicted to flood 30 percent of the city, leading to the displacement of 1.5 million people or more, the loss of 195,000 workplaces, and estimated damages to land and property reaching US$30 trillion.

Sudan. Droughts, flooding, and desertification lead to increased competition over environmental resources and contribute to human conflicts such as those in Darfur, one of the greatest humanitarian crises of this century.

Jordan Basin, West Bank–Jordan. The year 2008 marked the fifth consecutive year of drought for the Jordan River basin. As a result, many in the West Bank do not have access to water for most of the day.

Desert of Sinai Peninsula, Egypt. Rainfall in this region has decreased 20 to 50 percent over the past 30 years. Droughts and flash floods threaten the lives of the local Bedouin.

Nile Basin, Egypt-Ethiopia-Sudan. In 2006, flooding of the Nile River caused the deaths of 600 people, left 35,000 others without homes, and interrupted the lives of another 118,000 individuals. Some 3,000 houses were destroyed in Sudan from these floods.

Photograph by Dorte Verner

Source: Universitat Autònoma de Barcelona 2010.

knowledge may no longer be reliable without the support of forecasting technology and additional climate information. The impacts of climate change vary across the Arab region, and each country must respond with customized approaches and policies. These approaches depend in part on the perceived impacts of climate change and in part on a country's economic capacity to respond to those impacts.

This report addresses the Arab region as a whole, while acknowledging that the region and countries are heterogeneous and that adaptive policies must be flexible. Four subgroups of countries have been identified: (a) the least developed countries (LDCs; the Comoros, Djibouti, Mauritania, Somalia, Sudan, and the Republic of Yemen); (b) the Maghreb (Algeria, Libya, Morocco, and Tunisia); (c) the Mashreq (the Arab Republic of Egypt, Iraq, Jordan, Lebanon, the Syrian Arab Republic, and West Bank and Gaza); and (d) the Gulf region (Bahrain, Kuwait, Oman, Qatar, Saudi Arabia, and the United Arab Emirates). Some of the topics that are addressed in this report are uniform across the Arab world—such as the need for better climate information to improve policy making and meet development objectives—whereas others are more local. Moreover, this report provides strategic guidance on adaptation with a short time horizon—until about 2030—while taking into account longer-term climate change projections.

Many poor people are already being forced to cope with the impacts of climate change. In the Mashreq, in northeast Syria, the current multiyear drought has forced hundreds of thousands of people to move to the outskirts of major cities, leaving their livelihoods and social networks behind (see box 1.3). In countries such as the Republic of Yemen, one of the LDCs, women and children travel farther and farther distances to fetch dwindling water supplies. This additional labor often forces girls in rural areas to drop out of school, which deprives them of lifelong skills. Oman, located in the Gulf region, is experiencing significant movements of saline water into freshwater aquifers, which reduces people's access to drinking water and other resources. Small farmers in all Arab countries, particularly in the Maghreb, are experiencing reduced crop yields and lost outputs as a result of climate variability and change.

Climate change is a threat to short-, medium-, and long-term development. It restricts human potential and freedom and reduces the ability of people to make informed choices regarding their well-being and livelihoods (Mearns and Norton 2009; Verner 2010). As the *Stern Review* argues, the paramount aims are for climate change issues to become fully integrated into development policies and for international support of those policies to increase (Stern 2007). Social, economic, and human development are key to efforts to reduce potential conflicts, migration and displacement (see box 1.3), losses to livelihood systems, and damage to or declines in infrastructure. All of these issues will help enable people and communities to cope with climate change.

BOX 1.3

Effects of Severe Droughts on Rural Livelihoods in Syria

The recent prolonged drought in northeast Syria is having devastating effects on farmers and herders. Within these governorates, the livelihoods of more than 1 million people have been affected. Hundreds of thousands have been forced to leave their traditional ways of life and migrate to urban areas to find work.

In the Badia steppe rangeland of Syria, droughts have been especially damaging for small-scale herders. Interviews with communities in this region suggest that households with 200 sheep or fewer have been forced to give up herding and move to large towns and cities. Lower rainfall and the drying-out of riverbeds have caused farmers and herders to rely increasingly on groundwater for irrigation and livestock production. However, in some farming communities near Deir Ezzor, dried-up wells have led to the emigration of 8 out of every 10 families.

Rainfall in Syria is predicted to decline by more than 20 percent over the next 50–70 years. Already, groundwater levels are falling rapidly—by more than one meter per year in some regions—and lower rainfall levels will further accelerate this process through slower recharge rates. Given that about three-fourths of all farmland in Syria is rainfed, substantial decreases in yields can be expected as a direct consequence of lower precipitation rates.

Households with diversified sources of income such as remittances and off-farm activities are more resilient. Still, the impacts are present in all households and communities because of the continuous depletion of assets, reduced nutritional levels, lower school attendance rates, and reduced mobility.

Photograph by Dorte Verner

Source: World Bank 2011a.

TABLE 1.1

Socioeconomic Information for Arab Countries

	Land area (sq. km)	Agricultural land (% of land area)	Forestland (% of land area)	Population	Population growth (annual %)	Urban population (% of total)	Urban population growth (annual %)
Maghreb							
Algeria	2,381,740	17.4	0.6	35,468,208	1.5	66.5	2.4
Libya	1,759,540	8.8	0.1	6,355,112	1.5	77.9	1.7
Morocco	446,300	67.3	11.5	31,951,412	1.0	56.7	1.6
Tunisia	155,360	63.0	6.5	10,549,100	1.0	67.3	1.6
Mashreq							
Egypt, Arab Rep.	995,450	3.7	0.1	81,121,077	1.8	42.8	1.8
Iraq	434,320	20.1	1.9	32,030,823	3.0	66.4	2.8
Jordan	88,780	11.5	1.1	6,047,000	2.2	78.5	2.3
Lebanon	10,230	67.3	13.4	4,227,597	0.7	87.2	0.9
Syrian Arab Republic	183,630	75.7	2.7	20,446,609	2.0	54.9	2.6
LDCs							
Comoros	1,860	83.3	1.6	734,750	2.6	28.2	2.8
Djibouti	23,180	73.4	0.3	888,716	1.9	88.1	2.3
Mauritania	1,030,700	38.5	0.2	3,459,773	2.4	41.4	2.9
Somalia	627,340	70.2	10.8	9,330,872	2.3	37.4	3.5
Sudan	2,376,000	57.5	29.4	43,551,941	2.5	45.2	4.5
Yemen, Rep.	527,970	44.4	1.0	24,052,514	3.1	31.8	4.9
Gulf							
Bahrain	760	9.2	1.3	1,261,835	7.6	88.6	7.6
Kuwait	17,820	8.5	0.3	2,736,732	3.4	98.4	3.4
Oman	309,500	5.9	0.0	2,782,435	2.6	71.7	2.6
Qatar	11,590	5.6	0.0	1,758,793	9.6	95.8	9.7
Saudi Arabia	2,149,690	80.7	0.5	27,448,086	2.4	83.6	4.0
United Arab Emirates	83,600	6.8	3.8	7,511,690	7.9	78.0	8.0

Sources: Authors, based on UNDP 2011; World Bank 2011b.

Note: HDI = Human Development Index; PPP = purchasing power parity; — = not available. Data are those most recently available; most apply to 2010 or a few years prior.

Climate Change Affects People and the Economy

Climate change puts additional stress on the economies and people of Arab countries. Climate variability and change can lead to, and add to, disruptions in social, infrastructural, environmental, or productive systems and resources, which in turn can slow economic growth and increase poverty. Countries that rely heavily on climate-sensitive sectors, such as agriculture and fisheries, or that have low income levels; high poverty rates; lower levels of human capital; or less institutional, economic, technical, or financial capacity will be the most vulnerable.

The Arab region varies greatly in climate, culture, education, literacy, and access to resources, and thus in vulnerability. The region includes six

Urban population > 1 million (% of total)	Population in areas with elevation < 5 m (% of pop.)	GDP per capita, PPP (constant 2005 international $)	HDI value 2011	Poverty ratio at US$1.25/day (PPP) (% of pop.)	Labor force, total	Employed in agriculture (% total employment)	Agriculture, value added (% of GDP)
7.9	3.5	7,521	0.698	6.8	14,844,724	20.7	11.7
17.4	4.7	15,361	0.760	—	2,304,613	19.7	1.9
19.3	3.8	4,227	0.582	2.5	11,845,622	40.9	15.4
0.0	9.5	8,566	0.698	2.6	3,821,103	25.8	8.0
19.0	25.6	5,544	0.644	2.0	26,536,263	31.6	14.0
22.9	6.5	3,195	0.573	4.0	7,275,555	23.4	8.6
18.3	4.2	5,157	0.698	0.4	1,817,540	3.0	2.9
45.8	9.1	12,619	0.739	—	1,444,266	—	6.4
34.3	0.3	4,741	0.632	1.7	6,311,814	19.1	22.9
0.0	14.0	984	0.433	46.1	327,568	—	46.3
0.0	7.6	2,087	0.430	18.8	389,680	—	3.9
0.0	20.4	1,744	0.453	21.2	1,415,617	—	20.2
16.1	2.2	—	—	—	3,531,417	—	65.5
11.9	0.2	2,023	0.408	—	13,246,649	—	23.6
9.7	1.8	2,267	0.462	—	6,053,143	54.1	14.3
0.0	66.6	23,755	0.806	—	591,876	1.5	0.9
84.2	22.8	49,542	0.760	—	1,334,479	2.7	0.5
0.0	5.5	24,226	0.705	—	1,092,450	6.4	1.9
0.0	23.1	73,196	0.831	—	1,158,535	2.3	—
38.9	1.0	20,374	0.770	—	10,109,647	4.8	2.6
20.9	7.3	42,353	0.846	—	4,455,171	4.2	1.0

LDCs with annual gross domestic product (GDP) per capita as low as US$1,000 (lower in the case of Somalia).[10] Those countries, with the exception of Djibouti, contain large rural populations that account for over 50 percent of the total population and that maintain agricultural sectors worth 20 percent or more of total GDP. By comparison, the annual GDP of largely urban countries such as Kuwait, Qatar, and the United Arab Emirates is more than US$50,000 per capita (see table 1.1). Literacy rates of adult women are below 50 percent in Morocco and the Republic of Yemen; in Lebanon and Libya over 50 percent of young women are enrolled in tertiary education, more than for men of similar ages.

The populations in Arab countries are growing; in some places they are increasing at unprecedented rates. Qatar and the Republic of Yemen have an annual population growth rate of 9.6 and 3.1 percent, respectively (see table 1.1). The total population of the Arab world is likely to hit

700 million people by 2050, which is roughly twice the size of today's population. This growth will increase the demand for scarce resources, including water and land. Climate change and the increased demand for food will affect food prices and will therefore affect Arab countries, because the region is heavily reliant on food imports (see chapter 4).

Climate stress, combined with better social and infrastructural services in cities, has already led to the rapid urbanization of many Arab countries. As a result, millions of people have left their rural homes to settle in urban centers. The most recent data show a large variation among countries in terms of the urban-rural divide. Some countries are almost completely urbanized (98 percent of Kuwaitis live in urban areas); others are still largely rural (Comoros, Egypt, Mauritania, Somalia, and the Republic of Yemen have 28, 43, 41, 37, and 32 percent of their respective populations residing in urban areas; see table 1.1). Thus, climate change adaptation must occur in both rural and urban areas. Although different adaptation options will be deployed in these different environments, in both settings, local women and men, especially the poor, must play an integral role in building the resilience of their livelihoods and well-being.

All people of the Arab region are vulnerable to the impacts of climate change and variability on water availability, food security, and their health. Cities have their own vulnerabilities, which are made worse by rapid growth that is partly driven by migration of the rural poor. Tourism is an increasingly important contributor to many economies of the region, but it too could be threatened if climate change is ignored. Women are both a most vulnerable group and the main agents for the social changes necessary for economies to cope with a changing climate. These topics are addressed in the following chapters of this report.

Climate Change Is Likely to Reduce Household Well-Being in the Short and Long Term

Projections show that the economies of Arab countries will be increasingly affected by climate change as time passes. These challenges are illustrated by background case studies prepared for this report on income, livelihoods, well-being, and poverty in Syria, Tunisia, and the Republic of Yemen. Findings from these three diverse countries are based on qualitative analyses (World Bank 2011a) and quantitative modeling (the computable general equilibrium, or CGE model).

Although experts agree on climatic trends (see chapter 2), it is less clear what the socioeconomic impacts of climate change will be. The ability to assess these impacts is challenged by the generally complex relationship between meteorological, biophysical, and economic interactions; the expected diversity of local impacts within countries; and the relatively long time horizon of the analysis. The findings of a modeling suite, presented below, address these analytical challenges by linking the downscaling of selected global circulation models (GCMs), crop modeling, global economic modeling, and subnational-level CGE modeling with microsimulation modeling. This approach estimates the potential economic impacts of climate variability and change at both global and local levels.

Arab countries will be affected by climate change at the national and local levels and will also suffer from the effects experienced by other countries. At the local level, an increase in temperatures, and in some cases a reduction in precipitation, is projected to reduce agricultural yields: wheat yields, for example, may decrease by about 60 percent by 2050 in some parts of the Arab world. In addition, because climate change will likely reduce agricultural yields globally, world market prices for major food commodities are projected to rise. Given the high dependence of Arab countries on imported food (combined with relatively limited agricultural potential), these global dimensions are particularly important for the Arab world.

The long-term local and global implications of climate change in Syria, Tunisia, and the Republic of Yemen are projected to lead to a large total reduction in household incomes (figure 1.1). Income reductions accumulate over time. By 2020, household incomes in Syria are projected

FIGURE 1.1

Cumulative Impacts of Climate Change on Household Income for Syria, Tunisia, and the Republic of Yemen

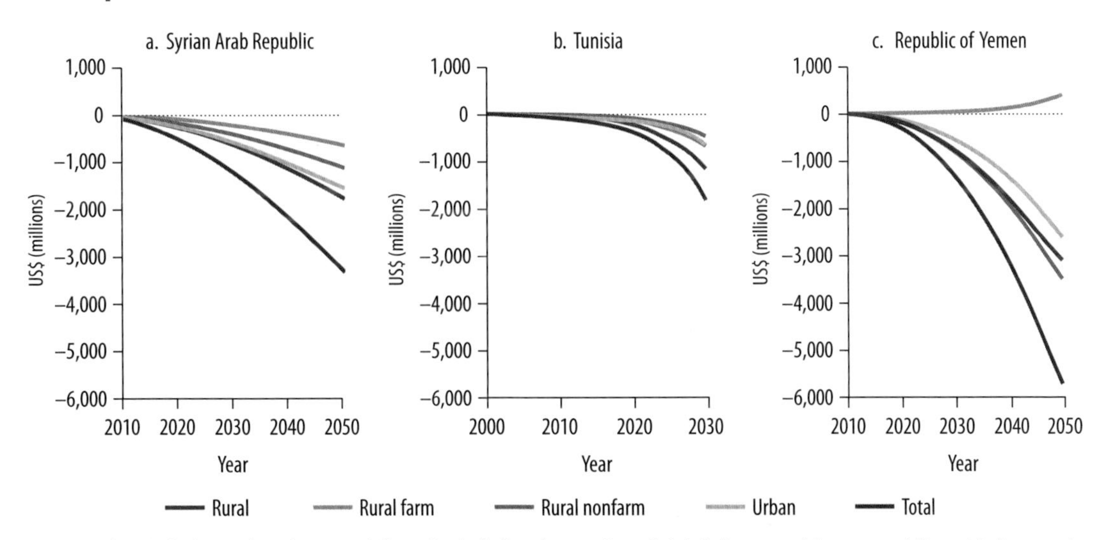

Source: Authors' calculations based on a modeling suite, including downscaling of global climate models, crop modeling, global economic modeling, and subnational-level economic modeling.

Note: Graphs represent results from the global climate model MIROC A1B scenario. Exchange rates are US$1 = 203 Yemeni rial, 48 Syrian pounds, and 1.44 Tunisian dinars. Results for Syria and the Republic of Yemen cover 40 years and Tunisia 30 years.

to be US$527 million (1.1 percent of GDP) lower than in a perfect mitigation scenario. By 2030, these losses are projected to increase by US$1.2 billion (2.5 percent of GDP), and by 2050 losses will reach US$3.4 billion (6.7 percent of GDP).

Climate change also has a negative impact on well-being in the Republic of Yemen. By 2020, household incomes are projected to be US$314 million lower (1.3 percent of GDP) compared with a perfect mitigation scenario; by 2030 these losses will reach US$1.3 billion (5.6 percent of GDP), and by 2050 losses will reach US$5.7 billion (23.9 percent of GDP). Tunisian households will also increasingly suffer from climate change. Initial income losses are estimated at US$100 million (0.4 percent of GDP) over a 10-year period and then accumulate to US$393 million and US$1.8 billion (1.4 and 6.7 percent of GDP after 20 and 30 years, respectively). These are rather optimistic estimates, considering that the chosen model (MIROC A1B, the model for interdisciplinary research on climate) is among the GCMs with medium-level impact and that the country-level CGE models for Syria, Tunisia, and the Republic of Yemen assume a large degree of endogenous adaptation. For example, MIROC A1B takes into account that people can freely adapt to a changing climate by switching crop patterns or moving out of agriculture and into other sectors of the economy that have greater development potential.

There are indications that the number of droughts has increased in some regions and will become more frequent in the future (see spotlight 1 on disaster risk management, following chapter 2).[11] Lower rainfall leads to a reduction in crop yields or, in extreme cases, to the complete loss of harvests, especially in rainfed agricultural systems. Droughts also affect livestock, particularly animals that rely on pastures for feeding. It is expected that normally occurring dry periods will last longer, which exacerbates these impacts. In addition to these direct impacts on the agricultural sector and the families that rely on it, droughts also directly affect other sectors of the economy and, indirectly, nonfarm households. In Syria, for example, droughts have occurred almost every other year during the past half-century. The average drought reduces growth in economic output by about 1 percentage point of GDP nationally compared to nondrought years. Food security in Syria is significantly reduced during droughts, and the poor are particularly hard-hit, mainly through the loss of capital, lower incomes, and higher food and feed prices. Nationwide, poverty levels increase by about 0.3 to 1.4 percentage points—depending on the year and household group—and stay above nondrought levels even when the drought is over. Poor farm households are the most affected, followed by rural nonfarm and urban households.

Floods may also become more frequent as a result of climate variability and, as a result, induce heavy economic losses and spikes in food insecurity. The Republic of Yemen is a natural disaster–prone country that faces a number of hazards every year, with floods being the most common and serious of these events. Although regular flooding can be beneficial to agricultural practices in dry lands like those in the Republic of Yemen, high-magnitude flooding leads to the loss of productive land, the uprooting of fruit trees, the deaths of animals caught in floodwaters, and the destruction of infrastructure, such as irrigation facilities and rural roads. Impact assessments of the October 2008 tropical storm and floods in Wadi Hadramout in the Republic of Yemen showed that agriculture was the sector hardest hit by floods. Industry and service sectors tend to be relatively more resilient. Estimates put the total cumulative loss in real income over the period 2008–12 at 180 percent of preflood regional agricultural value added. As a result of direct losses from flooding, farmers' incomes in these areas suffer most during the year of the flood. For example, the number of hungry people rose dramatically by about 15 percentage points as an immediate result of the 2008 floods. Spillover effects have led to increases in hunger even in regions where the flooding had no direct impact.

Poor Rural Communities Are among the Most Vulnerable to Climate Change

Rural populations that depend heavily on agriculture are especially vulnerable to climate change (see Diaz 2008; Gerritsen 2008; Smith 2008; Sulyandziga 2008; Tauli-Corpuz and Lynge 2008). Qualitative data analyses on vulnerable communities, such as the Bedouin in northeast Syria and farmers in Tunisia, show that these groups are largely dependent on natural resources for their well-being and livelihoods.[12]

Historically, the ability of Bedouin communities to adapt to climate impacts has been limited. The Bedouin define themselves as pastoralists, but this livelihood strategy has been severely affected by climate change. In Syria's current multiyear drought, livestock herds have been reduced by 80–100 percent, which has led individuals and households to adopt new coping measures. These adaptation mechanisms include temporary and permanent migration and agricultural wage labor. Most of the thousands of Bedouin people who migrate settle informally on the outskirts of urban areas. These changes mark a departure from the traditional Bedouin way of life.

The Bedouin have experienced declines in their social and cultural assets as well as their social and cultural cohesion, which are all important characteristics of their adaptive capacity. The Bedouin's dependence on natural resources and cultural assets makes these communities especially vulnerable to climate variability. To maintain their livelihood strategies,

FIGURE 1.2

The Bedouin's Assets to Cope with Drought, 1990 and 2010

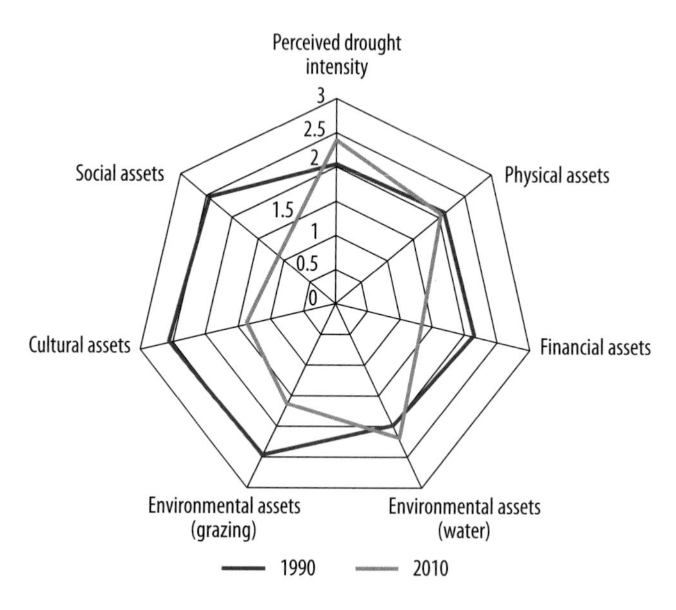

Source: World Bank 2011a.

Note: Data are from interviews conducted with 15 communities of the Bedouin in the Badia region, in Northeast Syria, December 2010.

they depend heavily on cultural, human, and social assets, including traditional knowledge systems and institutions that are now under increased stress. In interviews, the Bedouin said that their access to key livelihood assets is worse now than it was two decades ago (see figure 1.2).

The Bedouin perceive that the drought intensity in 2010 was higher than it was in 1990. The impact of the consecutive dry years starting in 2007 is clearly represented by decreased access to financial assets, such as loans and cash; environmental assets, such as available grazing areas; social assets, including trust and social networks; and cultural assets, such as the cultural leadership, capable institutions, and a strong sense of identity. Physical assets, such as infrastructure and means of transportation (including transportation of water) were not perceived to have worsened in 2010 compared to 1990. These findings show that the types of assets that the Bedouin possess are highly sensitive to climate change impacts, particularly drought. Therefore, adaptation initiatives that protect these assets would increase climate resilience.

Analytical work done for this report suggests that the poor people in Syria, Tunisia, and the Republic of Yemen suffer more than the nonpoor from climate change, and that rural households (farm and nonfarm) have been the group hardest hit by these adverse effects. In Syria, the lowest two income quintiles are projected to lose an accumulated US$1.2 billion (59 percent of 2010 GDP) as a consequence of climate change by 2050. Rural nonfarm households, the poorest group, are projected to lose US$3.5 billion in the Republic of Yemen (15 percent of 2010 GDP). In Tunisia, farmers are the group hardest hit by climate change, with losses of US$700 million (3 percent of 2010 GDP). This high level of vulnerability of the poor can be explained by the joint effect of (a) being net food buyers—those who spend a high percentage of their income on food—and (b) earning incomes from climate-sensitive productive strategies, namely unskilled farm labor. Although in general the poor suffer most, the three case studies show interesting differences between countries, depending on levels of household income and consumption structures. The most climate-affected household group in the Republic of Yemen is rural nonfarm households, whereas the hardest-hit household group in Tunisia is the farmers. However, urban households are also negatively affected by climate change in all three countries, and in Syria they are almost as hard-hit as rural households.

Climate Change Adaptation Is about Reducing Vulnerability

Definitions and Framework

Adaptation is about reducing vulnerability of countries, societies, and households to the effects of climate variability and change. Vulnerability depends not only on the magnitude of climatic stress, but also on the sensitivity and capacity of affected societies and households to cope with such stress (OECD 2009; also see box 1.4).

One conceptual framework for defining vulnerability and adaptation comes from Fay, Block, and Ebinger (2010). That framework is based on the Intergovernmental Panel on Climate Change's (2007) definition of vulnerability and seeks to capture the essence of the different concepts in the literature by defining vulnerability as a function of exposure, sensitivity, and adaptive or coping capacity (see figure 1.3).[13] As described in Fay, Block, and Ebinger (2010, 15), "the advantage of this approach is that it helps distinguish between what is exogenous, what is the result of past decisions, and what is amenable to policy action." This approach can be applied to communities, regions, countries, or sectors—for example, the Australian government applied this framework to agriculture.

Vulnerability is the degree to which a system is susceptible to, or unable to cope with, adverse effects of climate change, including climate variability and extremes. Vulnerability is a function of the character, magnitude, and rate of climate change, and the degree to which a system is

BOX 1.4

Definitions of Climate Change Adaptation

The Intergovernmental Panel on Climate Change defines adaptation as any "adjustment in natural or human systems in response to actual or expected climatic stimuli or their effects, which moderates harm or exploits beneficial opportunities."

The Organisation for Economic Co-operation and Development's Development Assistance Committee (OECD-DAC) defines climate change adaptation projects as those that "reduce the vulnerability of human or natural systems to the impacts of climate change and climate-related risks, by maintaining or increasing adaptive capacity and resilience. This encompasses a range of activities from information and knowledge generation, to capacity development, planning and implementation of climate change adaptation actions." Adaptation reduces the impacts of climate stress on human and natural systems and consists of a multitude of behavioral, structural, and technological adjustments. OECD highlights that timing (anticipatory versus reactive, ex ante versus ex post); scope (short-term versus long-term, localized versus regional); purposefulness (autonomous versus planned, passive versus active); and agent of adaptation (private versus public, societies versus natural systems) are important concepts when addressing adaptation.

Photograph by Dorte Verner

Sources: Authors' compilation based on OECD 2009 and IPCC 2001.

FIGURE 1.3

Conceptual Framework for Defining Vulnerability

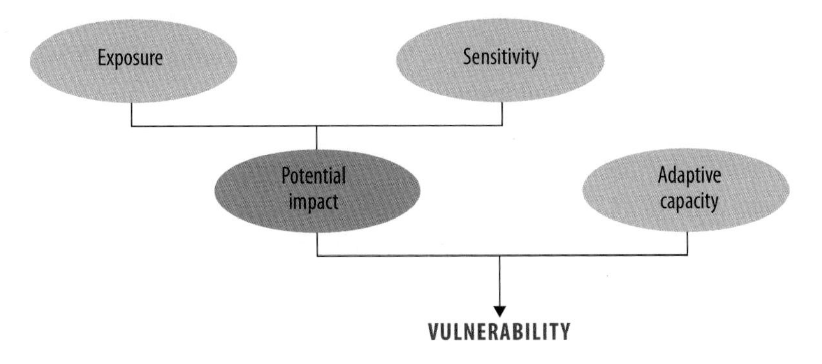

Source: IPCC 2001 (as presented in Fay, Block, and Ebinger 2010).

exposed, along with its sensitivity and adaptive capacity. Vulnerability increases as the magnitude of climate change exposure or sensitivity increases, and decreases as adaptive capacity increases (IPCC 2007).

The potential impact of climate variability and change on a community or sector depends on exposure and sensitivity (see figure 1.3). Exposure is determined by the type, magnitude, variability, and speed of the climate event, such as the changing onset of rains, minimum and maximum winter and summer temperatures, heat waves, floods, and storms. These are the exogenous factors.

Sensitivity is the degree to which a system can be affected by changes in the climate, and depends in part on how stressed the system already is. Poor people and communities will be more affected than the nonpoor, which may already face stresses before a climate event. With limited assets, the poor are inherently more sensitive to even minor climate events. These are the endogenous factors.

Vulnerability depends on the severity of the potential impact and the adaptive capacity of an affected community.[14] The capacity of a system or community to adapt is determined by access to information, technology, economic resources, and other assets. A system's adaptation capacity depends, moreover, on the community's having the skills to use this information, the institutions to manage these assets, and the equitable distribution of resources. In general, societies with relatively more equitable resource distribution will be better able to adapt than societies with less equitable distribution. This difference is because adaptive capacity avoids resource capture, corruption, and clientelism. The level of adaptive capacity tends to be positively correlated with levels of development: more developed societies tend to have more adaptive capacity (OECD 2009).

Climate Change Awareness Varies within and among Countries

Governments and policy makers in Arab countries are aware that climate change is happening, and some countries have taken action to address the issue. For example, many countries have prepared a National Adaptation Programme of Action and included climate change issues in their Country Partnership Strategy (see chapter 9). In 2009, a pan-Arab survey on climate change awareness was carried out through the region's media outlets and online. Responses from highly educated populations showed that 98 percent believe that the climate is changing and 89 percent believe that it is the result of human activity (Tolba and Saab 2009). Moreover, 84 percent reported that climate change poses a serious challenge to their country. The findings also revealed that people obtain most of their information from international media. Of the respondents, 51 percent answered that they disagree with the statement: "My government is acting well to address climate change." The responses varied by subregion: 59 percent disagreed in the Mashreq, 49 percent in Arab Africa, 44 percent in the Gulf, and 38 percent in the Republic of Yemen. However, because these surveys were conducted only in media sources that tend to be accessed by the most educated segments of the population, the results do not reflect the views of the less educated and impoverished.

Climate Change Adaptation Should Be an Integrated Part of Public Sector Management for Sustainable Development

Many countries, particularly the poorest and most exposed, will need assistance in adapting to the changing climate. They urgently need help in preparing for drought, managing water resources, addressing the impacts from rising sea levels, improving agricultural productivity, containing disease, and building climate-resilient infrastructure.

How to adapt to climate change is mostly a sovereign decision of individual countries, which includes governments, the private sector, and civil society. It is in each country's own best interest to build climate resilience and be as prepared as possible for the known and unknown consequences of climate change. This section provides a simple sketch of a framework for an integrated government adaptation process. This resiliency-building approach, which is based on findings from chapters 2 through 8 and takes into account regional characteristics, is developed in greater detail in chapter 9.

The prospect of climate change adds another element to be integrated into national planning. Governments, with assistance from the private sector and civil society, can ensure that a country's development policies, strategies, and action plans build resilience to a changing climate. As this report shows, an integrated approach to climate change adaptation at the country level calls for leadership, action, and collaboration and requires sound strategies for identification, integration, and implementation. Moreover, strategies need to be supported by legislation and action plans, including the necessary frameworks and a strong domestic policy. If they are not, these strategies can result in incoherent outcomes and maladaptation.

This report aims to provide guidance to policy makers on how to address climate change adaptation. The Framework for Action on Climate Change Adaptation introduced in this chapter highlights that adaptation is a long-term, dynamic, and iterative process that will take place over decades. Governments will need to make decisions despite uncertainty about how both society and climate will change and revise their adaptation strategies and activities as new information becomes available. Many standard decision-making methodologies are inappropriate, and alternative, evidence-based methods for selecting priorities within an adaptive management framework will be more effective.

The elements in an adaptive management model include (a) management objectives that are regularly revisited and revised accordingly, (b) a model or models of the system being managed, (c) a range of management choices, (d) the monitoring and evaluation of outcomes, (e) a mechanism for incorporating learning into future decisions, and (f) a collaborative structure for stakeholder participation and learning. This model is commonly used in many fields, but is particularly used in environmental policy.

In addition, and complementary to an adaptive management approach, the OECD (2009) has highlighted five enabling conditions that support successful integration of climate change adaptation into development processes. These conditions help to ensure that multiple perspectives are brought into the policy decision process, thereby ensuring that the policy solutions that are tried are based on evidence and are in line with an inclusive management approach. These enabling conditions are as follows:

- A broad and sustained engagement with, and participation of, stakeholders, such as government bodies and institutions, communities, civil society, and the private sector
- A participatory approach with legitimate decision-making agents
- An awareness-raising program on climate change, designed for households, civil society organizations, opinion leaders, and educators
- Information gathering to inform both national and local adaptation decisions

- Response processes to short- and long-term climatic shocks

The Adaptation Pyramid (figure 1.4) provides a framework to assist stakeholders in Arab countries in integrating climate risks and opportunities into development activities. It is based on an adaptive management approach but also highlights the importance of leadership, without which adaptation efforts are unlikely to achieve the actions necessary to minimize the impacts of climate change.

The base of the pyramid represents the four iterative steps that form the foundation for sound decision making related to climate change adaptation:

1. Assess climate risks, impacts, and opportunities for action.
2. Prioritize policy and project options.
3. Implement responses in sectors and regions.
4. Monitor and evaluate implementation, then reassess the climate risks, impacts, and opportunities.

The arrows on the four sides of the pyramid highlight the iterative nature of adaptation decision making. Adaptation is a continuous process that takes place over time, and adaptation activities will be subject to revision as new information becomes available. To this process is added the apex of leadership. The base of the pyramid is elaborated in the next sections.

FIGURE 1.4

The Adaptation Pyramid: A Framework for Action on Climate Change Adaptation

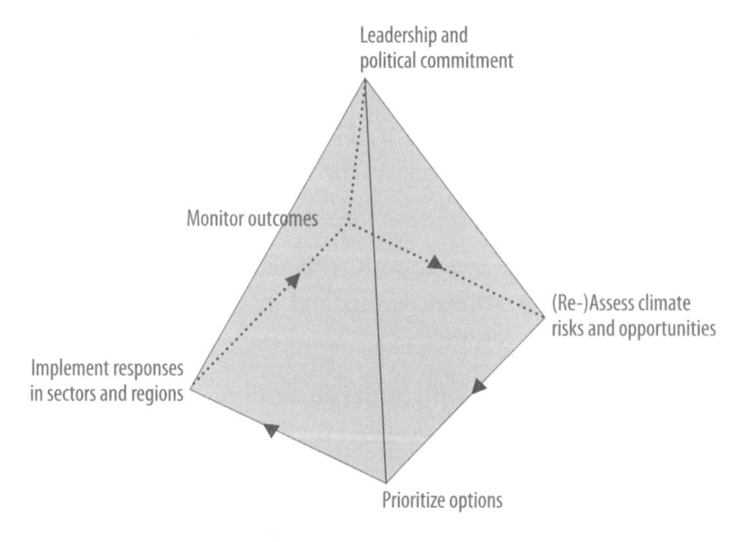

Source: Authors' representation.

Step 1: Assess Climate Risks, Impacts, and Opportunities

In this first step, a wide range of quantitative and qualitative analyses could be used.[15] All of these tools rely on access to climate and socioeconomic data for information on climate impacts, including on vulnerable groups, regions, and sectors. Understanding the risks and impacts requires that practitioners have data on current climate variability and change as well as projections and uncertainty about the future climate. Similarly, information on past adaptive actions and on coping strategies needs to be gathered and evaluated in light of the changing climate. Chapter 9 goes into more detail about this adaptive process. Approaches to this step should be adjusted to suit the issues, locality, and circumstances.

Step 2: Prioritize Options

The second step is to identify and prioritize adaptation options within the context of national, regional, and local priorities and goals and, in particular, the financial and capacity constraints. Expectations of climate change make the consideration of longer-term consequences of decisions more important, as short-term responses may miss more efficient adaptation options or even lead to maladaptive outcomes, such as the further development of highly vulnerable locations. An effective approach is robust decision making (see chapter 9), which allows practitioners to identify choices that provide acceptable outcomes under many feasible scenarios of the future.

Step 3: Implement Responses in Sectors and Regions

Adaptive responses will often be somewhat at odds with immediate, local priorities, and thus the third step of implementing the agreed-on responses requires cooperation and understanding at national, sectoral, and regional and local levels (often jointly). At the national level, adaptation needs to be integrated into national policies, plans, and programs and into financial management systems, including the five-year plans prepared in a number of Arab countries. Moreover, it includes sustainable development and poverty reduction strategies and plans; policies, regulations, and legislation; investment programs; and the budget. In addition, national adaptation strategies can guide the mainstreaming of adaptation into other national policies as well as implementation at sectoral and local levels. This integration could involve the formation of an interministerial committee at various levels with participation of the private sector, academia, and civil society.

Climate change needs to be considered in all sectoral activities, particularly in climate-sensitive sectors such as water, agriculture, tourism, and health (see chapters 3, 4, 6, and 8). Sectoral plans and strategies must take intersectoral effects into account. For example, water is one of the key sectors for successful adaptation in Arab countries, but any policies concerning water management will affect agriculture and energy (irrigation), city planning (drinking water and wastewater), gender (time spent by women and girls collecting water), and health (waterborne diseases).

The local level is ultimately the level at which climate change impacts will be felt and responded to. This report presents key issues and possible policy solutions at two levels: rural and urban (see chapters 4 and 5). Rural livelihoods tend to be anchored in climate-sensitive sectors such as agriculture, whereas urban livelihoods tend toward service sectors. Changes in agricultural productivity attributable to climate change, together with population growth, may increase food prices for those who do not produce food and may even disrupt food supplies. Changes in agricultural productivity may also accelerate the rural to urban migration, often initiated by men, from rural areas. This out-migration creates challenges for rural women left behind and puts pressure on services in rapidly expanding urban areas.

Step 4: Monitor Outcomes

Monitoring is essential to ensure that adaptation-related strategies and activities have the intended adaptation outcomes and benefits. Comprehensive qualitative and quantitative indicators can help proponents of adaptation projects recognize the strengths and weaknesses of various initiatives and adjust activities to best meet current and future needs. The monitoring framework should explicitly consider the effects of future climate change, particularly for projects with a long time horizon.

Monitor and Evaluate While Reassessing the Climate Risks, Impacts, and Opportunities

The Adaptation Pyramid is an iterative process; hence, the next step will be to reassess activities, taking into account new and available information, such as revised projections about future climate change or the effectiveness of previously applied solutions.

Make Leadership Central to Successful Adaptation

Effective climate change adaptation will not occur without strong leadership. International experience shows that leadership needs to be initiated at the national level by a prominent ministry or senior government champion, such as the prime minister, minister of planning or economy, or state planning commission. This champion will also require the support of a strong team composed of representatives of relevant ministries, governorates, local authorities and institutions, the private sector, academia, and civil society organizations, and ideally will involve members from opposition parties to ensure continuity through a changing government. Clearly, this leadership team would reflect the context of individual Arab countries and circumstances (see box 1.5).

Leaders are also needed at other levels of government, from all sectors, and within civil society and private sector organizations. Those leaders need support through information access and educational opportunities, and they must be treated as legitimate agents in decision-making processes. For example, in 2009, the Republic of Yemen created an interministerial panel for climate change adaptation chaired by the deputy prime minister, with ministers from 13 key ministries and other relevant actors. Moreover, the leadership must interact with other states with regard to intergovernmental issues (for example, riparian states sharing the water flow of the Euphrates, Khabur, or Nile Rivers). Thus, regional and international organizations, such as the United Nations Framework Convention on Climate Change (UNFCCC), will also play an important role.[16]

This book is about climate change, its impacts on people and on the systems on which they depend, and ways people might adapt to climate change. The book highlights a number of issues and areas that are being affected by climate change. One important message of the book is that climate change should be taken into account in all activities, but the book cannot provide solutions or options for all issues. For example, transboundary water issues are already being addressed by international task forces, and this book can deal only with how climate change might affect their decisions. However, anticipation of climate change can be the stimulus for improving interventions and speeding action as has been seen in other countries such as in Australia, where water laws and management were extensively changed in response to a prolonged drought and the anticipation of further climate change.

By nature, adaptation to climate change is a dynamic process, and so is the governance of adaptation. Political change, including changes originating from the Arab Spring, can provide an opportunity to increase civil society participation in adaptation governance and a move toward a more inclusive approach to addressing climate change issues and building climate resilience.

Notes

1. The members of the League of Arab States are Algeria, Bahrain, the Comoros, Djibouti, the Arab Republic of Egypt, Iraq, Jordan, Kuwait, Lebanon, Libya, Mauritania, Morocco, Oman, Qatar, Saudi Arabia, Somalia, Sudan, the Syrian Arab Republic, Tunisia, United Arab Emirates, the West Bank and Gaza, and the Republic of Yemen.

BOX 1.5

Jordan Desert Ecosystems and Livelihoods Project

The goal of a partnership between the World Bank's Middle East and North Africa Region and the Jordan Desert Ecosystems and Livelihoods Project (DELP) is to contribute to the enhancement of livelihoods in desert ecosystems by harnessing their value in an environmentally and socially sustainable manner so that the flow of desert goods and services can be optimized. The program, which includes projects in Algeria, Egypt, Jordan, and Morocco, will promote deserts in the region as ecosystems of major importance, offering a full range of intrinsic values, through unique and highly adaptive services.

The Jordan DELP is one of four projects under the umbrella of the program. Its goal is to sustain ecosystem services and livelihoods in four poverty pockets in the Badia region through diversification of community income sources; preservation and sustainable use of natural and rangeland resources; and capacity enhancement of target stakeholders and beneficiaries. Main project activities will include (a) supporting sustainable rangeland rehabilitation and livelihoods in the southern Badia and (b) promoting community-centered ecotourism and resource use in the northern Badia, along with complementary capacity-building, awareness-raising, and knowledge management activities.

The Jordan DELP will focus on four poverty pockets in southern and northern Badia. In the south, an area that comprises the poverty pockets of Al Jafr, Al Husseinieh (subdistricts within the Ma'an governorate), and Deisa (Aqaba governorate) will be targeted; in the north, ecotourism activities will be implemented along a route that begins in Al Azraq (southeast of Amman) and ends in the Burqu protected area in the east, targeting the communities in the Ar Ruwayshid poverty pocket. To enhance the returns on investment, the route will link to Al Azraq and Shaumari—with important and already established protected areas and ecolodge facilities—to offer an ideal itinerary. Jordan's Badia region makes up 80 percent of the country's territory and has about 6.5 percent of the population. Bedouin communities live and practice seasonal livestock browsing in the Badia, which also contains significant and unique habitats of global importance and supports a number of endangered species.

During the past 20 years, anthropogenic pressures, mainly overgrazing and speculative agricultural and mining initiatives, as well as climate change impacts, have severely degraded the land and the Badia's unique biodiversity. The Bedouin in Jordan are the main custodians of the Badia ecosystem and also the main resource users. Therefore, restoration and preservation of the Badia's degraded ecosystem services need to go hand in hand with improvements in the Bedouin's livelihoods. The Jordan project is responsive to Global Environment Fund strategies and priorities under the Biodiversity and Land Degradation focal areas.

Source: Authors' compilation.

2. IPCC (2007) points out that there are "sharp differences across regions, and those in the weakest economic position are often the most vulnerable to climate change and are frequently the most susceptible to climate-related damages, especially when they face multiple stresses. There is increasing evidence of greater vulnerability of specific groups." IPCC (2007) makes specific mention of traditional peoples and ways of living only in the cases of polar regions and small island states.
3. In 2003, the total population of the region reached 305 million (4.7 percent of the world's population). The population grew at an annual average rate of 2.6 percent in the past two decades (Tolba and Saab 2009). In the Arab countries, around 85 million people live on less than US$2 per day—that is, 30 percent of the region's total population in 2000.
4. The Arab countries have contributed less than 5 percent of the total accumulated carbon dioxide in the atmosphere.
5. The yearly median value is 403 cubic meters per capita.
6. The highest temperature was measured in Pakistan (53.5°C, or 128.3°F); the other four Arab countries were Iraq (52.0°C), Saudi Arabia (52.0°C), Qatar (50.4°C), and Sudan (49.7°C). See http://www.wunderground.com/blog/JeffMasters/comment.html?entrynum=1831. Oman, specifically Khasab Airport, recorded a new world high minimum temperature with a scorching 41.7°C (107.0°F) low on June 23, 2011.
7. Only Cyclone Gonu of 2007, a category 5 storm, was a stronger Arabian Sea cyclone, killing about 50 people in Oman, with damage estimated at roughly US$4.2 billion.
8. This report does not address in detail topics that are being addressed extensively in other ongoing work in the Middle East and North Africa Region of the World Bank, such as climate migration.
9. The Western term *Bedouin* is actually a double plural. In the Arabic language, the people known as Bedouin refer to themselves as *Bedu* (also plural). This report uses Bedouin because it is the recognizable English language term. The definition of who is and who is not a Bedouin has become somewhat confused in recent times, as circumstances have changed and the desert herders have had to adapt their traditional nomadic life. Generally speaking, a Bedouin is an Arab who lives in one of the desert areas of the Middle East and raises camels, sheep, or goats. The Bedouin traditionally believe they are the descendants of Shem, son of Noah, whose ancestor was Adam, the first man (according to the book of Genesis of the Bible). Bedouin are considered by some to be the "most indigenous" of modern Middle Eastern peoples (http://www.everyculture.com/wc/Rwanda-to-Syria/Bedu.html, March 5, 2011), meaning that they lived in the region before anyone else. The first appearance of nomadic peoples in the Arabian Desert can be traced back as far as the third millennium BCE.
10. Somalia's annual GDP per capita is US$600 (purchasing power parity, 2010 estimates) according to the CIA World Factbook (http://ciaworldfactbook.us/home).
11. A drought is defined as an extended period (months to years) during which a region receives consistently below-average precipitation, leading to low river flows, reduced soil moisture, and thus adverse effects on agriculture, ecosystems, and the economy.

12. These analyses were conducted by the authors on the basis of a modeling suite, including downscaling of global climate models, crop modeling, global economic modeling, and subnational-level economic modeling.
13. An overview of adaptation frameworks is given in Füssel 2009.
14. Sensitivity and adaptive capacity is usually inversely correlated, as shown for Eastern European countries and Central Asia in Fay, Block, and Ebinger (2010).
15. For an overview of available tools to assist in climate risk analysis, see http://climatechange.worldbank.org/climatechange/content/note-3-using-climate-risk-screening-tools-assess-climate-risks-development-projects.
16. The UNFCCC is the ideal political forum for reaching agreement on international action on climate change. Fully meeting the challenges of climate change will require action at many levels and through many channels.

References

Allison, Ian, Nathaniel Bindoff, Robert Bindschadler, Peter Cox, Nathalie de Noblet-Ducoudré, Matthew England, Jane Francis, Nicolas Gruber, Alan Haywood, David Karoly, Georg Kaser, Corinne Le Quéré, Tim Lenton, Michael Mann, Ben McNeil, Andy Pitman, Stefan Rahmstorf, Eric Rignot, Hans Joachim Schellnhuber, Stephen Schneider, Steven Sherwood, Richard Somerville, Konrad Steffen, Eric Steig, Martin Visbeck, and Andrew Weaver. 2009. *The Copenhagen Diagnosis: Updating the World on the Latest Climate Science*. Sydney: University of New South Wales Climate Change Research Centre.

Chomiz, Kenneth. 2011. "Climate Change and the World Bank Group: Climate Adaptation." Independent Evaluation Group, World Bank, Washington, DC.

Diaz, Estebancio Castro. 2008. "Climate Change, Forest Conservation, and Indigenous Peoples' Rights." Paper presented at the International Expert Group Meeting on Indigenous Peoples and Climate Change, Darwin, Australia, April 2–4.

Fay, Marianne, Rachel I. Block, and Jane Ebinger. 2010. *Adapting to a Climate Change in Eastern Europe and Central Asia*. Washington, DC: World Bank.

Füssel, Hans-Martin. 2009. "An Updated Assessment of the Risks from Climate Change Based on Research Published Since the IPCC Fourth Assessment Report." *Climatic Change* 97 (3): 469–82.

Gerritsen, Rolf. 2008. "Constraining Indigenous Livelihoods and Adaptation to Climate Change in SE Arnhem Land, Australia." Paper presented at the International Expert Group Meeting on Indigenous Peoples and Climate Change, Darwin, Australia, April 2–4.

IPCC (Intergovernmental Panel on Climate Change). 2001. *Climate Change 2001: Impacts, Adaptation, and Vulnerability—Working Group II Contribution to the Third Assessment Report of the Intergovernmental Panel on Climate Change*. Cambridge, U.K.: Cambridge University Press.

———. 2007. *Climate Change 2007: The Fourth Assessment Report of the Intergovernmental Panel on Climate Change*. Cambridge, U.K.: Cambridge University Press.

Kronik, Jakob, and Dorte Verner. 2010. *Indigenous Peoples and Climate Change in Latin America and the Caribbean*. Washington, DC: World Bank.

Mearns, Robin, and Andrew Norton. 2009. *The Social Dimension of Climate Change: Equity and Vulnerability*. Washington, DC: World Bank.

OECD (Organisation for Economic Co-operation and Development). 2009. "Integrating Climate Change Adaptation into Development Co-operation." Paris: OECD.

Parry, Martin L., Oswaldo F. Canziani, Jean P. Palutikof, Paul J. van der Linden, and Clair Hanson, eds. 2007. *Climate Change 2007: Impacts, Adaptation, and Vulnerability—Working Group II Contribution to the Fourth Assessment Report of the Intergovernmental Panel on Climate Change*. Cambridge, U.K.: Cambridge University Press.

Smith, Kimberly. 2008. "Climate Change on the Navajo Nations Lands." Paper presented at the International Expert Group Meeting on Indigenous Peoples and Climate Change, Darwin, Australia, April 2–4.

Stern, Nicholas. 2007. *Stern Review on the Economics of Climate Change*. Cambridge, U.K.: Cambridge University Press.

Sulyandziga, Rodion. 2008. "Indigenous Peoples of the North, Siberia, and Far East and Climate Change: From Participation to Policy Development and Adaptation Measures—Challenges and Solutions." Paper presented at the International Expert Group Meeting on Indigenous Peoples and Climate Change, Darwin, Australia, April 2–4.

Tauli-Corpuz, Victoria, and Aqqaluk Lynge. 2008. "Impact of Climate Change Mitigation Measures on Indigenous Peoples and on Their Territories and Lands." United Nations, Permanent Forum of Indigenous Peoples, Seventh Session, New York.

Tolba, Mostafa K., and Najib W. Saab. 2009. "Arab Environment Climate Change: Impacts of Climate Change on Arab Countries." Arab Forum for Environment and Development, Beirut.

United Nations Development Programme (UNDP). 2011. *Human Development Report 2011: Sustainability and Equity—A Better Future for All*. New York: Palgrave Macmillan. http://hdr.undp.org/en/reports/global/hdr2011/.

Universitat Autònoma de Barcelona. 2010. "Experts Will Be Studying Political and Social Conflicts Caused by Climate Change and the Fight for Water in Eleven Regions of Europe and Africa." Press release, February 24. http://www.alphagalileo.org/ViewItem.aspx?ItemId=69107&CultureCode=en.

Verner, Dorte, ed. 2010. *Reducing Poverty, Protecting Livelihoods, and Building Assets in a Changing Climate: Social Implications of Climate Change Latin America and the Caribbean*. Washington, DC: World Bank.

Weiss, Harvey, and Raymond S. Bradley. 2001. "What Drives Societal Collapse." *Science* 291 (5504): 609–10.

World Bank. 2010. *World Development Report 2010: Development and Climate Change*. Washington, DC: World Bank.

———. 2011a. "Syria Rural Development in a Changing Climate: Increasing Resilience of Income, Well-Being, and Vulnerable Communities." Report 60765-SY-MENA, World Bank, Washington, DC.

———. 2011b. *World Development Indicators 2011*. Washington, DC: World Bank. http://hdr.undp.org/en/media/HDR_2010_EN_Table1_reprint.pdf.

CHAPTER 2

Ways Forward for Climatology

Although mostly arid or semiarid, the Arab region encompasses a diversity of climates, including temperate zones in the northern and higher elevations of the Maghreb and Mashreq, tropical ocean climates in the Comoros, and varied coastal climates along the Arabian, Mediterranean, and Red Seas; the Gulf of Arabia; and the Indian Ocean. Climate change is expected to have different impacts in these different climate zones.

To date, the Arab region has not been addressed as a discrete region in climate change research assessments, such as in the Intergovernmental Panel on Climate Change (IPCC) reports. Typically, information must be inferred from analyses carried out in other regions. Recent literature from regional studies confirms the broad conclusions of the IPCC Fourth Assessment Report (AR4) regarding increasing temperatures and mostly reduced rainfall, but sometimes differs regarding the details. Although models generally agree on rainfall decrease in the Maghreb and Mashreq and on rainfall increase around the Horn of Africa, they agree less frequently on the already dry central part of the Arabian Peninsula. The future climate in this region will depend in large part on the position of the Inter-Tropical Convergence Zone (ITCZ) (see box 2.1). All models project that it will move further northward but disagree on the precise displacement and location of that shift, although at least two-thirds of the models used in the IPCC AR4 indicate that the Horn of Africa and the southernmost part of the peninsula is projected to receive more rainfall. Much of the region also falls into the transitional zone between areas with projected decreases in rainfall and those with projected increases. Because models differ as to the precise location of that transition, it is difficult to project the exact magnitude of rainfall changes.

Photograph by Dorte Verner

BOX 2.1

Some Basic Definitions

Climate scenario: A climate scenario is a plausible and often simplified representation of the future climate, based on an internally consistent set of climatological relationships that has been constructed for explicit use in investigating the potential consequences of anthropogenic climate change, often serving as input to impact models. Climate projections often serve as the raw material for constructing climate scenarios, but climate scenarios usually require additional information, such as information about the observed current climate. A climate change scenario is the difference between a climate scenario and the current climate.

Downscaling: The accuracy and representativeness of climate model data, among other factors, depend on using a high enough resolution to be able to represent features of interest adequately. Global circulation models (GCMs) cannot deliver this resolution because of computer constraints. Therefore, high-resolution limited-area regional climate models (RCMs) are used to generate climate change scenarios at a higher resolution. They are driven by boundary conditions taken and interpolated from the GCM. In a "nudging" zone of typically a few grid points in the RCM grid, the GCM values are relaxed toward the RCM grid. Inside this transition zone, the meteorological fields are RCM generated. Every few (typically six) hours, new boundary conditions are generated from the GCM data. In other words, the RCM can generate its own structures inside the domain, but the large-scale circulation depends on the driving GCM. Many modeling centers around the world follow this approach; however, it requires substantial computer resources and is comparatively slow.

North Atlantic Oscillation (NAO): The North Atlantic Oscillation consists of opposing variations of barometric pressure between areas near Iceland and near the Azores. It affects the strength and position of the main westerly winds across the Atlantic into Europe and the Mediterranean. When the pressure difference is high (NAO+), the westerlies are stronger and track more to the north, leading to cool summers and mild wet winters in Europe, but causing drier conditions in the Mediterranean. In the opposite phase, the westerlies and the storms they bring track farther south, leading to cold winters in Europe, but more storms in the Mediterranean and more rain in North Africa.

Inter-Tropical Convergence Zone (ITCZ): The Inter-Tropical Convergence Zone is an equatorial zonal belt of low pressure near the equator where the northeast trade winds meet the southeast trade winds. As these winds converge, moist air is forced upward, resulting in a band of heavy rainfall. This band moves seasonally. In Africa, it reaches its northernmost position in summer and also interacts with the Indian monsoon, bringing rains to the southern part of the Arab region (southern Sahel), but the northward extent varies from year to year, making both conventional weather forecasting and climate modeling difficult.

Storm surge: Storm surge is a rise of the seawater above the normal level along a

BOX 2.1 *Continued*

shore associated with a low-pressure weather system, typically tropical cyclones and strong extratropical cyclones. It is the result of both the low pressure at the center of the storm raising the ocean's surface, as well as the wind pushing the water in the direction the storm is moving. The storm surge is responsible for most loss of life in tropical cyclones worldwide.

Flash floods: Flash floods usually refer to rapid flooding that happens very suddenly, usually without advance warning. They are different from regular floods, in that they often last less than six hours. Flash floods, with intense rainfall, normally occur in association with the passage of a storm or tropical cyclone, especially when the rain falls too quickly on saturated soil or dry soil that has poor absorption capacity. Flash floods may also refer to a flooding situation when barriers holding back water fail, such as the collapse of a natural ice or debris dam or a human-made dam.

Tropical cyclone: A tropical cyclone is a storm system characterized by a large low-pressure center and numerous thunderstorms that produce strong winds and heavy rain. Tropical cyclones strengthen when water evaporated from the ocean is released as the saturated air rises, resulting in condensation of water vapor contained in the moist air. They are fueled by a different heat mechanism than other cyclonic windstorms such as nor'easters and European windstorms. The characteristic that separates tropical cyclones from other cyclonic systems is that at any height in the atmosphere, the center of a tropical cyclone will be warmer than its surroundings, a phenomenon called *warm-core storm systems*. The term *tropical* refers to both the geographic origin of these systems, which form almost exclusively in tropical regions of the globe, and their formation in maritime tropical air masses.

Climate extreme: Usually, both extreme weather events and extreme climate events are referred to as climate extremes. In particular, an extreme heat event (also referred to as a *heat wave*) is a prolonged period of excessively and uncomfortably hot weather, which may be accompanied by high humidity. Following the general definition of an extreme event, there is no universal definition of a heat wave; the term is relative to the usual weather in the area. Temperatures that people from a hotter climate consider normal can be termed a heat wave in a cooler area if they are outside the normal climate pattern for that area. The term is applied both to routine weather variations and to extraordinary spells of heat, which may occur only once a century. Severe heat waves have caused catastrophic crop failures, thousands of deaths from hyperthermia, and other severe damages. Rainfall events of high intensity are often considered extreme, although natural variability may actually account for the event as regularly occurring. Often it is the impact of the event that implies that the event is considered extreme; that is, the occurrence of a value of a weather or climate variable above (or below) a threshold value or near the upper (or lower) end of the range of its observed values.

Source: IPCC 2012.

Although several efforts are under way to improve the availability of downscaled climate change information for the Arab region based on existing global and regional circulation models and scenarios, the international climate change modeling community is moving ahead with new global modeling approaches and climate change scenarios. This effort will include a new set of improved and consistent modeling results that will be available for all regions, with the first results appearing within the next one to two years. How these new scenarios will compare to existing knowledge is still unclear, although preliminary analysis suggests that the results are broadly in line with the results provided in this chapter.

Regional and local climate projections depend on the availability of a good set of observational data to determine current trends and to translate outputs from global climate models to regional scales (downscaling). Unfortunately, many observational and modeling gaps exist in the study of climate change in the Arab region. Under these circumstances, it is important not to fall into the trap of interpolating from global information and applying it to the region, or of jumping from interpretations of long-term projections to statements about the near term. This chapter brings together various pieces of work communicated through the international scientific literature that have been done for the Arab region, which provide an insight into the region's future climate. From that material, this chapter finds that the Arab region will remain predominately arid, with some areas becoming even drier and hotter, but rainfall patterns will change, and the increase in flooding events already being observed is likely to continue in the future.

Despite Sparse Observational Data, the Projections Are That Most of the Arab Region Is Becoming Hotter and Drier

The climate in Arab countries ranges from the Mediterranean, with warm and dry summers and wintertime rainfall, through subtropical zones, with variable amounts of summer monsoon rains, to deserts with virtually no rain. During winter, variability in the North Atlantic Oscillation influences the position of storm tracks; annual variations in rainfall in western and central North Africa (the Maghreb), most of the Mashreq, and the northern part of the Arabian Peninsula are largely governed by this NAO effect. The eastern part of the region (the eastern parts of the Mashreq, the Gulf, and central part of the Arabian Peninsula), where it rains mainly during the winter, is almost without rainfall in summer. The southeast of the region (the Republic of Yemen and Oman) is influenced by the Indian monsoon system, which is largely controlled by the position of the ITCZ (see box 2.1), and therefore has a secondary summer maximum of rainfall. Occasionally, these countries also experience serious consequences of tropical cyclones.

There Is a Scarcity of Meteorological Surface Observation

With a few exceptions, the availability of climate-station data to establish baseline climate across the Arab region is very limited compared to most other parts of the world (map 2.1). This scarcity of data hampers detection of climate change as well as the interpretation of projected changes, because changes must be compared to a verifiable current climate. The rescue of existing but undigitized climate data and the establishment of well-chosen, permanent, high-quality observational sites will be necessary to establish more rigorous models in the future. Map 2.1 does not show all of the station data available in the region because not all data are available through open-access databases. Most of data are withheld, even between government departments, for military or other reasons. As an example, map 2.2 shows rainfall stations in 2007 run by the Republic of Yemen's Ministry of Irrigation and Agriculture. Only a few of these stations are available through open-access databases.

MAP 2.1

Spatial Distribution of Stations with at Least 10 Years of Monthly Rainfall

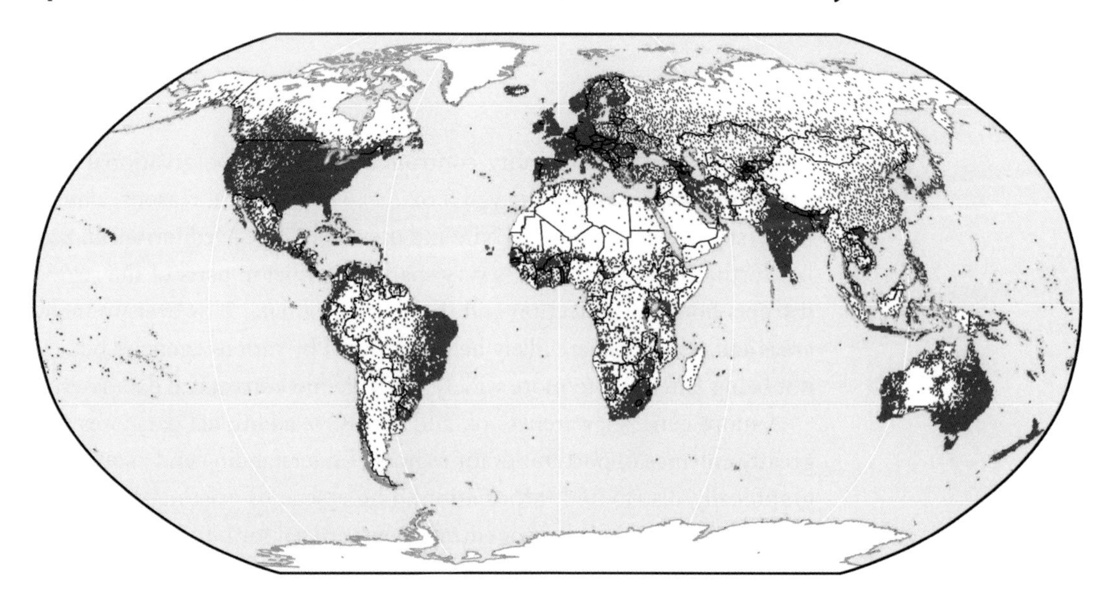

Source: Schneider et al. 2010.

Note: Available in the Global Precipitation Climatology Centre (GPCC) database (global number of stations in June: 64,471).

MAP 2.2

Rainfall Stations in the Republic of Yemen Run by the Ministry of Irrigation and Agriculture, 2007

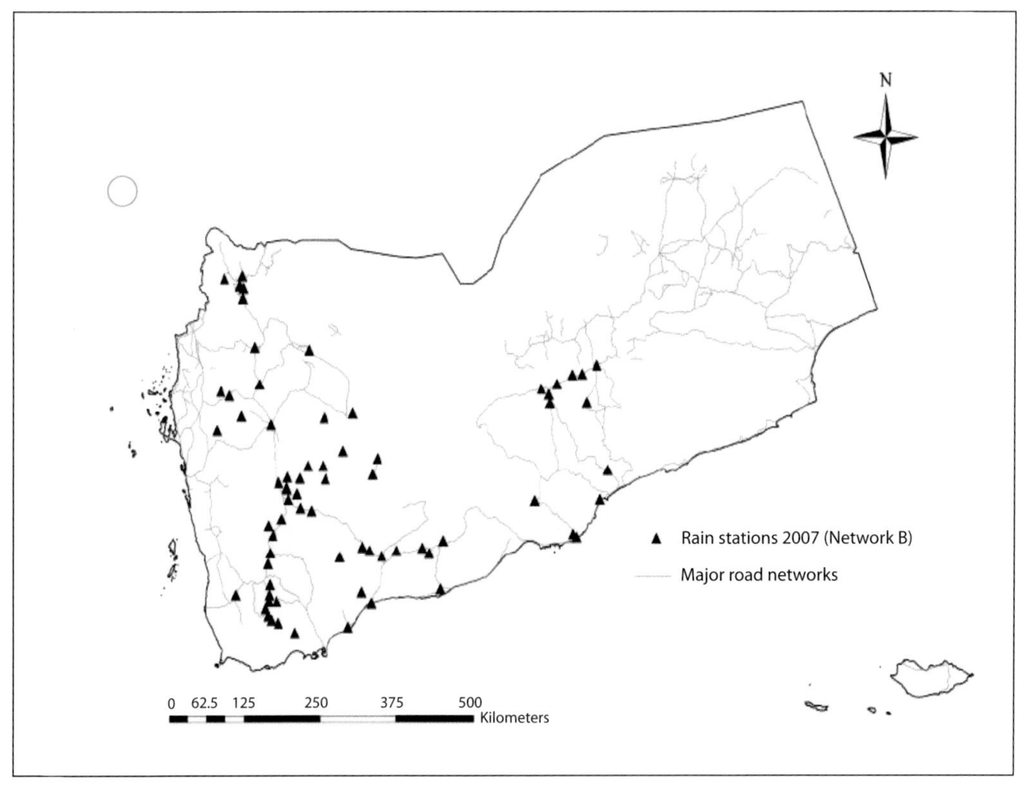

Source: Rob L. Wilby, personal communication with author, 2011.
Note: Data were provided to Wilby after extensive discussions with local authorities.

The distribution of quality-controlled, long-term observational sites within the Arab region is uneven (box 2.2). For historical reasons, a number of stations exist along the Nile and the coast of the Mediterranean Sea, but further inland coverage is very sparse. Conflict in parts of the region disrupts both the collection and the sharing of data. However, in many areas additional data are likely being gathered by various agencies but are not being entered into more widely available meteorological databases.

A more general awareness of, and access to, additional databases will greatly enhance opportunities for regional understanding and use of geographically distributed information. The rescue of existing data, their digitization, and their homogenization are crucial for building capacities to further strengthen mapping and understanding of the baseline climate and ongoing changes within Arab countries (box 2.3).

BOX 2.2

Observational Networks

Meteorological services throughout the region should consider extending and should secure the spatial coverage of their observational networks. This action is necessary for ensuring a minimal station density, which is required to reflect climate variability and possible change. Furthermore, ensuring an adequate network will also be beneficial to the quality of weather forecasting and early warning systems.

Source: Authors' compilation.

BOX 2.3

Access to Data

Meteorological and hydrological institutes or services: Services should provide access to quality-controlled weather and climate data. Often, meteorological services are under the governance of the Ministry of Defense, and the most recent meteorological data might be regarded as sensitive and restricted information; therefore, a civil authority should be in charge of making less recent data (for example, older than one year) available at daily or subdaily frequencies. Along with this, it is important to provide and regularly update a compilation of the data available, the conditions for their use, and the procedures to access the data.

Data rescue and digitization: Promotion of data rescue and digitization of manually archived meteorological data will enhance access and understanding of climate change information at national and regional levels. Digitization must include documentation about the data source, such as station location and possible relocations, instrumentation, environment, and all changes. Such an initiative is also important to safeguard documents, which otherwise may be lost.

Source: Authors' compilation.

Aridity Predominates, but the Arab Region Contains a Wide Range of Climates

As in most of the world, the critical climate variable for human settlement patterns in the Arab region is rainfall. All desert regions receive annual rainfall totals of less than 200 millimeters. The result of combining the available observations through geographic information system (GIS) mapping is presented in map 2.3. The central parts of the Sahara receive less than 50 millimeters. In a region ranging from southwestern Algeria to western Egypt, no rain at all was observed during the 20th century. Mediterranean zones in the north and subtropical regions in the south typically receive above 500 millimeters of rainfall per year, whereas the annual rainfall in southern Sudan and the Comoros is more than 1,000 millimeters. Consequently, most of the Arab region is classified as hot arid desert. As in other arid regions, rainfall varies greatly between years, with the coefficient of variation exceeding 100 percent in the deserts (see map 2.3 and annex 2A). This means that there can be years with little or no rainfall and years in which the rainfall greatly exceeds the average, but also that it is difficult to identify trends in the amount of rainfall.

Mean annual temperature is between 20°C and 25°C in the desert regions, up to 28°C on the Arabian Peninsula, between 15°C and 20°C in the Mediterranean and subtropical zones, and close to 25°C in the tropi-

MAP 2.3

Rainfall in Arab Countries and Year-to-Year Variations

a. Mean annual rainfall

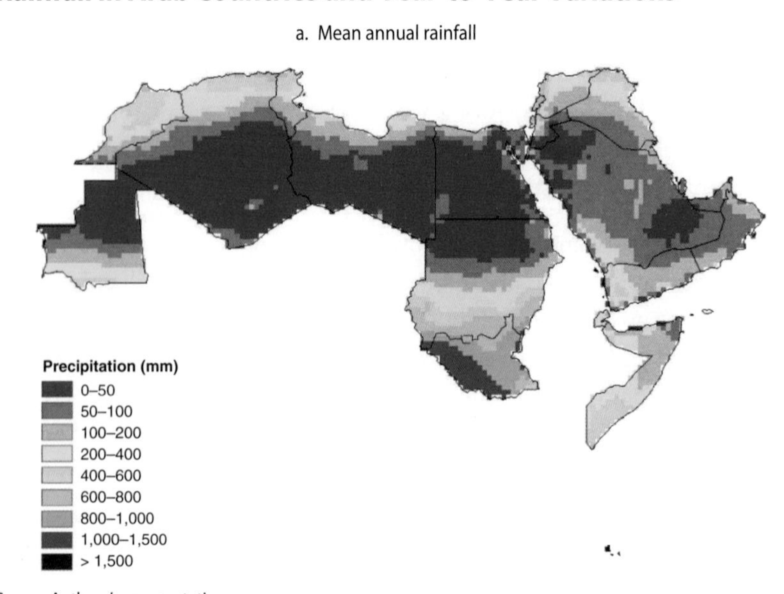

Source: Authors' representation.

Note: Areas with mean annual rainfall of 0 millimeters are shown in white in panel a.

cal regions. In tropical zones, the annual amplitude is very small, whereas variability increases further north. In the mountainous regions of Iraq, Lebanon, and Morocco, it is sometimes even cold enough for occasional snowfall. Often, monthly mean summer temperatures exceed 30°C and in a few places even 35°C (see annex 2A).

Rainfall in most of the Arab countries depends partly on the state of the North Atlantic Oscillation, which is the dominant source of climate variability in the Atlantic-European-Mediterranean and Middle East regions. Its influence can be seen in weather patterns, streamflows, and subsequent ecological and agricultural effects. Cullen et al. (2002) identified two components of Middle Eastern streamflow variability. The first reflects rainfall-driven runoff and explains 80 percent of the variability in river flows from December to March. This component is correlated, on interannual to interdecadal time scales, to the NAO phase (in a positive NAO phase, the climate in the Middle East is cooler and drier than average and vice versa for the negative phase). The second principal component (the so-called Khamsin) is related to spring snowmelt in the mountains and explains more than half of the streamflow variability from April to June. A prevailing positive NAO phase, as in the 1990s and 2000s, can therefore result in drought conditions in the region, including in the Euphrates-Tigris and Jordan River basins.

b. Interannual variability

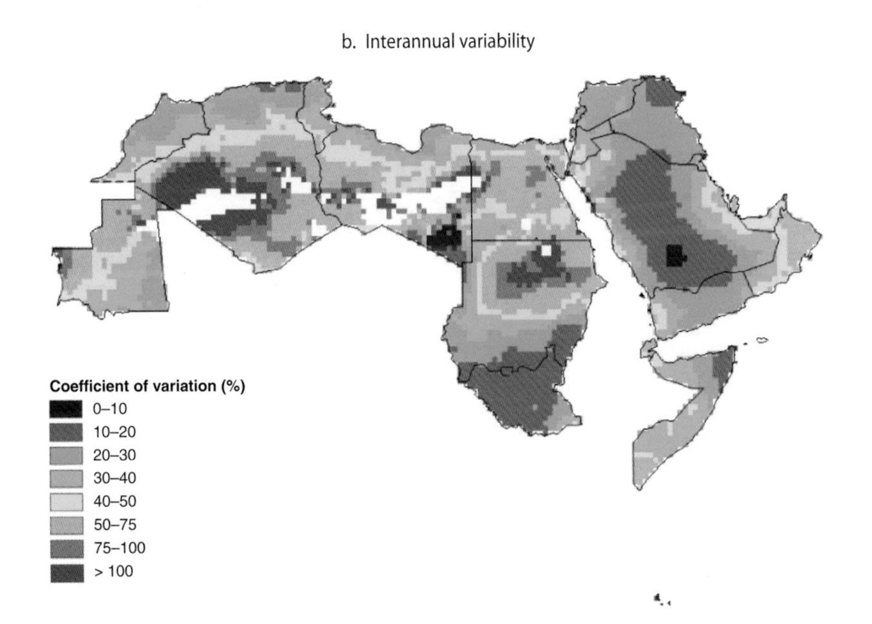

Arab Countries Have Warmed and Most Have Become Drier

Despite some local deviations, the available evidence clearly indicates a warming trend within the past 100 years or more. In a publically available data set (GHCN2), there are 119 stations in the Arab region that can be regarded as having reliable data for the period 1961–90 (map 2.4, panel a). Most of them show an overall temperature increase of 0.2°C to 0.3°C per decade, mainly in the Maghreb and in Sudan. In the Mashreq, a number of stations, often in larger cities, show a temperature decrease for 1961–90, notably Damascus (Syrian Arab Republic) and Kuwait City (Kuwait). For the more recent period, 1991–2011, however, almost all 91 stations with data have a significant positive trend of 0.3°C to 0.4°C per decade (map 2.4, panel b). Inclusion of stations with larger data gaps (map 2.4, panels c and d) changes this result only slightly. The fact that temperatures rose only slightly or even decreased prior to 1990 may be related to the gener-

MAP 2.4

Positive Temperature Trends Seen in the Majority of Available Ground Stations

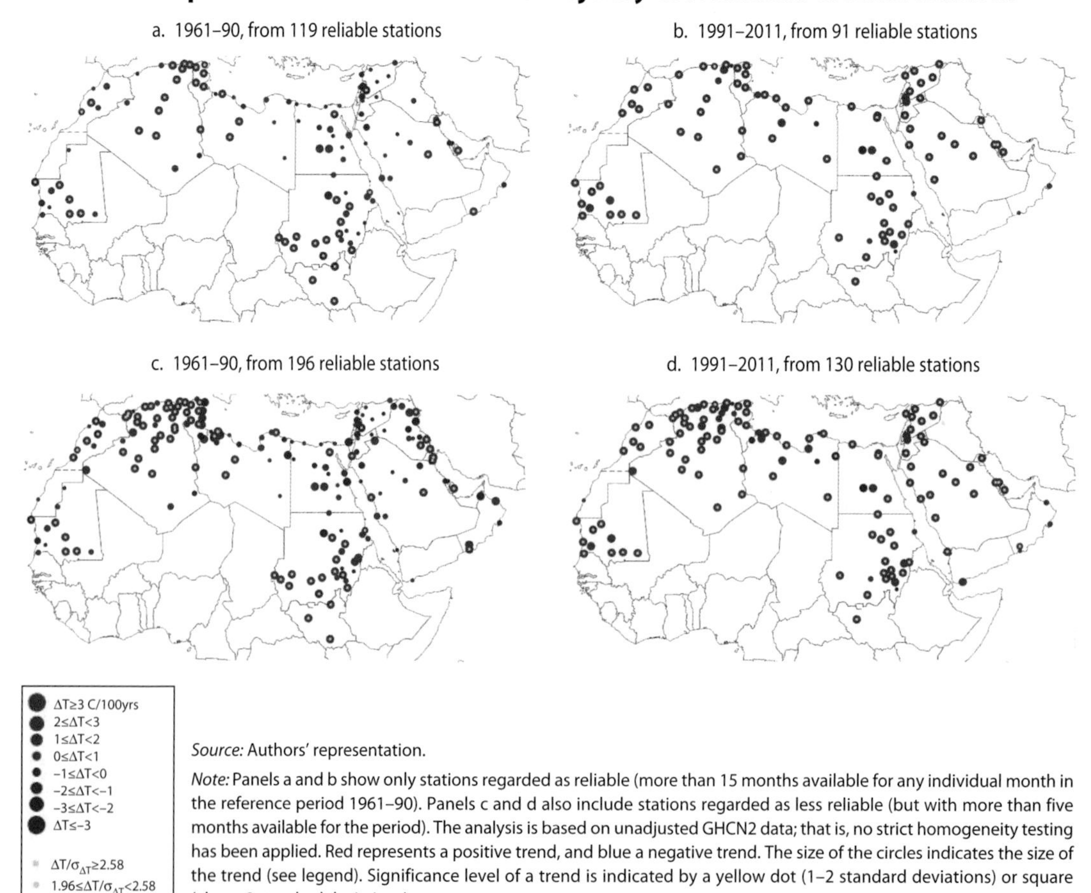

Source: Authors' representation.

Note: Panels a and b show only stations regarded as reliable (more than 15 months available for any individual month in the reference period 1961–90). Panels c and d also include stations regarded as less reliable (but with more than five months available for the period). The analysis is based on unadjusted GHCN2 data; that is, no strict homogeneity testing has been applied. Red represents a positive trend, and blue a negative trend. The size of the circles indicates the size of the trend (see legend). Significance level of a trend is indicated by a yellow dot (1–2 standard deviations) or square (above 2 standard deviations).

ally positive NAO phase during this period. The few stations with longer records suggest that rainfall has decreased over the past century.

During recent decades, a wealth of satellite information that can augment ground observations has become available (map 2.5). Improved capacity to use existing international programs and take part in the design of future programs for satellite retrievals and data products therefore appears to be a necessity (box 2.4). Even with all available ground-based observations at hand, the spatial coverage will remain limited throughout the region. As a result, no systematic analyses or verification of the present climate as it can be interpreted from satellites have been provided from within the Arab region.

Map 2.4 shows large regions with almost no observations in the Arab world. To obtain climate information from regions otherwise void of data, reanalyses can be used. They are designed to use as much relevant information as possible from an incomplete, possibly error-prone observational database and can therefore serve as an approximate ground truth. Several of these products are available, covering the past several decades. Reanalyses are useful not only to derive climatologies for regions without conventional data, but also, for example, to construct drought-monitoring indexes (see Villholth et al. 2012). Again, enhanced capacity to make best use of these data on a national and regional scale would be helpful.

Three Case Studies from Different Regions of the Arab World

A case study from the Maghreb: Morocco

Annual and seasonal rainfall varies greatly in Morocco, with coefficients of variation ranging from 25 percent in coastal areas to more than 100 percent in the Sahara (Knippertz, Christoph, and Speth 2003; see also map 2.3). Nevertheless, rainfall has declined since the 1960s by as much as 40 percent in spring. This trend coincides with the change from a generally negative NAO phase to a more recent dominant positive phase. The maximum length of dry spells has also increased by more than two weeks over the same period (Wilby 2007a), and the Atlas Mountains have experienced less rainfall (Chaponniere and Smakhtin 2006).

A case study within the Mashreq

Recently, Wilby (2010) prepared climate change observations and projections for the Mashreq countries of Jordan, Lebanon, and Syria in the eastern Mediterranean. Here the data availability is less favorable than in Morocco. The World Meteorological Organization (WMO) lists 11 stations for Jordan, 7 for Lebanon, and 24 for Syria. Önol and Semazzi (2009) have gathered data from a number of additional stations. However, regions without any data also exist, particularly in western Iraq, east-

MAP 2.5

Spatial Distribution of Rainfall from a Study Conducted in the Region, 1998–2009

a. RegCM_B50 annual rainfall

b. TRMM annual rainfall

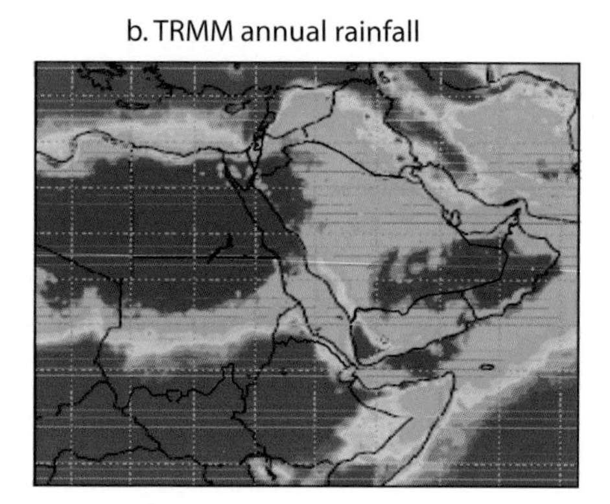

1 20 30 40 50 100 150 200 250 300 350 400
Millimeters

Source: Almazroui 2011.

Note: Rainfall data (millimeters) in panel a were simulated by a regional climate model (RegCM); data in panel b were obtained by averaging TRMM satellite data; all are averaged for the period 1998–2009.

BOX 2.4

Capacity Building Using Regional Climate Information

An important aim for all governments is to exploit international programs and data products that are useful on a national or regional level. These available programs include the use of satellite retrievals and derived products as well as the use of comprehensive data sets such as reanalysis products. Since the late 1970s, a wealth of satellite information has become available, which in many cases can be used to augment surface observations. Reanalyses make best use of all available data for the region and can therefore serve as a ground truth where no ground observations are available. Further, capacity building is needed in almost all Arab countries to make use of and visualize climate data information, for example, by means of geographic information systems. Therefore, developing a stronger regional collaboration between government agencies, research institutions, and universities is important. Success is contingent on the capacity to coordinate and promote a broad awareness and wide dissemination of generated knowledge.

Source: Authors' compilation.

ern Jordan, and northern Syria. The present-day climate in this region is quite different from countries further west, with relatively cool, wet winters and hot, dry summers, generally without any rain. Temperature and rainfall are strongly affected by altitude and the distance from the sea (see, for example, map 2.5). In the Lebanon Mountains, up to 1,400 millimeters per year of rain are observed (in winter it is often snow), but the deserts of southeast Syria and southern Jordan receive less than 100 millimeters per year. Temperatures above 50°C have been observed near the Dead Sea. To explore the time-space characteristics of rainfall, daily rescaled data from the TRMM satellite observations have been used (Kummerow et al. 2000; Simpson, Adler, and North 1988).

A temperature rise since the 1970s has been observed in all three countries. Mahwed (2008) considered meteorological records at 26 stations in Syria; Freiwan and Kadioğlu (2008a, 2008b) examined monthly rainfall data from Jordan; and Shahin (2007) looked at several stations throughout the Middle East. The greatest warming has occurred for summer minimum temperatures, which have risen at a rate of 0.4°C per decade. Consequently, a decrease in the diurnal temperature range has been observed, which is consistent with earlier studies (Nasrallah and Balling 1993; Zhang et al. 2005). There is no clear indication whether rainfall has changed in recent decades, but estimates from TRMM data (map 2.5) suggest a slight decrease in winter and spring, probably related to shifts in cyclone tracks. However, the trends are small compared to the interannual variability (Wilby 2010). Map 2.5 illustrates that a regional climate model (RCM) as used within the region is able to faithfully simulate rainfall climatology to a large extent.

A case study from the Arabian Peninsula

Wilby (2008) compiled an assessment of climate and climate change for the Republic of Yemen, despite the lack of reliable data. Obvious errors and missing data make it particularly difficult to apply statistical downscaling procedures. The lack of obvious trends in rainfall averages or extremes may be, in part, due to bad data. The only station in the Republic of Yemen with a reasonably long and reliable time series for rainfall is Aden, for which monthly means exist since 1880. However, that station shows no significant trends in annual rainfall.

More Extreme Events Are Being Observed

From a climate change point of view, changes in extremes are more interesting than changes in average values. Unfortunately, researchers do not always use the same definitions of extremes (Bonsal et al. 2001), making it difficult to make global or even regional comparisons. Frich et al. (2002)

tried to standardize definitions of extreme indexes, but they focused on areas with ample data, which meant the indexes were difficult to apply in large parts of the world, including the Arab region.

Figure 2.1 shows time series anomalies averaged over the Middle East for the hottest days (that is, days with a maximum temperature higher than 90 percent of those observed over the baseline period 1971–2000) and the coolest days (that is, days in which the maximum temperature is in the lowest 10th percentile). Both a decrease in the number of the coolest days and a more recent abrupt increase in the number of the hottest days are visible. Trends have also become more coherent and more significant in recent periods (Zhang et al. 2005). For rainfall, the results are much more variable.

FIGURE 2.1

Change in Days with Maximum and Minimum Temperatures

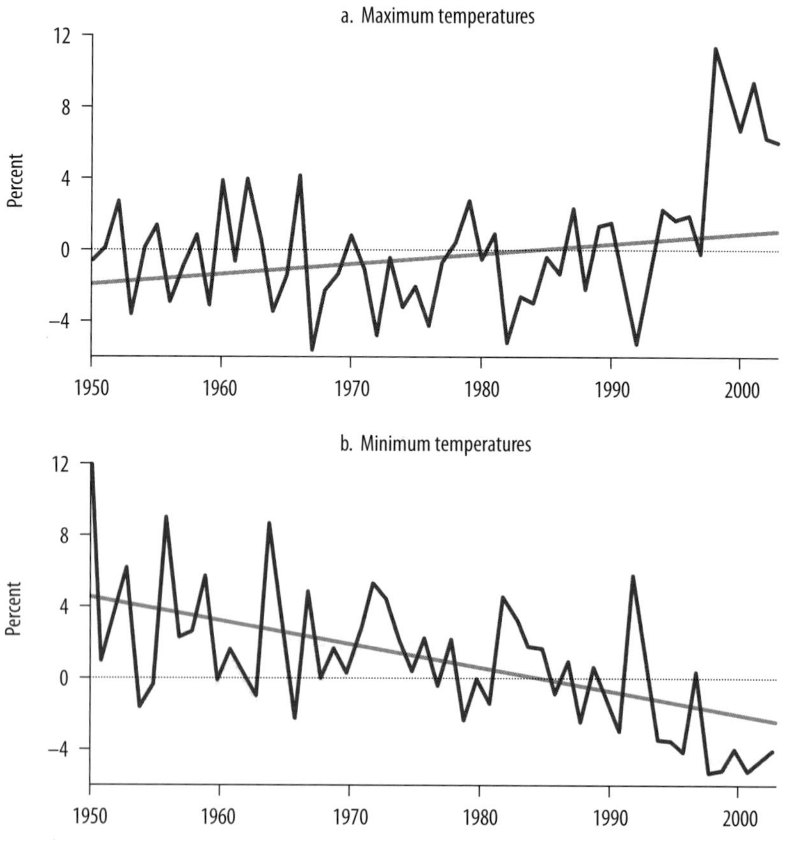

Source: Zhang et al. 2005.

Note: Time series anomalies, averaged over stations in the Middle East, of the percentage of days with maximum temperature above the 90th (maximum) and below the 10th (minimum) percentile with respect to the period 1971–2000, with the linear trend for the period 1950–2003.

IPCC AR4 Projects Warming and Aridity

For the IPCC AR4, a broad range of climate models were coordinated to perform a large number of simulations with GCMs for the known historical forcings (anthropogenic and natural) and various future emission scenarios in order to assess climate change projections. Taken together with information from observations, these coordinated simulations (referred to as a *multimodel data set*, or MMD) provide a quantitative basis for assessing many aspects of future climate change. All models assessed in the IPCC AR4 have projected increases in global mean surface air temperature continuing throughout the 21st century and driven by increases in anthropogenic greenhouse gas concentrations, with warming proportional to the associated radiative forcing. The best estimated projections indicate that decadal average warming by 2030 is insensitive to the choice among the three nonmitigated scenarios from the IPCC Special Report on Emission Scenarios (SRES), that is, the B1, A1B, and A2 (map 2.6). Furthermore, the projected warming is very likely to exceed the natural variability observed during the 20th century (about 0.2°C per decade in the global average). By the end of the 21st century (2090–99), projected global average air warming relative to 1980–99, under the SRES emission scenarios, will range from a best estimate of 1.8°C (likely range 1.1°C to 2.9°C) for the low scenario (B1) to a best estimate of 4.0°C (likely range 2.4°C to 6.4°C) for the high scenario (A2) and to a best estimate of 2.8°C (likely range 1.7°C to 4.4°C) for the moderate scenario (A1B) (IPCC 2007). Warming is very likely to be greatest over land, with a maximum over the high northern latitudes, and least over the Southern Ocean and parts of the North Atlantic Ocean.

Projected Changes in Climate over the Next Few Decades Do Not Depend on the Scenario Chosen

Much of the uncertainty in climate projections for the end of the century arises from the particular emission scenario (emission pathway) selected. However, in the near term (until about 2050), the scenario choice is not very important. This range is depicted by map 2.6. The left column represents the projected model mean temperature change for the near term, and the right column shows the projected change at the end of the century, both for three quite different emission scenarios.

IPCC AR4 Projects Drying and Warming for the Arab Region

The Arab countries lie within three neighboring subregions used by the IPCC AR4 SRES. For the purpose of this review, the African do-

MAP 2.6

Projected Surface Temperature Changes for the Early and Late 21st Century Relative to the Period 1980–99

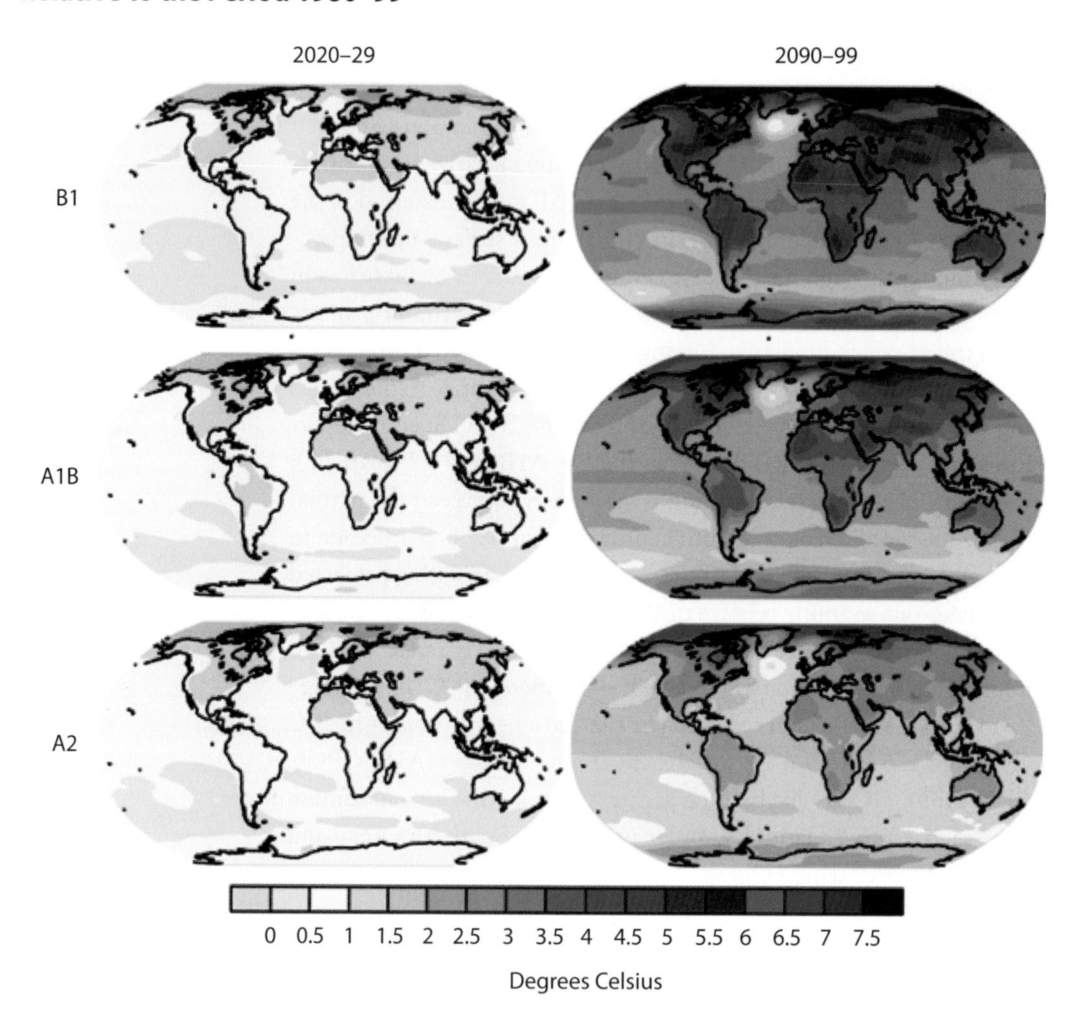

Source: Meehl et al. 2007.

Note: The two panels show the AOGCM multimodel average projections for the B1 (top), A1B (middle), and A2 (bottom) SRES scenarios averaged over the decades 2020–29 (left) and 2090–99 (right). The dependency on the scenario is insignificant for the near future but will become increasingly important toward the end of the century.

mains provide the most complete coverage of the areas of interest (map 2.7).

- The Arab countries are expected to warm by between 3°C and 4°C by the late 21st century under the IPCC A1B emission scenario, which is roughly 1.5 times the global mean response. Warming is evident in all seasons, with the greatest increase in summer (map 2.7, upper panel; J. H. Christensen, Hewitson et al. 2007).

MAP 2.7

Temperature and Rainfall Changes over Africa (and the Arabian Peninsula) Based on 21 IPCC AR4 Models under the IPCC A1B Scenario

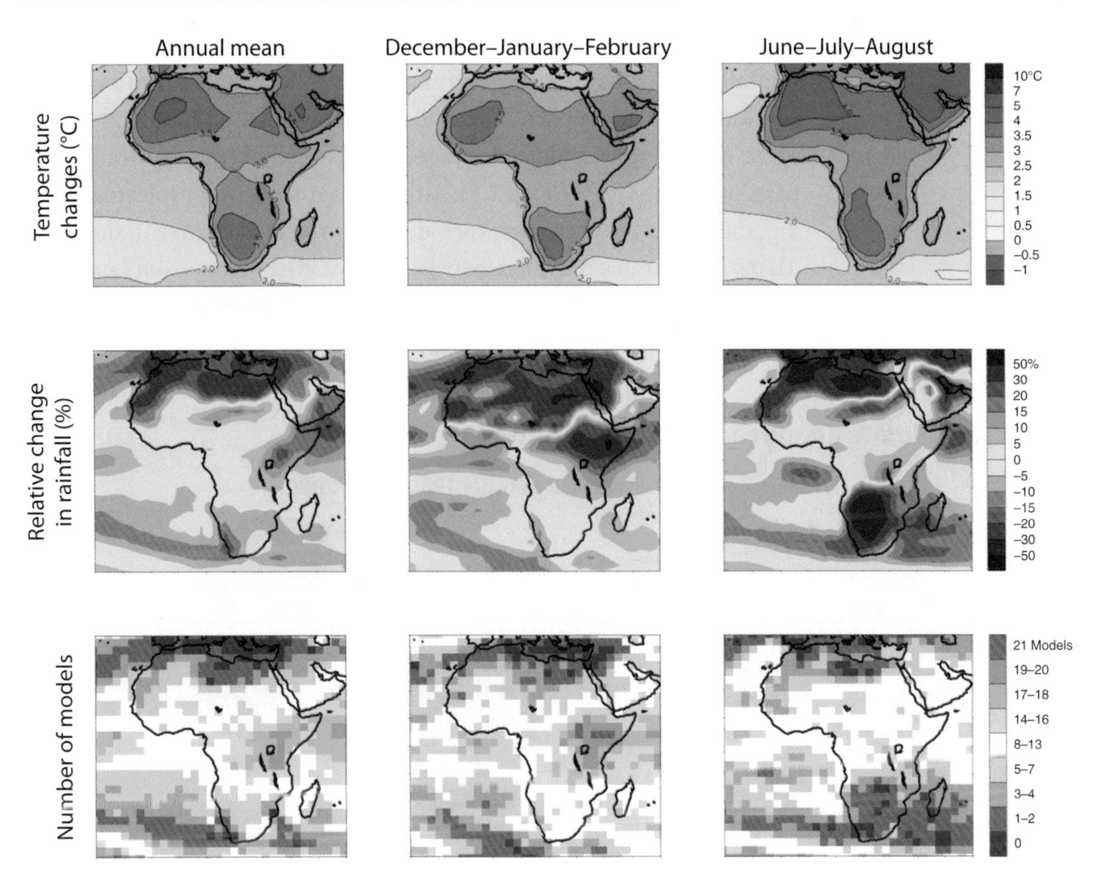

Source: J. H. Christensen, Hewitson et al. 2007.

Note: Top row: annual mean, northern hemisphere winter, and summer, temperature (°C) changes over periods 1980–99 and 2080–99 averaged over 21 models. Middle row: same as top row, but for relative change (percentage) in rainfall. Bottom row: number of models out of 21 that project increases in rainfall.

- Models project a northward displacement of the ITCZ but disagree on how far. Therefore there is general agreement on an increase in rainfall around the Horn of Africa, but less agreement about the change over the already dry central part of the Arabian Peninsula (map 2.7, middle and lower panels; J. H. Christensen, Hewitson et al. 2007).
- Annual rainfall is expected to decrease in much of Mediterranean Africa but increase in East Africa and the southern half of the Arabian Peninsula. A 20 percent drying in the annual mean is typical along the African Mediterranean coast in the A1B scenario by the late 21st century in nearly every climate model, with drying extended down the

west coast. The annual number of rainy days is very likely to decrease, and the risk of summer drought is likely to increase in the Mediterranean basin (see also map 2.10, later in this chapter).

All the GCM outputs of the scenario projections for temperature, rainfall, and other relevant extreme indexes for the Arab region were combined for this report. The evidence shows that although all models agree on strong warming, they agree on rainfall changes for only some parts of the region (see map 2.8). Although agreeing on a projected warming, the models show a large spread in the projected change in the mean. Map 2.9 illustrates the range between a low (MRI model from Japan) and high (ECHAM5 from Germany) model response compared to the overall

MAP 2.8

Projected Climate Change for Late This Century

a. Mean annual temperature change, 1980–99 to 2080–99

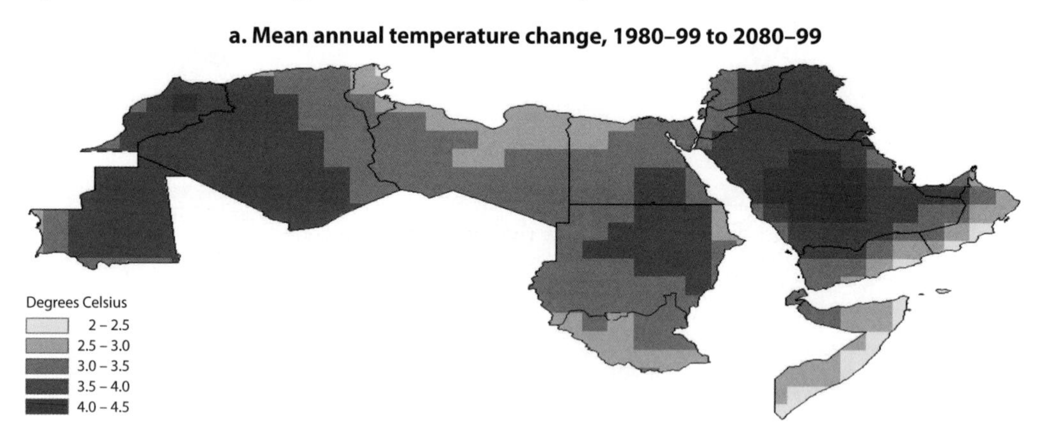

b. Mean annual rainfall change, 1980–99 to 2080–99

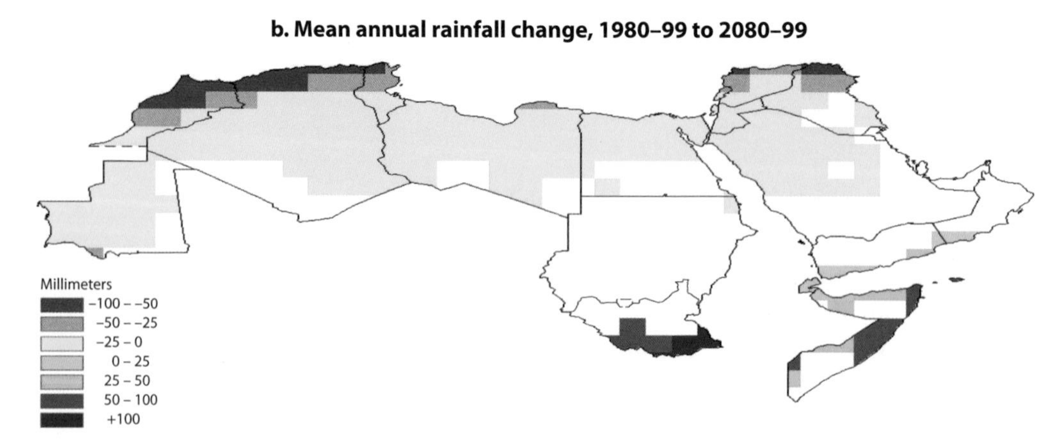

Source: Authors' compilation.

Note: Panel a: mean annual temperature change (2080–99 versus 1980–99) based on an average of the 24 IPCC AR4 GCMs. Panel b: mean annual rainfall change (2080–99 versus 1980–99) based on an average of 23 GCMs. White areas indicate where fewer than two-thirds of the models agree on the sign of the change.

model mean. For rainfall, consensus does not exist regarding the sign of rainfall change (white areas in map 2.7, lower panel, and map 2.8). This disagreement is to be expected because current rainfall is low and very variable, and the models suggest that this will continue, with sometimes a little less or a little more rainfall, but still largely dry.

Interpreting Climate-Related Extreme Events Needs to Consider Many Factors

Extremes occur even without climate change

Extreme events are not interpreted purely from a meteorological point of view. A weather or climate event, although not necessarily extreme in a statistical sense, may still have an extreme impact, either because it crosses a critical threshold in a social, ecological, or physical system, or because it occurs simultaneously with another event, which, in combination, leads to extreme conditions or impacts. Conversely, not all extremes necessarily lead to severe impacts. The impact of a tropical cyclone depends on where and when it makes landfall. Changes in phenomena, such as monsoons, may affect the frequency and intensity of extremes in several regions simultaneously, which indicates that the severity of an event may also depend on the overall geographical scale being affected. A critical or even intolerable threshold defined for a large region may be exceeded before many, or any, of the smaller regions within it exceed local extreme definitions (for example, local versus global drought).

MAP 2.9

Annual Mean Temperature Response in Africa in 2 out of 21 IPCC AR4 Models

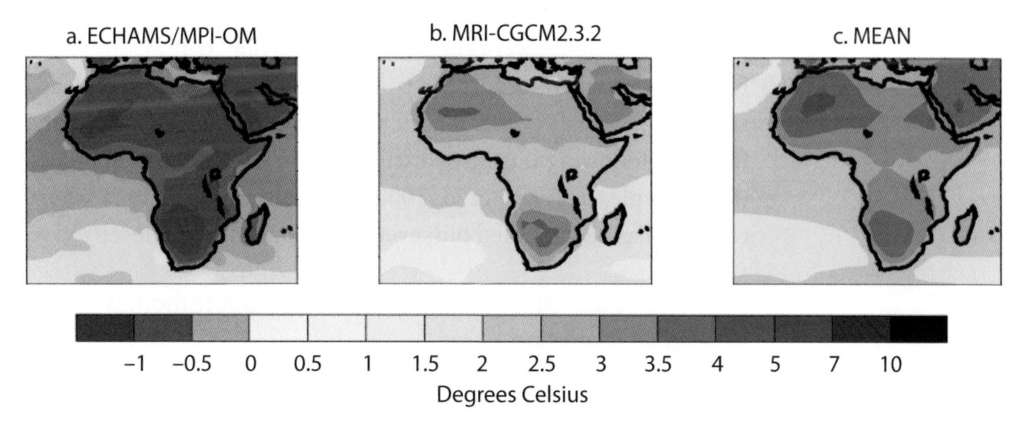

Source: J. H. Christensen, Hewitson et al. 2007.

Note: Shown is the temperature change from the periods 1980–99 to 2080–99 under the A1B scenario for the ECHAM5 model (left), the MRI model (center), and the mean of all 21 available models (right).

Many weather and climate extremes are the result of natural climate variability (including phenomena such as El Niño). At the same time, natural decadal or multidecadal variations in the climate provide the backdrop for possible anthropogenic changes. Even if no anthropogenic changes in the climate were to occur over the next century, a wide variety of natural weather and climate extremes would still occur. Projections of changes in climate means are not always a good indicator of trends in climate extremes. For example, observation and modeling show that rainfall intensity may increase in some areas and seasons even as total rainfall decreases. Thus, an area might be subject to both drier conditions and more flooding.

This is illustrated on a global scale (map 2.10). Based on a multimodel analysis they found simulated increases in rainfall intensity for the end of the 21st century (upper panel), along with a somewhat weaker and less clear trend of increasing dry periods between rainfall events for the A1B scenario (lower panel). For the Arab region, the statistical signal is weak but with an indication of increasing risk of both increased rainfall intensity and increased length of dry day spells.

Extreme impacts do not require extreme climate events

Events that may be perceived as extreme may actually be due to the compound effect from two or more extreme events, combinations of extreme events with amplifying events or conditions, or combinations of events that are not in themselves extreme but that lead to an extreme event or impact when combined. The contributing events can be similar (clustered multiple events) or very different. There are several varieties of clustered multiple events, such as tropical cyclones with the same path generated a few days apart. Examples of compound events resulting from events of different types are varied: for instance, high sea level coinciding with tropical cyclone landfall, or a combined risk of flooding from sea-level surges and rainfall-induced high river discharge (van den Brink et al. 2005). Compound events can even result from "contrasting extremes," for example, the projected near-simultaneous occurrence of both droughts and heavy rainfall events mentioned earlier or, more anecdotally, flash flooding following bushfires attributable to fire-induced thunderstorms (for example, Tryhorn et al. 2008). Overall, this is an area where little research has been carried out, even at the international level.

Climate-related extreme events as seen by climate models depend on resolution

Extreme events are often localized in both space and time (for example, the track of an extreme thunderstorm), so the coarse-resolution climate models are not suited (nor designed) to capture the many extreme events that are so important from the point of view of impacts. This factor has been one of

MAP 2.10

Changes in Rainfall Extremes Based on Multimodel Simulations from Nine Global Coupled Climate Models

a. Precipitation intensity

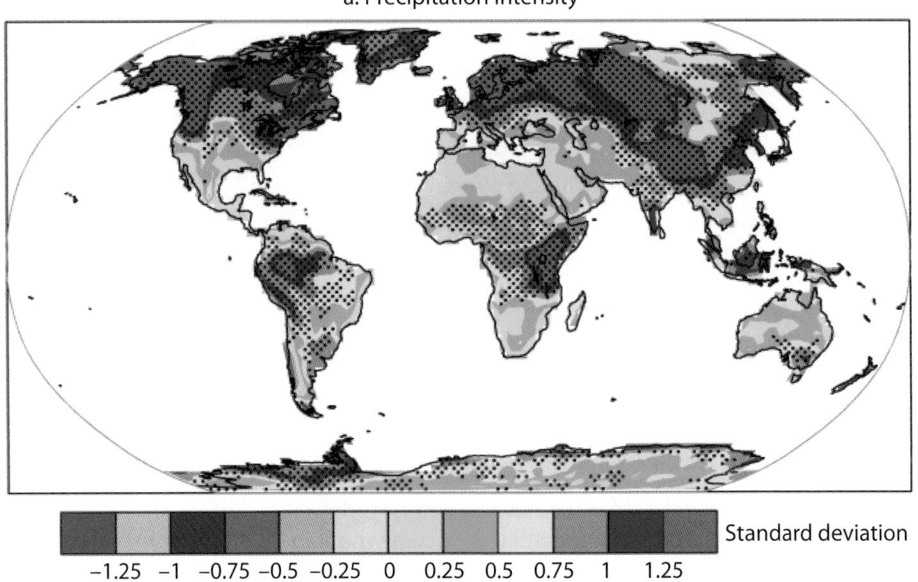

b. Dry days

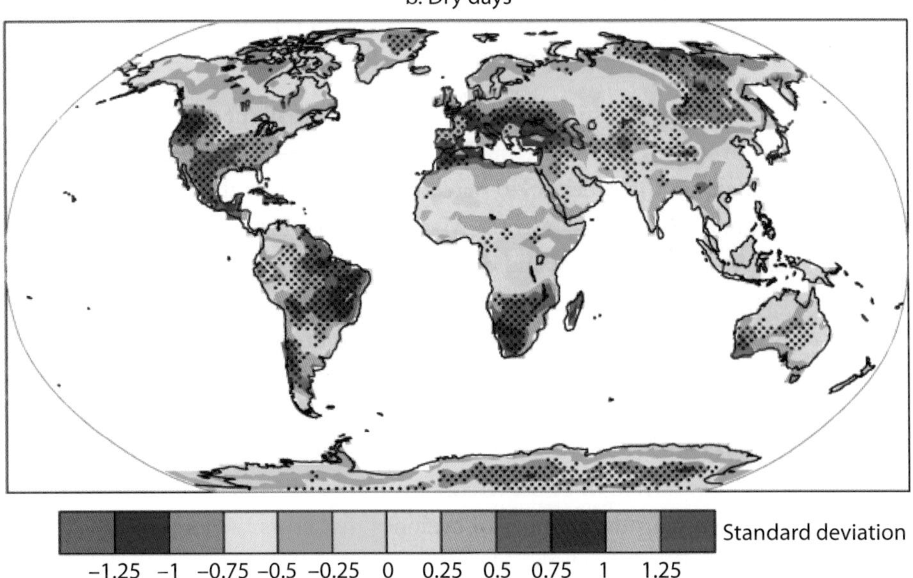

Source: Meehl et al. 2007.

Note: Panel a: changes in spatial patterns of simulated rainfall intensity between two 20-year means (2080–99 minus 1980–99) for the A1B scenario. Panel b: changes in spatial patterns of simulated dry days between two 20-year means (2080–99 minus 1980–99) for the A1B scenario. Stippling denotes areas where at least five of the nine models concur in determining that the change is statistically significant. Each model's time series was centered on its 1980–99 average and normalized (rescaled) by its standard deviation computed (after detrending) over the period 1960–2099. The models were then aggregated into an ensemble average at the grid-box level. Thus, changes are given in units of standard deviations.

the major rationales for dynamical downscaling using RCMs (see box 2.5). Using Europe as an example, Christensen and Christensen showed that by using a high-resolution RCM nested in a GCM, a more realistic pattern of rainfall change can be simulated (J. H. Christensen and O. B. Christensen 2003; O. B. Christensen and J. H. Christensen 2004). Map 2.11 illustrates this finding. For changes in the mean (left and middle panels), the increase in rainfall in southern Europe is confined to the southern slopes of the resolved topography. In the coarse-resolution GCM, this precipitation increase is distributed over a much wider area than in the real world, whereas in the RCM, it is confined to much smaller regions. Therefore, projected changes in extreme rainfall (right panel) appear more credible when stemming from a high-resolution model.

Downscaled Projections Are Available for the Region

Several regional assessments of future climate in the Middle East and North Africa region have been conducted, building on new modeling and downscaling exercises (box 2.5). Most of them concentrate on the Mashreq and the Arabian Peninsula. A smaller number of studies have also been conducted for the Maghreb.

Eastern Mediterranean and the Arabian Peninsula Are Projected to Become Drier, Especially in the Rainy Season

Black (2009) investigated the projected change in the monthly mean rainfall over the eastern Mediterranean and found a significant decrease in rainfall, on the order of 40 percent, at the peak of the rainy season (December and January) over the Mashreq. This change is due to a reduction in both the frequency and duration of rainy events. Before and after the rainy season, the situation is less clear, with some areas projected to get wetter and others drier. These results are broadly consistent with wider surveys of global models included in the IPCC AR4 (for example, see Dai 2010; Evans 2009; Kitoh, Yatagai, and Alpert 2008; Lionello and Giorgi 2007) identified to be induced by the northward displacement and reduction in the strength of the Mediterranean storm track and, consequently, a reduction in the number of cyclones that cross the eastern Mediterranean basin and reach the Mashreq (Bengtsson and Hodges 2006).

Higher Resolution Modeling Is Needed to Improve Projections in the Complex Topography of the Mashreq

Evans (2010) found that the climate change signal in rainfall differed between the driving GCM and a higher resolution RCM nested therein. It

BOX 2.5

Climate Models and Downscaling

The accuracy and representativeness of climate model data depend, among other factors, on the horizontal resolution. Features smaller than the distance between two grid points cannot be resolved; therefore it is important to use a high enough resolution to be able to represent features of interest adequately, such as coastlines, islands, lakes, and mountain ranges.

A straightforward way of adding spatial detail to GCM-based climate change scenarios could be so-called perturbation experiments, where the GCM data are interpolated to a finer resolution, and then these interpolated changes are combined with observed high-resolution climate data. In essence, however, this is only an interpolation exercise, because it does not add any meteorological information beyond the GCM-based changes; furthermore, the method implies that the spatial patterns of present-day climate are assumed to remain constant in the future.

A second option, dynamical downscaling, is more costly. A higher resolution limited-area model (a regional climate model, or RCM) is used to generate climate change scenarios at a higher resolution. State-of-the-art RCMs have a typical resolution of 25 kilometers, although a few investigations have been conducted at higher resolution on the 5- to 10-kilometer scale. Because such a model covers only part of the globe, it needs boundary conditions generated by a global model. These boundary conditions are interpolated from the GCM to the RCM. In a "nudging" zone of typically a few grid points in the RCM grid, the GCM values are relaxed toward the RCM grid. Inside this transition zone, the meteorological fields are RCM generated. Every few (typically six) hours, new boundary conditions are generated from the GCM data. In other words, the RCM can generate its own structures inside the domain, but the large-scale circulation depends on the driving GCM. Many modeling centers around the world follow this approach, which, however, requires substantial computer resources and is comparatively slow.

A third approach can be termed *statistical downscaling*. Compared to the straightforward interpolation discussed above, additional meteorological knowledge can go into these models. One example is the dependence of rainfall on height and wind directions. From box map 2.5B, it is clear that the GCM always underestimates the height or steep slopes of mountain ranges. Rainfall will therefore generally be underestimated in the GCM compared to higher resolution approaches. Box maps 2.5A and 2.5B also show that there generally is a directional dependence of rainfall (as in the case of the Atlas Range); that is, the increase in rainfall with increasing resolution will not be uniformly distributed but will actually depend on wind direction. At least three general methods of statistical downscaling exist: those based on regression approaches, circulation-type schemes, and stochastic weather generators, which are used to construct site-specific scenarios. A weather generator is calibrated on an observed daily weather series over some appropriate period, usually for a site, but possibly for a catchment or a small grid box. It can then be used to generate any number of series of daily weather for the respective spatial domain. Such a stochastic time

BOX 2.5 *Continued*

series will in theory have the correct (that is, observed) climate statistics for that domain. The parameters of the weather generator can then be perturbed using GCM output, allowing the generator to yield synthetic daily weather for the climate change scenario. However, to derive the appropriate parameter perturbation from the GCM data, other downscaling methods must be used. It is also important to ensure that low-frequency variations (for example, multidecadal variations) are adequately captured. Statistical downscaling approaches always require a very good observational database from which to derive the statistical properties. Therefore, the weather generator is not necessarily cheaper or faster than running an RCM for dynamical downscaling. Also worth noting is that downscaling methods generally cannot easily be transported from one region to another. Just as with an RCM, the derived regional scenarios will depend on the validity of the GCM data.

As discussed above, there are considerable uncertainties in deriving climate projections for the Arab countries. Nevertheless, some general conclusions can be drawn by taking into account the different sources of uncertainties. Apart from observational uncertainties, discussed earlier in this box, uncertainties may reflect future emissions, natural climate variability, and differences between models. Future emissions are of course uncertain, so a number of representative emission scenarios have been defined (Nakicenovic and Swart 2000; see also figure 2.3 later in this chapter).

For the next 30–40 years, differences between these scenarios generally are smaller than the other uncertainties. This

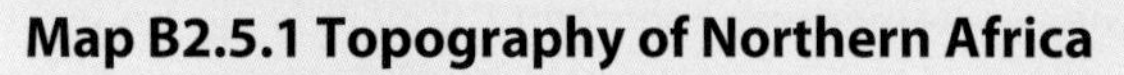

Map B2.5.1 Topography of Northern Africa

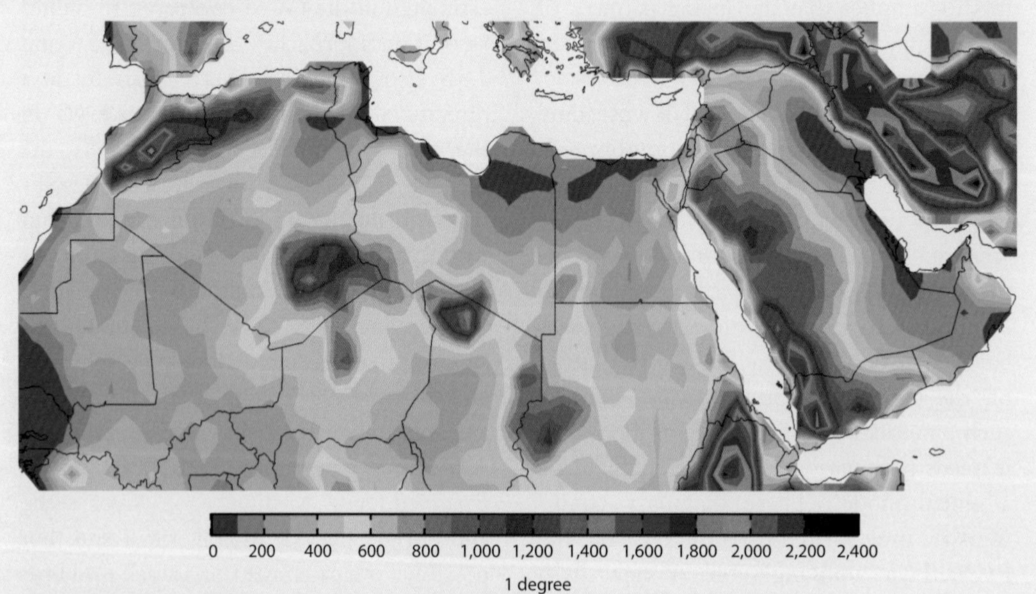

Source: Authors' representation.

Note: The map is based on the GTOPO30 data set from the EROS Data Center (EDC) Distributed Active Archive Center (DAAC), Sioux Falls, South Dakota. Available online at http://edcdaac.usgs.gov/gtopo30/gtopo30.html. Also see Gesch, Verdin, and Greenlee (1999). Color coding is directly related to topographic height. The resolution is 1 kilometer.

BOX 2.5 *Continued*

Map B2.5.2 Model Topography at a Horizontal Resolution of 1 Degree
a. Resolution of 1.0 degree (about 100 km) b. 0.2 Degrees (about 20 km)

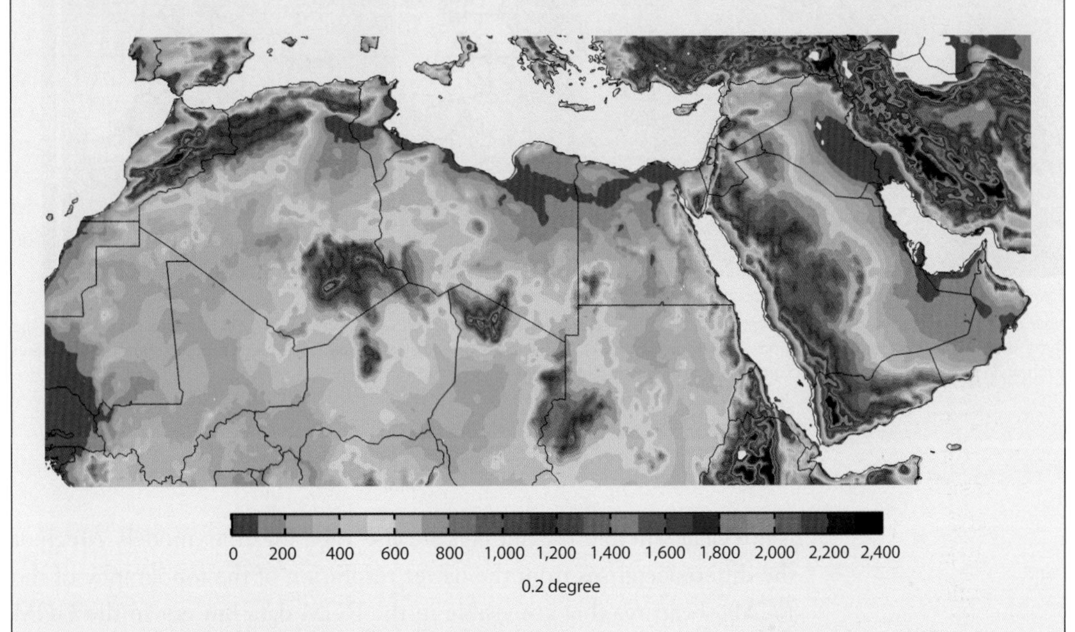

Source: Authors' representation, based on the GTOPO30 data set.
Note: In a comparison to the high-resolution map in map 2.5A, a resolution on the order of 20 kilometers is necessary to resolve features such as the Iranian Plateau, the Anti-Atlas, or the Al-Sarat in the Republic of Yemen.

issue is discussed further in Prudhomme et al. (2010) and Wilby and Dessai (2010). The ratio between natural climate variability (that would have occurred anyway regardless of human activities) and human-caused climate change can be quantified by running ensembles of simulations, that is, by using more than one GCM and more than one RCM.

Several large international programs exist that have followed this approach, and it has been applied, for example, in the IPCC AR4. Finally, differences between models are caused by different mathematical formulation, resolution (horizontal, vertical, and temporal), and, consequently, parameterization of subgrid-scale processes. However, most models have been validated extensively, so climate scientists generally know for past and present climate how a model behaves in different parts of the world and how sensitive it is to changes in concentrations of greenhouse gases or other forcings. From these considerations, scientists can assess an envelope of probable climate changes under future conditions. This has been taken into account in the following paragraphs, but is, for the sake of brevity, not mentioned every time.

Source: Authors' compilation.

MAP 2.11

Example of the Benefit from High-Resolution Modeling

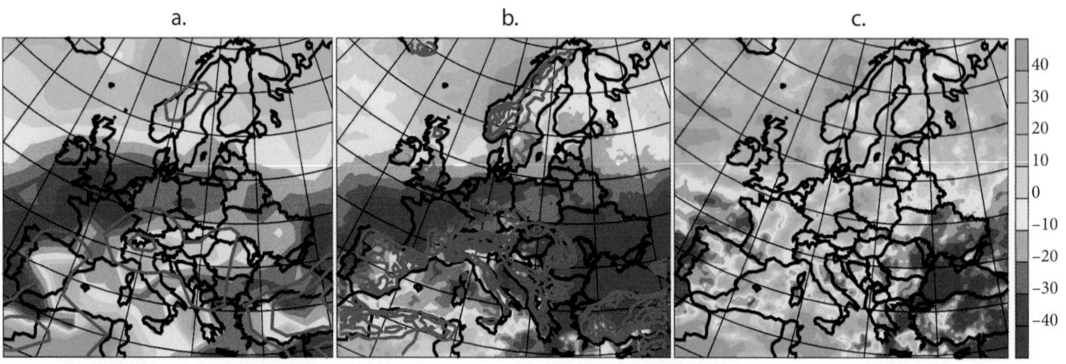

Sources: J. H. Christensen and O. B. Christensen 2003; O. B. Christensen and J. H. Christensen 2004.

Note: Panel a: simulated percentage change in rainfall at the end of the century with a coarse-resolution GCM (250-kilometer grid). Panel b: the same result for the RCM (50-kilometer grid). Panel c: percentage change in extreme rainfall for the same RCM. Blue contours show the 500-meter contours in the model topography according to the model resolution.

is not clear whether the findings are also valid for other models. Much of the difference arises from the better resolution of the topography of the RCMs. Features that are visible in the RCM data but not in the GCM data affect the large-scale circulation, even in regions quite far away, by causing upslope air mass movement and subsequent rainfall in the RCM, whereas rain is dumped in the wrong locations in the GCMs because they cannot resolve these topographic structures.

In the RCM simulation, temperature changes to the end of the century (map 2.12) are smallest in winter (around 2°C) and largest in summer (on the order of 5°C, but up to 10°C on the Iranian Plateau).

According to this study, rainfall is projected to decrease over the Mashreq, whereas an increase is simulated in the central part of Saudi Arabia in summer and autumn (map 2.13). This reflects a change away from the direct dependence on storm tracks toward a greater amount of rainfall triggered by the lifting of moist air along the mountains, a feature totally absent in the coarse-resolution simulations of Evans (2009). As a consequence, the rest of the Mashreq is projected to receive considerably less rain, mainly in winter and spring, which is also related to the displacement in the storm tracks. The Saudi Desert, on the other hand, could receive more rain in late summer when it is projected that the ITCZ will be further north than in the present-day climate. An increase in autumn rainfall is projected for southeastern Iraq and Kuwait (not shown), attributable to the advection of moist air along the Zagros Mountains (figure 2.2).

MAP 2.12

MM5/CCSM Modeled Change in Seasonal Mean Temperature, 2095–99 compared with 2000–04

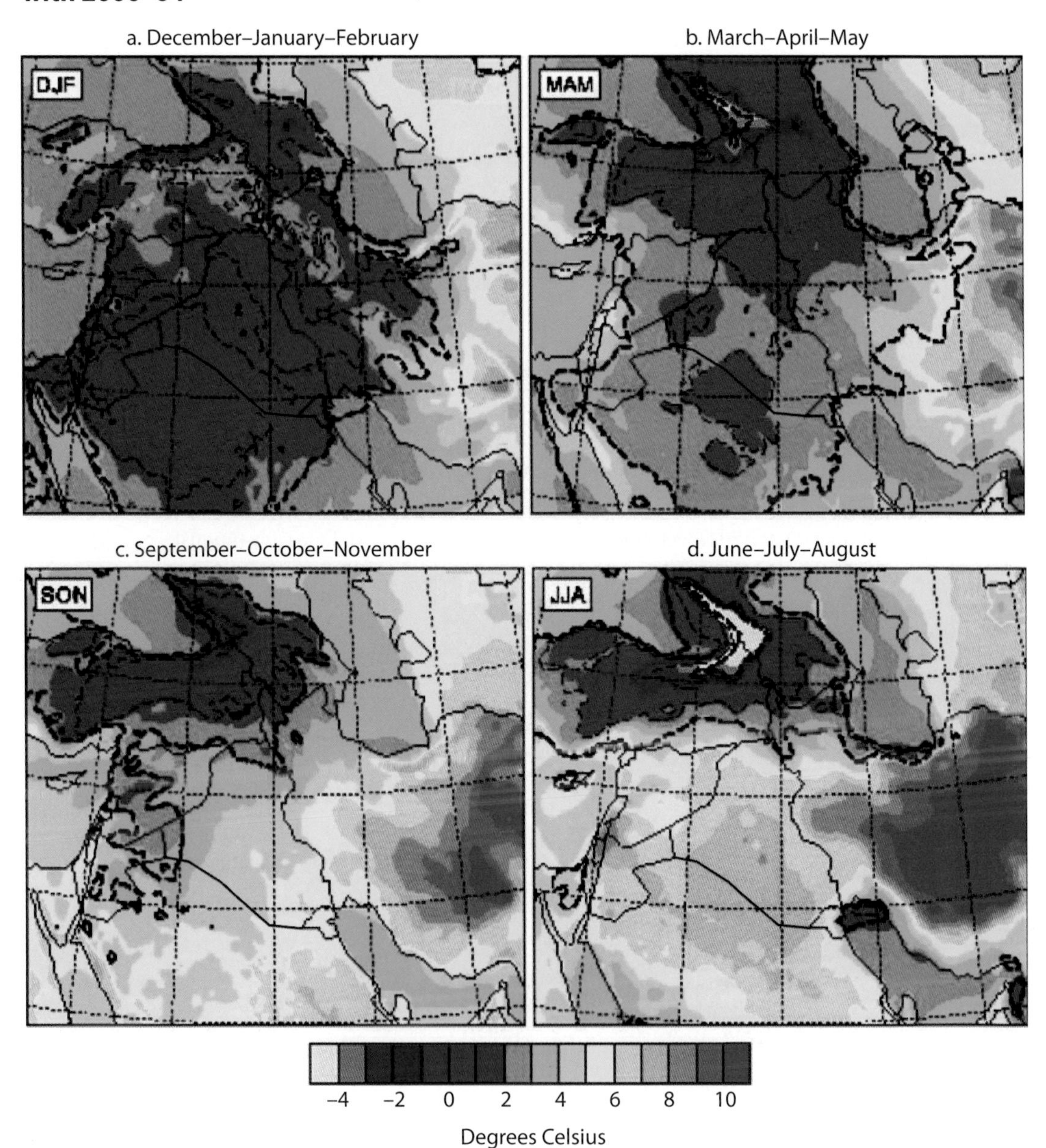

Source: Evans 2010.

Note: The 0.9 and 0.99 significance levels are indicated by the thin and thick dotted lines. IPCC scenario A2.

According to figure 2.2, a minor increase in rainfall is projected in late summer for the Fertile Crescent, including northeast Syria. For this region, the rainfall contribution from storms will be less than for the present-day climate, and upslope lifting processes will play a larger role.

MAP 2.13

MM5/CCSM Modeled Change in Seasonal Mean Rainfall, 2095–99 compared with 2000–04

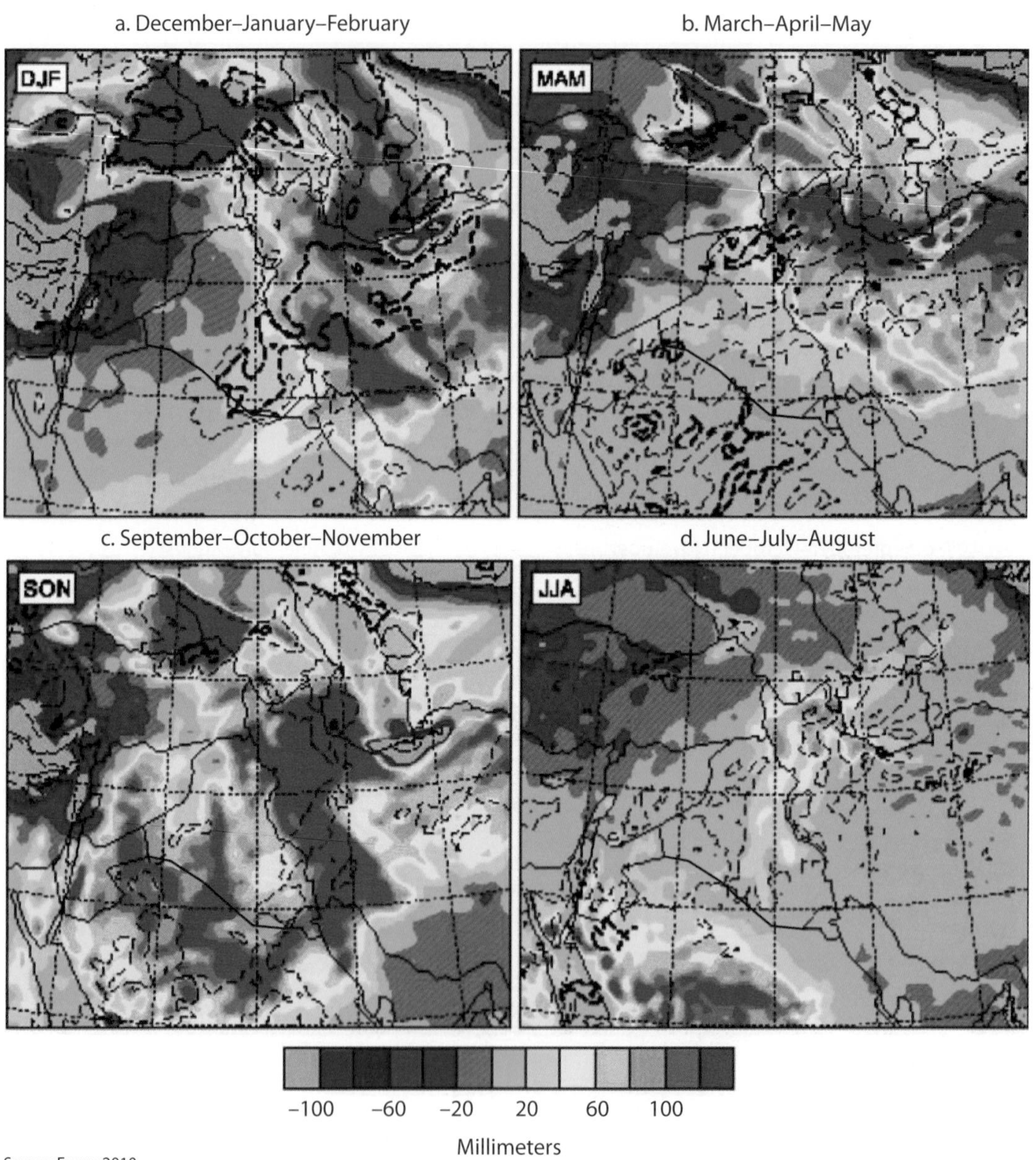

Source: Evans 2010.

Downscaled Models Project a Hotter and Drier Maghreb

For Morocco and Mauritania, a temperature increase of about 5°C is projected by the end of the century, with a maximum during the summer. This is related to a decrease in soil moisture attributable to decreasing rainfall and consequently an enhanced risk of droughts (J. H. Christensen,

FIGURE 2.2

MM5/CCSM Modeled Monthly Mean Rainfall for Early and Late 21st Century

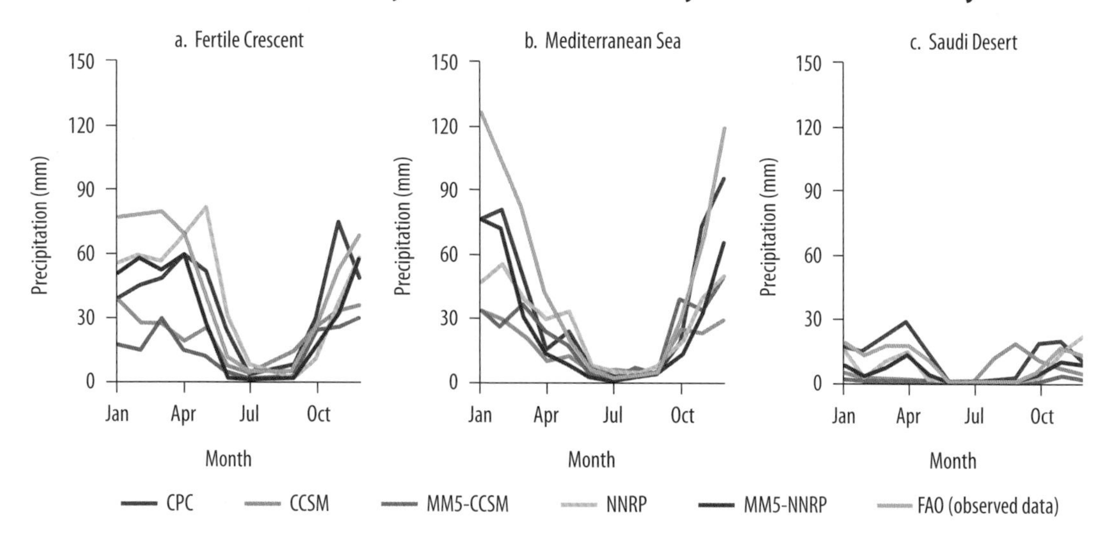

Source: Evans 2010.
Note: Scenario A2 = CCSM, MM5-CCSM, NNRP, MM5-NNRP.

Hewitson et al. 2007). Less clear is the effect of NAO trends and the subsequent possibility of enhanced droughts in a future climate, but a small tendency toward more positive NAO indexes and, therefore, less rainfall is backed up by two studies (Coppola et al. 2005; Rauthe, Hense, and Paeth 2004). Large differences exist between GCMs, with strong positive trends in ECHAM4 and no discernible trends at all in HadCM2 (Wilby 2007a). According to Krichak, Alpert, and Kunin (2010) and Raible et al. (2010), it could also be possible that increased polar intrusions following cyclone passages lead to no net winter rainfall change in the region.

In Morocco, the data density is high enough to allow for a statistical downscaling approach. With HadCM3 as the driving GCM and using the A2 and B2 scenarios, climate projections have been calculated for 10 locations in Morocco (Wilby 2007a) as shown in figure 2.3.

According to the study, temperature increases are smallest in Agadir (coastal station) and largest in Ouarzazate in the Atlas Mountains, where the summer temperature is projected to rise by more than 6°C by the 2080s. The frequency and severity of heat waves is projected to increase. According to Wilby (2007a), almost 50 days per year with a maximum above 35°C in Settat (near the coast) and Beni Mellal (in the Atlas foothills) are projected by the end of the century. Except in Agadir and Marrakech, less rainfall is projected, ranging from about a 25 percent decrease in the south to approximately a 40 percent decrease in the agroeconomic zone in the north.

Map 2.14 shows an assessment of drought in the Mediterranean region (including substantial parts of the Maghreb) for present-day climate and two scenarios, each at two different horizontal resolutions (Gao and Giorgi 2008). Both scenarios show the drought risk around the Mediterranean Sea increasing from west to east, which is worse in the Mashreq but notable also in other regions, including southern Europe.

Figure 2.4 shows rainfall statistics for the Mashreq. Most prominent are a statistically significant decrease in the number of rainy days and a general decrease in winter rainfall. According to the GCMs in J. H. Christensen, Hewitson et al. (2007) and RCM experiments by Önol and Semazzi (2009), temperatures in the region are projected to increase on the order of 2°C in winter and up to 6°C in the inland regions in summer. A reduction in winter rains on the order of 25 percent and an increase of drought duration by up to 60 percent are expected based on the A1B scenario (Kim and Byun 2009). The authors also predict a northward expansion of the Arabian Desert and an increase of autumn rainfall over the Fertile Crescent by up to 50 percent.

Agricultural Lands Are Threatened by Increasing Aridity in the Mashreq

Although less favorable than for Morocco, the data density is sufficient to allow for statistical downscaling in the Mashreq for stations in Amman (Jordan), Kamishli (Syria), and Kfardane (Lebanon). While for Kamishli,

FIGURE 2.3

Mean Seasonal Temperature Changes for the 2080s Downscaled from HadCM3 under the A2 Emission Scenario

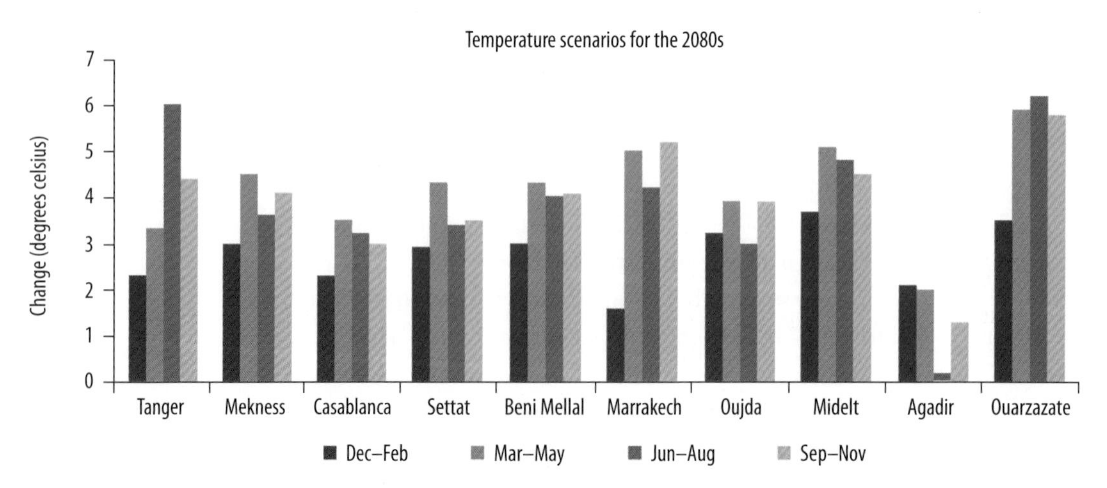

Source: Wilby 2007a.

MAP 2.14

Projections of Drought Index

a. Present day drought risks

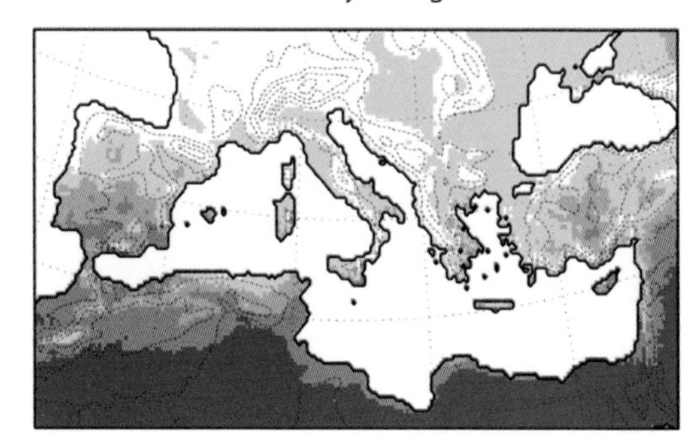

b. Future drought risks under A2 scenario

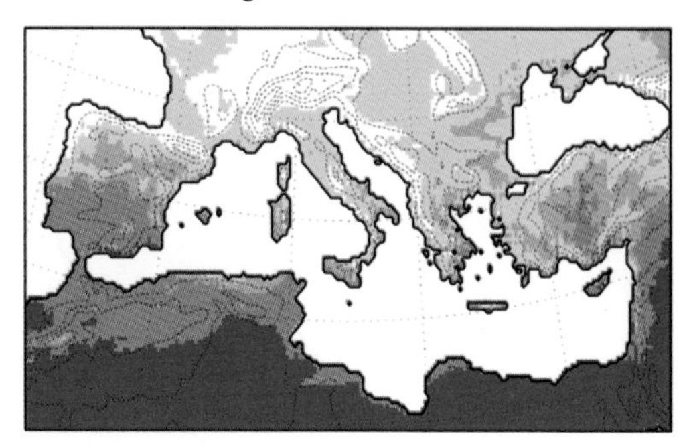

c. Future drought risks under B2 scenario

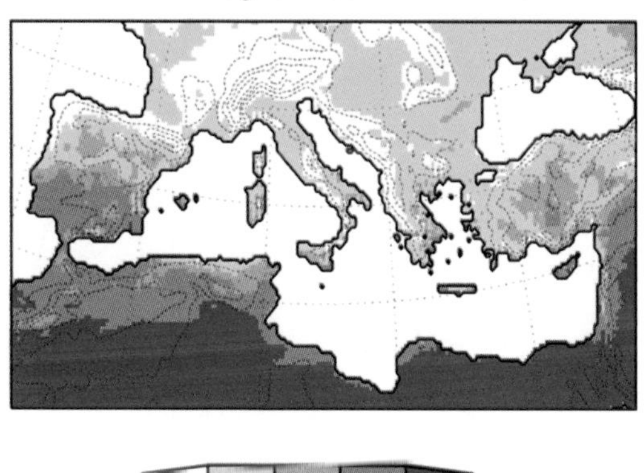

1.1 2.3 3.4 10

Source: Gao and Giorgi 2008.

Note: Panel a: reference simulation; panel b: difference in drought index between the A2 and reference simulations; panel c: difference for B2 and reference simulations. Aridity regimes are defined for index (Idx) values: 0 < Idx ≤ 1.1—humid (surplus moisture regime; steppe to forest vegetation), 1.1 < Idx ≤ 2.3—semihumid (moderately insufficient moisture; savanna), 2.3 < Idx ≤ 3.4—semiarid (insufficient moisture; semidesert), 3.4 < Idx ≤ 10—arid (very insufficient moisture; desert), 10 < Idx—hyperarid (extremely insufficient moisture; desert).

the present-day climate can be reconstructed quite well, results are less convincing for Amman. This may be the result of missing data and problems associated with station relocation and urban growth (Smadi and Zghoul 2006). Depending on scenario and location, the models project temperature increases of 3–4°C (A2) and 2–3°C (B2), respectively, and rather large decreases in rainfall by up to 50 percent near the coast (Lattakia, Syria) and about 15–20 percent inland (for example, in Palmyra, Syria) (Wilby 2010). This reduction in rainfall happens almost entirely during the winter, as generally no rain is observed at all during summer under present-day conditions.

As mentioned above, it has been suggested (Evans and Geerken 2004) that the limit for rainfed agriculture is close to an average annual rainfall

FIGURE 2.4

Seasonal Cycle of Rainfall Statistics for the Mashreq

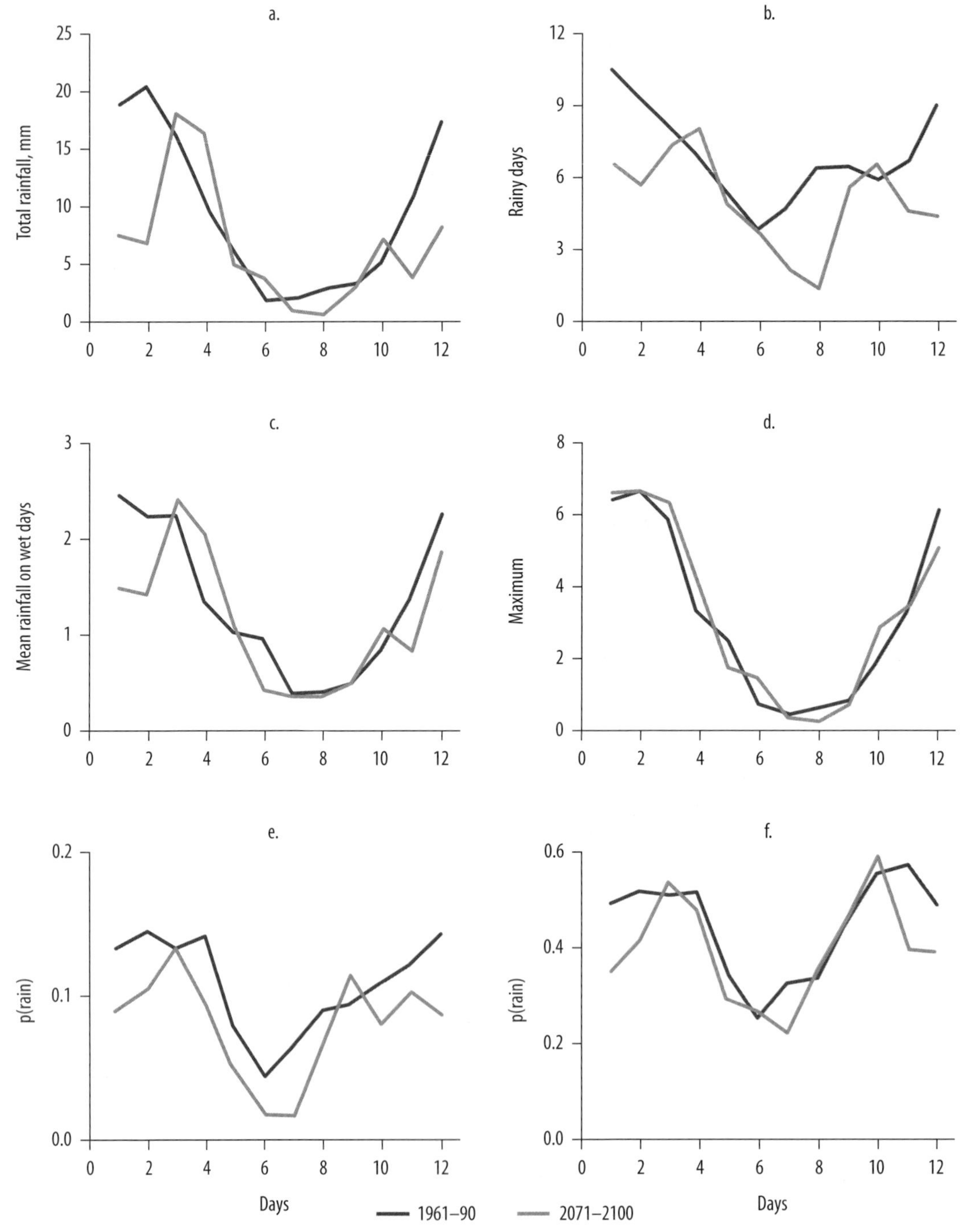

Source: Black 2009.

Note: Panels a–f: total rainfall, wet days, mean rainfall on wet days, maximum rainfall per wet day, probability of rainfall given no rainfall on the day before, and probability of rainfall given a wet day the day before as a function of the month. Red lines represent 1961–90, gold lines represent 2071–2100 for the A2 SRES scenario.

of 200 mm per year. Taking Amman as an example (average annual rainfall for the period 1961–90 was 260 mm), this suggests that agriculture would no longer be possible in 2080. However, because of the large interannual variability (coefficient of variation in Amman is 38 percent, which implies that there is a 1 in 3 chance for an annual rainfall sum below 200 mm even in present-day conditions), agriculture could cease to be economically viable considerably earlier. On the basis of the HadCM3 data and the A2 scenario, the chance of a dry year (< 200 mm) would be 50 percent around 2020 and 80 percent around 2060. This finding is consistent with previous studies that suggest that a 200 mm low annual rainfall limit could move on the order of 75 kilometers northward by the end of the century (Evans 2009). If rainfall is too low, agriculture is traditionally replaced by grazing. The same scenario projects an increase in the dry season from six to eight months by 2080 for Amman.

Another RCM simulation covering the Arabian Peninsula and neighboring regions has been conducted (Tolba and Saab 2009). Apart from the coastal plains in Oman, Somalia, and the Republic of Yemen, where the salt marshes remain comparatively humid and therefore experience less warming, relatively uniform warming is projected. Rainfall changes are small in areas that are already arid under present-day conditions, but major drying is projected for the Mashreq.

The Risk of Heat Stress Is Intensifying

Increases of mean temperature in the future, and in particular an increase of extreme warm days, may increase the risk of heat stress. A recent study (Diffenbaugh et al. 2007) showed that under the A2 scenario, the number of days with a dangerous or extremely dangerous heat index is dramatically increased along the Mediterranean coast of all Arabic countries by the end of the century in comparison with the present baseline period. The peak changes may be up to 65 days per year (see map 2.15). Willett and Sherwood (2012) used a heat index based on the wet-bulb temperature and found positive trends for the period 1973–2003 and an increase in future extreme heat events under the A1B scenario. Map 2.15 shows changes in the probability of another monthly heat index based on temperature, humidity, and physiology (Steadman 1979) in the Arab world under the new emission scenarios RCP4.5 and RCP8.5 (compare figure 2.6, later in this chapter). The model indicates an increase in the heat index for 1990–2009 in Sudan and Somalia, and a further increase of about 10 percent (more in Sudan and Somalia) is modeled for midcentury, increasing to more than 40 percent toward the end of the century. Under the RCP8.5 scenario, there is also a large increase in the heat index in the Maghreb region.

MAP 2.15

Probability of Monthly Heat Index above 30°C

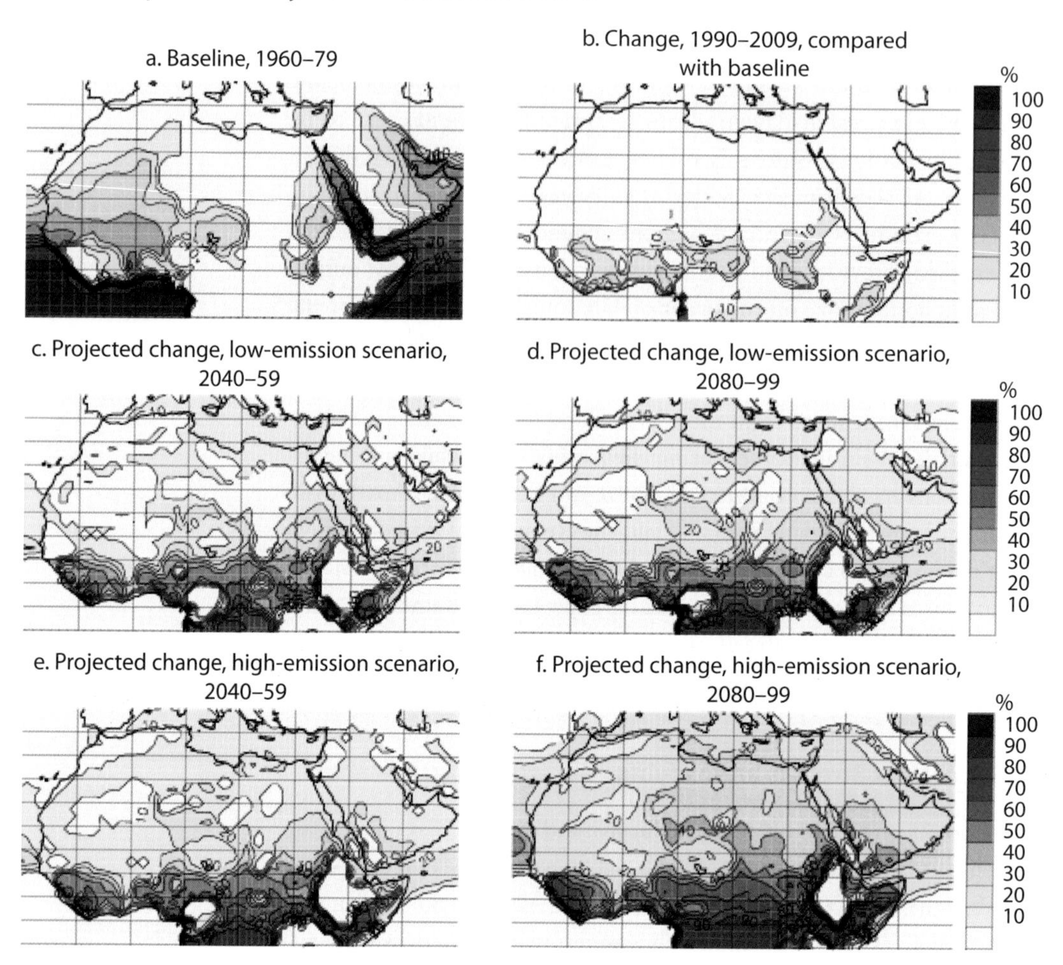

Source: Authors' representation.

Note: Panel a: heat index under 1960–1979 climatology. Panel b: changes in heat index for 1990–2009 with respect to the climatology (shown in panel a). Panel c: changes for 2040–59 under the RCP4.5 scenario. Panel d: changes for 2080–99 under the RCP4.5 scenario. Panel e: changes for 2040–59 for the RCP8.5 scenario. Panel f: changes for 2080–99 in the RCP8.5 scenario. The heat index is based on Steadman (1979).

Sea Levels Will Continue to Rise

Global Mean Sea Level Has Been Observed to Have Steadily Increased since 1870

The level of the sea at the shoreline is determined by many factors that operate over a great range of temporal scales: hours to days (tides and weather), years to millennia (climate), and longer. The land itself can rise and fall, so regional land movements must be accounted for when using tide gauge measurements for evaluating the effect of oceanic climate change on coastal sea levels. Coastal tide gauges indicate that the global average sea level rose during the 20th century. Since the early 1990s, sea level has also been continuously observed by satellites, with near-global coverage. Satellite and tide gauge data agree at a wide range of spatial scales and show that global average sea level has continued to rise during this period. Figure 2.5 illustrates the observed steady increase in global sea level since 1870 to the present. In total, the global increase has been approximately 18 centimeters. Sea-level changes show geographical variation because of several factors, including the distributions of changes in ocean temperature, salinity, winds, and ocean circulation.

The IPCC AR4 states that over the period 1961 to 2003, global ocean temperature has risen by 0.1°C from the surface to a depth of 700 meters. This temperature increase results in a thermal expansion of seawater that has contributed substantially to sea-level rise in recent decades. Over the period with data available, warming was greatest between 1993 and 2003 but was somewhat less afterward. The thermal expansion of the oceans, however, is not sufficient to explain the observed sea-level rise, which also occurs from cryospheric changes, including the increased melting of glaciers and ice caps, and the thawing and calving of Greenland's inland ice.

Recent expert meetings, such as at the Copenhagen Climate Change Conference in March 2009, discussed whether the IPCC AR4 had underestimated the amount of sea-level rise and that ocean warming could be about 50 percent greater than the IPCC had previously reported. Material presented at the conference suggested that the rate of sea-level rise increased during the period from 1993 to the present, mainly because of the growing contribution of ice loss from Greenland and Antarctica. Observations show that the area of the Greenland ice sheet at the freezing point or above for at least one day during the summer period increased by 50 percent during the period 1979 to 2008, particularly during the extremely warm summer of 2007 (K. Steffen and R. Huff, University of Colorado at Boulder, personal communication, 2009; Mote 2007).

Ice sheets may also lose mass through ice discharge, which is also sensitive to regional temperature. The best estimate is that the Greenland ice sheet has been losing mass at a rate of 179 gigatons per year since 2003, corresponding to a contribution to global mean–level rise of 0.5 millimeters per year (Dahl-Jensen et al. 2009). This is approximately twice the amount estimated by the IPCC in 2007.

There Is a Risk That Global Mean Sea Level May Increase 1.5 Meters by 2100

Climate models are consistent with the ocean observations and indicate that thermal expansion is expected to continue to contribute significantly

FIGURE 2.5

Annual Averages of the Global Mean Sea Level

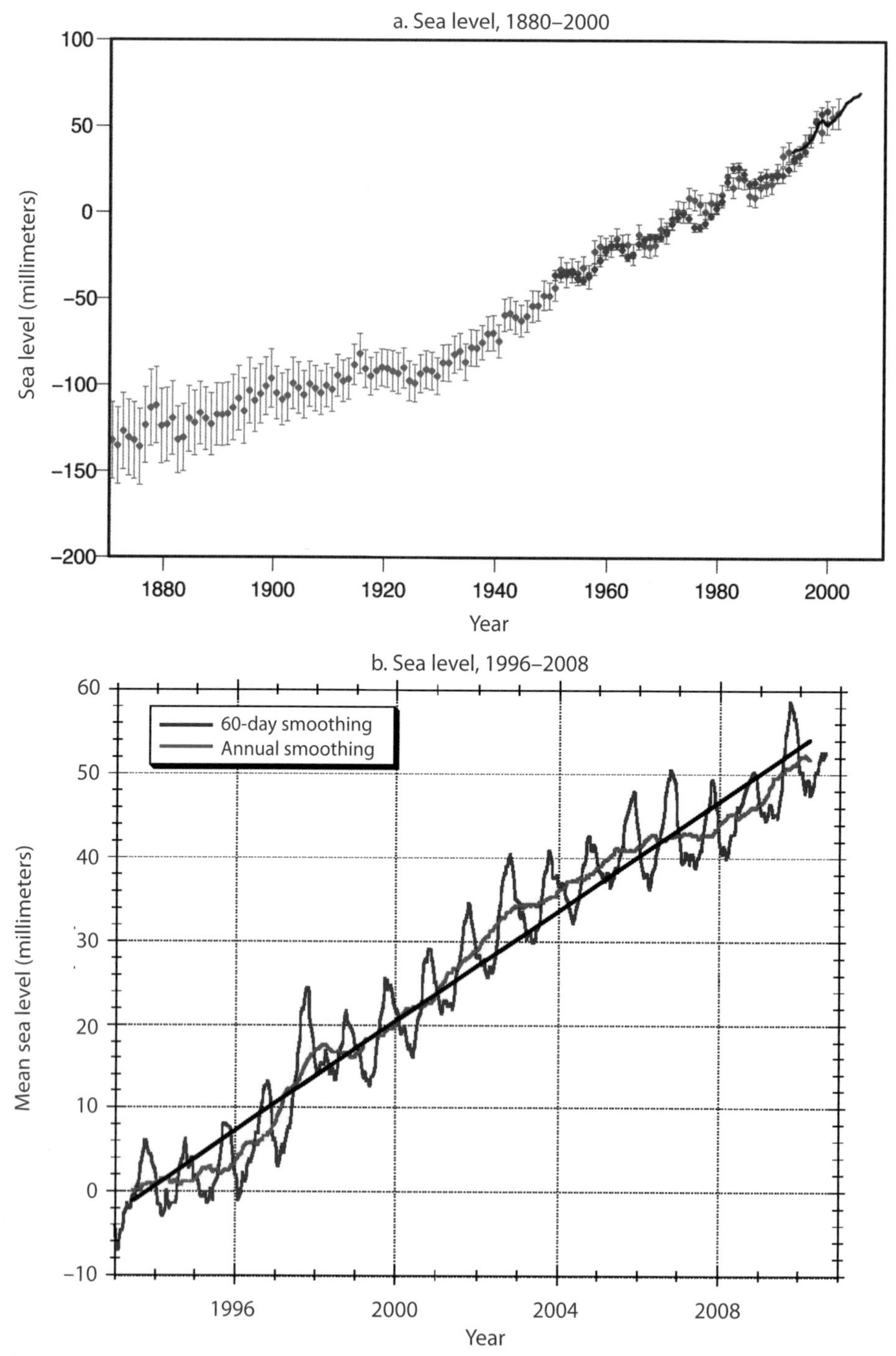

Source: IPCC 2007; Nerem et al. 2010.

Note: Panel a: Reconstructed sea-level fields since 1870 (red), Tide gauge measurements since 1950 (blue), and Satellite altimetry since 1992 (black). Panel b: Updated satellite estimates. Units are in millimeters relative to the average for 1961–90. Error bars are 90 percent confidence intervals.

to sea-level rise over the next 100 years. New estimates suggest a sea-level rise of around one meter or more by 2100 (IOP Science 2009). Because deep ocean temperatures change slowly, thermal expansion will continue for many centuries even if atmospheric greenhouse gas concentrations are stabilized today.

In a warmer climate, models suggest that the ice sheets could accumulate more snowfall, which would lower the sea level. However, in recent years, any such tendency has probably been outweighed by accelerated ice flow and greater discharge. The processes of accelerated ice flow are not yet completely understood, but they are likely to continue to result in overall net sea-level rise from the two large ice sheets of Greenland and Antarctica. Recent studies (for example, Grinsted, Moore, and Jevrejeva 2010) have demonstrated that a committed future global sea-level rise independent of emission scenarios could be as high as 1.5 meters.

The greatest climate- and weather-related impacts of sea level will be felt during extremes associated with tropical cyclones and midlatitude storms, on time scales of days and hours. Low atmospheric pressure and strong winds produce large local sea-level excursions called *storm surges*, which are especially serious when they coincide with high tide. Changes in the frequency of the occurrence of these extreme sea levels are affected by changes both in mean sea level and in the meteorological phenomena causing the extremes.

A New Round of Improved Projections Is Coming

To prepare for the upcoming IPCC report (AR5), a new, coordinated modeling effort is under way. The effort will be more focused on developing consistent downscaled outputs for all terrestrial regions of the Earth and on improving the projection of changes in extreme events. The first results of this effort have already appeared and will continue to emerge within the next one to two years. The projections are expected to considerably improve scientists' understanding of future climates.

IPCC AR5 Will Bring a New Framework for Modeling Climate

A new generation of climate models

Since the IPCC AR4 was published, increasing efforts by the climate modeling community have continued to address outstanding scientific questions that arose as part of the assessment process; improve understanding of the climate system; and provide estimates of future climate change. A new set of coordinated climate model experiments, known as phase 5 of the Coupled Model Intercomparison Project (CMIP5) (Taylor, Stouffer, and Meehl 2011), has become a high priority on the research agendas of most major climate modeling centers around the world. The results from this new set of simulations are expected to provide valuable information and knowledge that is particularly relevant to the next international assessment of climate science, such as the IPCC AR5.

Compared to the previous generation of models that contributed to the IPCC AR4, the climate models participating in CMIP5 have greatly improved through the adoption of new findings about parameterizations of subgrid-scale physical processes, inclusion or further development of aerosol schemes, carbon cycle models, variable vegetation cover, and more. In particular, a subset of the new models explicitly incorporates the global carbon cycle as one of the model components (subsequently referred to as *earth system models* or ESMs). These coupled carbon-climate model simulations provide a way of diagnosing the role of carbon-climate feedback and quantifying the allowable emissions for a given climate change target.

A new way of making climate change scenarios

A set of new emission scenarios has been adopted in the CMIP5 framework (Moss et al. 2008, 2010). Unlike the SRES scenarios used during the past two decades, which explore only emission pathways in the absence of climate policy, the new scenarios are defined by the radiative forcing levels (that is, the climate signal) of representative concentration pathways (RCPs) that are compatible with socioeconomic development, including adaptation and mitigation (van Vuuren et al. 2010). Four RCPs are selected for CMIP5 experiments: one nonmitigated (RCP8.5) and three that take into account various levels of mitigation (RCP6.0, RCP4.5, and RCP2.6), with labels according to the approximate target radiative forcing in Wm^{-2} at about 2100 (figure 2.6). For comparison, the figure also shows three SRES emission scenarios. In comparison with the SRES, the RCPs provide more regionally detailed scenario information, such as aerosol emissions, geographically explicit descriptions of land use and related emissions and uptakes, and detailed specifications of emissions by source type, which are needed by new advances in climate models.

A new focus on near-term changes

A new goal of the CMIP5 experiments is to provide decadal climate predictions for the near term (out to about 2035). These experiments will be carried out with atmosphere-ocean global climate models (AOGCMs) that are properly initialized for the ocean and perhaps also sea ice and land surface, using either observations or initialization methods that have been developed recently. Some of the decadal simulations are expected to be performed using higher resolution to better resolve regional climate and extremes. An enhanced resolution in such experiments may enable a global horizontal grid scale as high as 50 kilometers, which is equivalent to the scale of RCMs currently used in downscaling studies. These experiments will also be able to better separate natural climate fluctuations from those that are anthropogenic (see also Hawkins and Sutton 2009, 2010).

More Effort Is Needed to Develop Good Impact Models to Translate Climate Change Projections to Potential Impacts

Great uncertainty remains about the precise magnitude of changes in climate and weather patterns at any given location and time. Some of this uncertainty derives from the inherently chaotic nature of the weather and climate system and from limitations in the models, but much uncertainty is linked to the future of human actions that release greenhouse gases and change land cover. Despite these uncertainties, decisions still have to be made about water management, agricultural practices, land use, and infrastructure, with the consequences of the decisions extending well into what climatologists know will be significantly changed climates. These decisions are often guided by impact models that, among many other factors, take into account climate and climate variability (box 2.6). Impact models have typically been guided by climate observations from the re-

FIGURE 2.6

Representative Concentration Pathways Compared with the Older SRES Scenarios

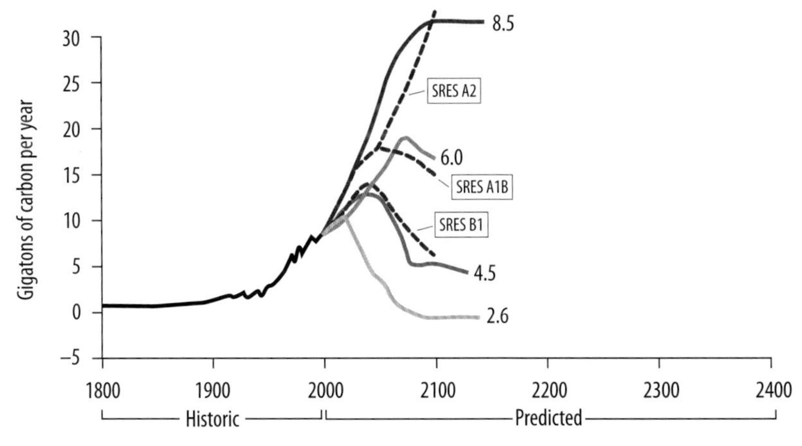

Source: Based on Moss et al. 2010.

Note: Energy and industry carbon dioxide emissions for the four RCP candidates for CMIP5 experiments. Three of the most commonly used Special Report on Emissions Scenarios (SRES) are shown for comparison.

cent past, such as 1961–90 weather data, but the past is no longer a good proxy for the future. The challenge is to find ways of representing both the understanding of future climate and the uncertainties in models of water use, decisions about land use, and infrastructure design.

A number of steps are involved in deciding how to incorporate future climate change and variability into impact models. The first step is to assess how the impact model considers current climate variability. Many models, in particular many economic models, treat climate as a constant, and they factor climate variability into a general error term or safety factor, as in bridge and building design. In such cases, the design of these decision processes will have to be reassessed to consider whether the safety margins can simply be extended, and if so, whether these margins will still lead to effective and efficient decisions. The challenge is to replace inputs from the recent past with inputs from well-selected models of the future. However, many of these future models require data at a

BOX 2.6

The Practice of Impact Modeling

National and regional collaboration: Meteorological services should enhance regional collaboration on early warning systems. This goal includes the use and dissemination of existing extended forecasts, which are available through the World Meteorological Organization (WMO) and other international institutions. Collaborative projects should exploit existing or new initiatives within international bodies of which Arab countries are members. Available products for the analysis of climate change risks and for impact assessment should be used. The Global Framework for Climate Services, launched by WMO, depends crucially on the active participation of all member states. Capacities to contribute to these efforts and to explore the outcome are much needed in most of the Arab countries.

International collaboration: International Centers of Excellence should be established, and existing ones, such as the King Abdullah University of Science and Technology in Saudi Arabia, should be better promoted on a regional scale. This expanded access can be achieved through staff exchange and enhanced collaboration at a regional and international scale.

Source: Authors' compilation.

much finer resolution than can be produced by GCMs, or even RCMs, and at time scales that are often not available from modeling runs, such as hourly or daily data.

Improving Climate Model Output Resolution in Both Space and Time Scales Is a Challenge

The horizontal resolution of GCMs is typically 100–200 kilometers, which is usually too coarse for direct use by impact models. Climate projections with higher spatial resolution can be obtained by dynamical downscaling, using high-resolution RCMs driven by initial and boundary conditions provided by GCMs (for example, Rummukainen 2010). This improved resolution makes it possible to include subgrid variability, such as small-scale topography (see box 2.1), and to resolve subgrid processes to improve the modeling of regional rainfall, for example. Indeed, a number of studies show that the results from impact models using downscaled climate information can be quite different from results directly based on GCM output (for example, Mearns et al. 2001; Olesen et al. 2007).

Different GCMs and RCMs use different physical formulations to describe atmospheric processes—that is, interactions between the atmosphere and the land and oceans—and thus provide different projections of future climate. The spread of results from these different models represents an uncertainty regarding future climate that must be accounted for by impact modelers; therefore the use of a single GCM or RCM should be avoided. However, the uncertainties are now being explored by ensemble simulations that apply different combinations of GCMs and RCMs (see, for example, J. H. Christensen, Carter et al. 2007; Kjellström and Giorgi 2010). One way of directly including this uncertainty is to calculate the probability distribution of important outputs, such as daily maximum temperatures, on the basis of information from the full set of model simulations (Déqué and Somot 2010).

In recent studies, the contribution of individual models to the probability distribution has been weighted according to each model's performance with respect to a given set of metrics (J. H. Christensen et al. 2010). The weighting procedure itself adds an extra source of uncertainty, and the method is still being explored. For example, the ability of a model to predict present-day climate could be seen as an indication that it will also be able to predict future climate. However, feedback mechanisms that were important in the past could be less important in a future climate with a different forcing, so the weightings themselves may not be appropriate (J. H. Christensen et al. 2008; Reifen and Toumi 2009). In addition, models may perform differently for different regions. The Coordinated Regional Climate Downscaling Experiment (CORDEX) initiative will study this function and offer new regional projections for a number of regions across the world. The initiative includes projections for the Arab region, although not as a single entity, but instead distributed across several proposed regional domains (Giorgi, Jones, and Asrar 2009).

Most downscaling methods combine climate model output with observed data. In this way, local features not captured by the climate model can be incorporated in the scenario data, allowing the impact model to be used with data in the form they were developed and tailored for. The most appropriate method of preparing climate scenario data depends on the local region and on the specific needs of the impact models.

Any dependence on time series data to calibrate and validate an impact model prevents the direct use of GCM or RCM data, because the climate models do not reproduce sequences of real weather events. The models are neither designed nor meant to be able to do that. Instead, a climate model, as the name suggests, is constructed with the aim of being able to reproduce climate, which, in a narrow sense, is usually defined as the average weather. More rigorously defined, climate is the statistical description in terms of the mean and variability of relevant quantities, such as temperature or precipitation, over a period of time, ranging from months to thousands or millions of years. Many climate variables in the model output are systematically offset and need to be bias corrected, which requires observational data.

Methods to transform climate model output into the impact application are therefore called for. The climate change scenarios can be further downscaled to a regional or local level by applying a weather generator (WG) approach. For example, a stochastic WG based on the series approach (Racskó, Szeidl, and Semenov 1991) has been proven to give a realistic representation of the duration of periods of wet and dry days and, thus, the duration of droughts, whereas the frequently applied Markov chain models (Semenov et al. 1998) are less successful in achieving realistic representation. The WG approach has the benefit of allowing for a good representation of current climatic conditions by calibrating the WG to observed data and allowing not only the inclusion of changes in mean climate from the RCMs (or GCMs), but also changes in climatic variability. This approach also has the flexibility to generate longer and multiple time series of synthetic weather data for use in impact models, thus allowing for better quantifying of the variability in response to climate change.

Even if suitable time series of weather data are generated to drive an impact model, the question remains as to what extent the impact model itself is calibrated and tested for weather conditions beyond those experienced in observed conditions. This information should be clarified when the model is used under climate change conditions, in particular when modelers are addressing values well outside past experiences or when extreme events are being assessed.

A Seamless Approach from Weather Forecasting to Seasonal-to-Decadal Projection Is a New Concept of Climate Prediction

As discussed earlier, a climate model is designed to be able to capture the statistical properties of the geographically and temporally varying weather that characterizes climate; it is not designed to predict the actual weather. A weather prediction or forecast, on the other hand, is typically formulated as a categorical statement about the state of the weather within a certain time frame, typically up to 10 or 15 days (medium-range forecasts). More recently, monthly and seasonal time-scale forecasts are also provided by many centers. These forecasts typically provide qualitative assessments of the most likely weather development, primarily based on long-term variations, such as expectation of droughts or moist conditions. A commonality between all forecasts is the dependency on a well-defined initial state that must have its origin in an observed state of the atmospheric (or, more broadly, climate) system. This assessment typically involves some kind of data assimilation system by which the forecast model is constrained toward the observed evolution of the weather or oceanic state, moist or dry land surfaces, and so forth. The techniques to adequately combine modeled information, typically represented by a model grid box, with point measurements at which most observations are made, comprise large research fields of their own.

In climate modeling, the model typically begins from an idealized initial state, which aims to represent conditions "typical" for the period it is intended to represent, often preindustrial conditions in which the levels of greenhouse gases and anthropogenic aerosol loads in the atmosphere were low. Then the known or estimated external forcings from the emission of greenhouse gases and aerosols, varying solar insolation, volcanism, and other drivers (for example, land use) are introduced into the freely running model. In this way, a modeled representation of the evolution of the climate since the industrial revolution is made. But the state of the climate in such a model for a particular decade, say 2001–10, is not meant to be compared with the real world for that decade. The model is designed to capture the changing characteristics of major phenomena such as El Niño's decadal fluctuations in monsoons or the North Atlantic Oscillation but not to forecast particular events.

This aspect of modeling is a problem when the challenge is to address the near-future climate changes, because natural fluctuations on the decadal time scale are major determinants. In essence, the probable evolution of climate over the next three to four decades is more difficult to depict than the climate at the end of the century under a prescribed emission scenario. This shortcoming is very often forgotten when impact assessments are called for.

Some predictability in the atmosphere on seasonal, interannual, and decadal time scales can arise from internally generated natural climate variability—often connected to oceanic variability—and certain types of external forcing (solar and volcanic eruptions). Internal variability that results in, for example, extensive, long-lasting upper ocean temperature anomalies has the potential to provide seasonal, interannual, and even decadal predictability in the overlying atmosphere, both locally and remotely through atmospheric "teleconnections." These teleconnections can guide predictions for the next few seasons. The question of whether or not skill extends to forecasts for the following decade and beyond is a scientific hot topic and currently a central issue in the upcoming AR5. A pioneer work of Smith et al. (2007) has demonstrated that the forecasting skill for surface temperature is substantially improved up to a decade with the global climate model system DePreSys, which takes into account the observed state in the atmosphere and ocean and thus predicts both internal variability and externally forced changes. Map 2.16 shows an example of the forecast of regional rainfall anomalies out to 2017 based on the DePreSys (Wilby 2007b). The model predicts strong reduction of rainfall over the lower Nile and Arabian Peninsula for the period of 2007–17. However, much more research is needed to understand whether the signals are robust, and if so, the underlying physical mechanism.

The level of predictability and apparent predictive skill arising from both internal and external forcing can vary markedly from place to place and from variable to variable. Recent research indicates that predictability is absent or very limited in some variables in most locations over the surface of the earth (see, for example, Hurrell et al. 2010; Latif et al. 2010; Meehl et al. 2009; Murphy et al. 2010). So climatologists cannot accurately predict all aspects of the climate over the coming decades even if they could perfectly resolve all the technical issues confronted when developing and conducting predictions. The nature of the climate system precludes the possibility of reliable decadal predictions of some climate variables in some locations. In such cases, the best estimate that can be provided for the quantities and regions in question for future decades is the information contained in historical climate records.

Nevertheless, there is some potential for seasonal prediction in the Arab world. As an example, there is a relationship between the phase of the Southern Oscillation Index (SOI, a measure of the phase of El Niño) in August–September and rainfall anomalies in Djibouti the following October–December. A negative SOI indicates El Niño episodes, wetter than normal conditions, and an increased probability for the outbreak of diarrhea and cholera, which are waterborne diseases (Rob L. Wilby, personal communication with author, 2011). Operational regional seasonal forecasts are issued by several institutions, including the Greater Horn of Africa Consensus Climate Outlook,[1] the International Research Institute for Climate and Society (IRI),[2] and the U.K. Met Office.[3]

MAP 2.16

Rainfall Anomalies for 2007–17

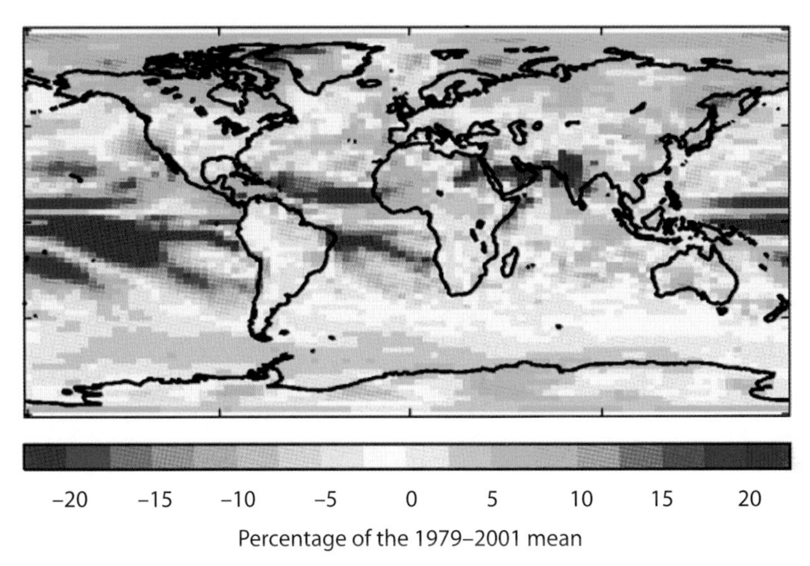

Source: Wilby 2007b.

Note: Predicted by the U.K. Met Office's DePreSys, given information on surface ocean heat content up to June 2005.

WMO has launched a new large-scale climate services initiative, the Global Framework for Climate Services (GFCS), with the goal to "enable better management of the risks of climate variability and change and adaptation to climate change at all levels, through development and incorporation of science-based climate information and prediction into planning, policy and practice" (WMO Secretariat 2009, 2). It will provide on-demand information such as regional weather and climate predictions to policy makers, businesses, and individual farmers and fishermen in particularly vulnerable parts of the world. GFCS will act as a tool to proactively reduce disaster risk across all levels of society and to support disaster risk management and climate risk management practices. This initiative should be of particular interest in the Arab countries, as it can provide access to climate data and help in preparing for the expected impacts of climate change.

The best choices for integrating climate modeling projections with impact modeling depend on location, time period, and application, and are best made through the cooperation of climate modeling and impact specialists. A discussion of the myriad options is not possible in this chapter, but annex 2B provides some guidelines for program leaders responsible for developing impact models. The guidelines will help with decisions that are likely to be affected by climate change, by stepping through the process of developing advice guided by the best available understanding of future climates.

ANNEX 2A

Monthly and annual rainfall and temperatures in the Arab World

	Rainfall (millimeters)						
Economy	January	February	March	April	May	June	July
Mean Monthly and Mean Annual Rainfall and Interannual Rainfall Variability							
Algeria	10	9	9	8	7	3	2
Bahrain	22	14	13	7	0	0	0
Comoros	320	238	204	227	135	133	107
Djibouti	9	14	28	23	11	2	27
Egypt, Arab Rep.	5	4	4	2	4	2	3
Iraq	35	31	31	28	11	1	0
Jordan	24	23	17	8	3	0	0
Kuwait	22	14	17	16	5	0	1
Lebanon	160	130	88	49	15	3	0
Libya	9	6	5	3	2	1	0
Mauritania	1	1	0	1	1	7	21
Morocco	39	41	45	36	21	8	2
Oman	7	11	12	13	6	8	11
Qatar	7	20	17	7	2	1	2
Saudi Arabia	6	6	12	16	9	2	4
Somalia	3	4	11	54	54	16	15
Sudan	1	2	8	21	44	56	91
Syrian Arab Republic	55	47	41	32	14	2	0
Tunisia	34	29	29	23	16	8	3
United Arab Emirates	12	14	16	7	2	0	0
West Bank and Gaza	129	99	60	20	3	0	0
Yemen, Rep.	7	8	16	24	18	10	27

	Temperature (°C)						
Economy	January	February	March	April	May	June	July
Mean Monthly Temperature							
Algeria	12.0	14.6	18.0	22.0	26.4	30.8	32.6
Bahrain	15.9	17.9	21.8	26.9	32.3	34.9	36.6
Comoros	26.0	25.9	26.2	25.9	25.0	23.5	22.8
Djibouti	23.5	24.4	26.0	27.7	29.4	31.6	32.4
Egypt, Arab Rep.	13.2	14.8	17.9	22.4	26.5	28.8	29.3
Iraq	9.0	11.1	15.0	20.7	26.5	30.8	33.3
Jordan	8.4	9.9	13.0	17.6	22.1	25.0	26.8
Kuwait	12.4	14.5	19.0	24.2	30.5	34.5	36.0
Lebanon	6.5	7.0	9.6	13.4	17.2	20.5	22.7
Libya	12.5	14.7	17.9	22.1	26.4	29.1	29.4
Mauritania	20.2	22.2	25.0	28.0	31.3	33.4	33.3
Morocco	9.4	10.8	13.1	15.3	18.8	22.7	26.0
Oman	19.9	20.7	23.3	26.6	29.4	29.9	28.5
Qatar	18.4	19.3	22.4	26.5	30.6	32.4	33.7
Saudi Arabia	15.4	17.1	20.4	24.5	28.9	31.5	32.0
Somalia	25.1	25.9	27.3	28.1	28.3	27.9	27.1
Sudan	22.1	23.7	26.4	28.9	30.2	30.1	28.9
Syrian Arab Republic	6.1	7.8	11.3	16.3	21.6	26.0	28.8
Tunisia	9.9	11.4	14.3	17.4	21.5	25.9	28.6
United Arab Emirates	17.1	18.3	21.6	26.2	31.2	33.9	35.0
West Bank and Gaza	10.8	11.6	13.7	17.3	21.0	23.5	25.4
Yemen, Rep.	18.4	19.3	20.9	23.6	26.0	27.9	27.0

Source: Authors' compiliation.

Note: CV = coefficient of variation

August	September	October	November	December	Mean annual	Interannual variability (CV, %)
4	6	8	12	10	88	75
0	0	0	5	18	79	51
82	33	51	72	171	1770	18
54	25	7	11	8	219	50
3	1	3	3	5	37	56
0	0	7	23	30	197	23
0	0	3	11	19	110	36
1	0	2	16	22	114	31
0	3	31	85	140	702	25
1	2	5	6	9	48	50
41	26	5	1	1	105	49
4	14	29	43	48	331	33
9	3	5	5	8	97	50
2	1	1	2	9	70	47
4	1	2	7	7	76	25
10	13	43	31	16	269	35
109	66	35	8	2	442	43
0	2	16	34	50	292	26
6	21	31	32	33	262	34
0	0	0	4	13	69	53
0	0	14	60	113	498	29
28	12	6	9	9	175	28

August	September	October	November	December	Mean Annual
31.9	28.7	23.5	17.5	13.0	22.6
36.1	33.3	28.7	23.0	18.0	27.1
23.1	23.9	25.1	26.0	26.1	25.0
31.7	30.2	27.6	25.4	23.8	27.8
29.4	27.5	24.6	19.3	14.7	22.4
33.1	29.5	23.9	16.3	10.5	21.7
27.1	25.0	21.1	15.1	10.1	18.4
35.8	32.7	27.1	20.0	14.1	25.1
23.3	21.4	18.1	12.8	8.0	15.0
29.2	27.4	23.5	18.2	13.8	22.0
32.2	31.2	29.4	25.0	20.6	27.7
26.1	22.6	18.2	13.7	10.3	17.3
27.4	27.3	26.1	23.7	21.4	25.3
33.5	31.6	28.2	24.0	20.3	26.8
32.1	30.2	25.8	20.9	16.8	24.6
27.1	27.7	26.8	25.8	25.1	26.9
28.4	28.5	28.0	25.2	22.6	26.9
28.7	25.3	20.0	13.0	7.7	17.7
28.7	25.7	20.8	15.3	11.3	19.2
34.7	32.5	28.9	24.1	19.3	26.9
25.9	24.5	21.9	17.1	12.6	18.8
26.8	26.1	23.1	20.4	19.2	23.2

Annex 2C: Dealing with Climate Risks—A Checklist

This checklist is designed to guide the leadership of a program or a large sectoral or cross-sectoral project that is assessed as being subject to climate risks. Examples include the construction of a new dam, a major rural road upgrade, the design of an agricultural irrigation program, or a major coastal tourism development. The checklist is designed to assist in incorporating climate risk, including climate change, in the program design. It is essentially a top-down approach to adaptation that unfortunately so often brings adaptation considerations rather late to the decision process (Wilby and Dessai 2010). This checklist is not the only—or necessarily the most effective—approach to adaptation planning.[4] However, the circumstances described here are common in development planning, especially for infrastructure. The checklist emphasizes opportunities and the need to consult widely with stakeholders throughout the process. Note that the order of the checklist is such that entries are *not* meant to reflect a purely sequential process.

Climate Risk Checklist for Policy Makers

Process	Comment
1. *Assess where you stand in the decision process.*	
a. If the decision process is in its early stages, then follow a process to identify the goals and needs of stakeholders, the risks from current and future climates, and the range of feasible adaptation options and criteria by which they may be evaluated.	Clearly, this is the best stage to begin considering adaptation actions. It allows the full range of viewpoints and climate-resilient options to be considered. This checklist cannot elaborate the open-ended and case-specific stakeholder engagement process in detail.
b. Are the core features of the design in place? If so, what information relating to future climate might challenge these existing decisions?	Here the assumption is that adaptation to climate risk and climate change is a component of a wider development objective (that is, this is not a stand-alone adaptation project).
c. What are the feasible options for modifying or fine-tuning the existing design?	These options need to be identified so that appropriate decisions can be made about how much assessment of the climate risks is needed. At this point, simple "rules-of-thumb" about climate change may need to be applied (for example, using expert judgment to assess the effect of a warmer climate with more variable rainfall on the proposed project).
d. What is the time horizon relevant to the decision process?	The time horizon is not necessarily limited to the longevity of a piece of infrastructure. For example, implementing improved flood resistance of a planned road system may be sufficient for the normal design horizon for roads, but it could also lead to denser settlement in inherently flood-prone areas, leading to unacceptable outcomes in the climates of the more distant future.
e. Consider the application of decision scaling (Brown 2011).	Decision scaling is a methodology that explores the climate sensitivities of a decision and then tailors the climate information that is gathered to focus on those sensitivities. This method can lead to quicker, more efficient, and more relevant input to the decision process.

Process	Comment
2. *Gathering the basic climate information.*	
a. What do the latest IPCC report and subsequent commentaries and updates say about your project location and sectors?	The IPCC reports will contain a summary of the current climate, observed changes, climate change projections, and possible impacts on your region. They are available at http://www.ipcc.ch.
b. Are more recent (since the most recent IPCC report) or more nationally or regionally explicit assessments available?	
i. Many countries have their own national assessments of climate change and its impacts.	Check national communications to the United Nations Framework Convention on Climate Change at http://unfccc.int and search for "National Reports."
ii. Assess the knowledge base and the validity of all assessments.	Many national and subnational reports are summaries from the IPCC with little additional information.
iii. Identify major discrepancies (if any) between the regional and explicit assessments of the IPCC and seek to establish an explanation.	Discrepancies may arise because of newer information, more regionally specific information, and different sources of information. Sometimes, however, they arise from errors and misinterpretations.
3. *Establish a climate baseline.*	
a. Seek cooperation and input from the national meteorological office and similar authorities (for example, for sea-level monitoring).	The climate baseline is a description of the current climate (often a 30-year period, such as 1961–90), including averages, variability, and trends. It is used to establish the current conditions, and most modeled projections (see section 4) are interpreted as changes to this baseline. Baseline climate information may exist in a form that is not publicly available. For example, data may be held by other government agencies, such as those responsible for agriculture or water management, or by the military. Many data may not exist in digitized form or, if data are digitized, basic quality control may still be lacking
b. Based on sections 1 and 2, decide on the variables that may affect the outcome and design of the project.	These variables are not only temperature and rainfall. They are more likely to be complex variables, such as runoff, dry spells, rainfall intensity, and wind and dust storms. Some of these variables are available from GCM modeling, whereas others will have to be derived through secondary modeling.
c. Explore the existing public databases for both station and gridded data.	See references in chapter 2 for coverage of the Arab region (for example, information provided to produce maps 2.1, 2.2, and 2.3). See also box 2.3.
d. Calculate averages, trends, and measures of variability for the variables of interest.	See the World Bank Climate Change Knowledge Portal, where much of this information is available (http://sdwebx.worldbank.org/climateportal/).
e. Share the baseline data with appropriate experts (sectoral and climate), and review and revisit section 1 if necessary.	These analyses may give insights to additional options in the project design.
4. *Establish the relevant range of climate projections.*	
a. Currently, 22 IPCC AR4 models form a core set and are available from the website listed opposite.	The IPCC Data Distribution Centre is located at http://www.ipcc-data.org/. During 2012 and 2013, results from a major new modeling effort will begin to become available. Consider how these results might affect or be taken into account in the decision-making process.

(table continues next page)

Climate Risk Checklist for Policy Makers

Process	Comment
b. Assess whether particular scenarios should be chosen (that is, A1B, A2, B1, B2). Note that for projections in the near term (next few decades), the choice of scenario makes little difference.	The current trajectory suggests A1B is close to being surpassed. A common pairing is B1 (most effective mitigation) with A1B (least effective mitigation).
c. Do not reject any GCM unless data are missing or an authoritative climatological reason indicates the data are inappropriate.	Some users seek a single "best" GCM for their region. *This approach is unwise because the criteria for choosing the "best" GCM are unclear.* One GCM might project current rainfall better in your region, whereas another matches temperatures better. Also, the ability to project current climate does not necessarily indicate the ability to project future scenarios.
d. Check the performance of each GCM selected for a broad match with the major observed meteorological phenomena for your location, such as rainy seasons, timing of monsoons, and so forth. If major discrepancies are found, seek further technical help before using that model.	Poor performance of a GCM for your region in projecting major patterns of seasonality, such as monsoon patterns or rainfall seasonality, is an a priori rationale for rejecting that GCM. But cases may exist where all models appear to be disqualified. In such cases, seek further technical assistance.
e. Check climate change projections for each of the selected models to see if any appear to be outliers (that is, they produce results very different from the other models or strong discontinuities between observed and near future projections). If major discrepancies are found, seek further technical help before using the model.	A very different climate projection for a single model may indicate that the model should be disregarded. However, the difference may also be due to internal variability, which simply indicates that the climate change signal for your region is not a robust feature in that model.
f. As a minimum requirement, explore a plausible range by using GCMs with a low, medium, or high global mean temperature or rainfall response to a particular scenario.	Use of a bracketing set of GCMs is a common approach when running the impact modeling or assessment is time consuming or expensive.
5. *Determine the climate projection data needed to drive the climate impact models and to assess impacts on the decision making.*	An enormous range of approaches is available for evaluating the impacts of climate change and variability, including well-established empirical damage formulas, watershed models, agricultural yield and production models, and so on. These "impact models" cannot be dealt with in detail here. Instead, some generic advice, applicable in most situations, is provided.
a. If climate was treated as a constant (that is, an *assumption of stationarity*) in the original design, then identify and focus on project performance criteria that might affect decision making and that are also likely to be affected by different climate projections.	For example, if standard hazard models (such as 1-in-100-year flood levels) were used, can they be updated for the different climate scenarios? If the hazard models are difficult to update, what would be the effect of raising the safety threshold (for example, from 1 in 100 years to 1 in 200 years)?
b. If qualitative methods were used to assess the project options (expert judgment, Delphi techniques, and so forth), climate risks may or may not have been incorporated. Go to section 8.	
c. If quantitative impact models were used, which climate variables were used as inputs?	These variables are usually well prescribed in the documentation of quantitative impact models.
i. Has the climate response of the impact model been sufficiently tested (validated) against observational data? If not, what was the justification for the impact model's selection?	This basic test determines whether the selected impact model is fit for its purpose.

Process	Comment
ii. Does this validation still apply to a changed climate?	For example, would a changed climate go beyond the range or domain of validation, especially for the impact of extreme events? Or might climate change introduce new phenomena not considered in the original model, such as salinization of cropland or even flooding in a location currently safe from flooding?
iii. If not, you may proceed, but with cautious interpretation of the modeling results. Also, seek additional ways of validating the models under the changed conditions.	
iv. How sensitive are the final outcomes of the impact models to the climate parameters?	The sensitivity of the model to variation in the input parameters should be a part of model development and testing. It may be that the important outputs from the impact model are sensitive to only some of the climate inputs. This step will help to focus on best describing the important climate variables. For example, a crop growth model may not be sensitive to seasonal changes in mean rainfall but very sensitive to the timing of the first rains of the wet season. Are your climate models able to provide the information that matters?
v. Do the models treat climate as a fixed (for example, use climatic means) or a stochastic (for example, use historical weather data) input?	If the model treats climate as fixed, you will need to design a series of model runs covering the range of climate projections being considered. Usually, this means running the impact model under the current climate scenario and, for example, a +1°C scenario, a +2°C scenario, and so on. If the model uses stochastic weather patterns, you will need to consider how to generate realistic weather sequences for the future climate projections (for example, via a "weather generator," which generates random weather sequences within a prescribed bound of means, variability, and cross correlation between variables, such as temperature and rainfall). *Do not use raw output from either GCMs or RCMs*. These are models of climate and not of "weather." Seek further advice if needed. Also see section 6.
vi. Might the modeling results under a changed climate affect other components of the modeling or the decision-making process?	For example, the current planning may have implicit or explicit assumptions about the availability of water for irrigation, constant trends in yields, or constant commodity prices. These assumptions may not be valid under a changed climate. If so, the impact modeling may need to be modified or extended to deal with these additional factors.
d. Are the climate data gathered in section 4 sufficient to rerun the impact models under changed climate scenarios?	
i. If not, can the necessary data be provided from the raw data repositories of the model runs?	The GCMs provide extremely detailed descriptions of the Earth's atmosphere and land surface for every few minutes of their run. Most of these data are not saved, but far more data are saved than appear in most compendiums and may be available from the modeling groups that conducted the experiments.

(table continues next page)

Climate Risk Checklist for Policy Makers

Process	Comment
ii. Can proxies for the missing data be found? Can the missing data be assumed constant, or can the impact models be run with a set of estimates of the missing data that bracket the expected range?	Some aspects of climate change are known to scale more or less linearly with global temperature. Assessing whether this may apply in a specific context requires technical assistance. Some guidance is also available in IPCC reports (for example, chapter 11 in the contribution from WG-I to AR4). In some cases, you can exchange place for time. For example, if data for a particular application in a region does not exist, analogue current climates may be identified, and pseudo data from such a place may be used as a proxy. In any case, further technical advice should be sought.
e. How important are other variables (such as population trends, commodity prices, increasing demand, and yield improvements) in determining the outcomes of the modeling or in guiding the decision? In other words, what sort of climate change might cause you to change or revise the project design. *Does a detailed treatment of climate matter in your decision process?*	This step is essentially a revisit of section 1 to ensure that you are still on track.
f. If the answers to these questions suggest that climate change may affect the project design and that the impact models can incorporate climate change effectively, then proceed to run the impact models under a representative range of climate scenarios determined in section 4b.	
6. *Consider how to apply the changed climate to the impact model.*	
a. Direct use of GCM, or even RCM, output is rarely applicable.	Most GCMs have well-documented biases, such as a tendency to be clearly too wet or too dry in a particular region, to exhibit too weak an annual cycle for certain climate parameters, or to systematically underrepresent extreme values (such as heavy rainfall). Such biases must be corrected for before using the GCM output for driving impact models.
b. The most common approach is to use changes in projection between a baseline period and some future climate (often called *deltas*).	Future climate projections are calculated as the difference between the model results for the time period of interest (for example, 2030–60) and the model results for the baseline period (for example, 1961–90)—that is, the delta. This difference is added to the observed climate data for the same baseline period to estimate the future climate.
c. If the decisions are sensitive to the impact model's treatment of climate variability, seek expert advice on the best downscaling methods.	Delta change in its most simple form will preserve the higher-order climate statistics of the *current* period (that is, its variability, extreme events, longer-term cycles). But climate variability and other factors may change in the future.
7. *Reassessing qualitative treatments of the impacts of climate, climate variability, and climate change on the decision process.*	
a. In most complex programs or projects, some elements of the design and decision process will be based on qualitative assessments. These elements should still be subject to scrutiny for climate impacts.	For example, the task may be a program for building schools in remote areas. Detailed climate modeling should not be necessary, but it would be useful to consider the environment in which these schools will operate over their lifespan (for example, higher temperatures, higher flood risk). Do such factors call for rethinking the building design or resiting?

Process	Comment
b. Involve climate experts in expert judgment exercises.	
c. Ensure that stakeholders, civil society representatives, and so forth are adequately briefed on climate risks and have the opportunity to apply this information to their inputs.	
d. Look at recent trends in extreme weather events and disasters for an indication of things to come.	The Centre for Research on the Epidemiology of Disasters database provides comprehensive disaster information (http://www.cred.be/), and a review of other sources can be found in Tschoegl, Below, and Guhar-Sapir (2006).
e. Apply sensible rules of thumb, such as the following:	
i. It will be a warmer world (the IPCC will give estimates for your location at future dates), but some places will also experience more extreme cold events because of changes in weather patterns.	Most impact specialists will be able to make an assessment as to the nature and extent of such changes in climate. They may advise that no changes in design are needed; be able to recommend adjustments (such as, in the school example, set the building further back from rivers, upgrade the water supply, and so forth); or conclude that further analysis and possibly modeling are needed.
ii. Rainfall will marginally increase in most regions (but check the IPCC reports), but most places will effectively be drier because of longer dry spells, higher temperatures, and increased evaporation.	Note that competition for water is increasing across all sectors. Climate change is likely to make that competition more intense and make allocation decisions more difficult.
iii. Rainfall probably will occur more erratically, with both extreme dry (drought) and extreme wet (flood) periods becoming more common.	
iv. Coastal regions will be subject to rising sea levels, but probably also to increased storm surges and wind damage, which are likely to be more important than direct sea-level changes in the near future.	
8. Cross-check the results of the impact modeling by whatever means available.	
a. Do the results scale sensibly with different levels of climate change?	For example, if you used GCMs representing low, medium, and high outcomes or scenarios representing low and high mitigation effectiveness, do the results of the impact modeling follow these trends in the way expected. If not, can plausible explanations be found, and can these explanations be further tested?
b. How do the results compare with previous studies?	
c. If a range of climate projections has been considered in the impact modeling, do any of the projections lead to unacceptable or undesirable outcomes?	Here you need to consider whether the option chosen is robust to the feasible range of climate outcomes. Consider whether you should apply robust decision making to the whole decision process (Lempert and Groves 2010; Lempert et al. 2004).
d. Have you left out any factors that you thought might be important but did not have enough information to include? If so, document the reasons and warn users of the omission. Could you make estimates that might bracket the scale or impact of these omitted factors?	For example, you may have good estimates of sea-level rise but no reliable estimates of changes in the height and frequency of storm surges and therefore omit such estimates. In terms of flooding impact, the storm surges are possibly far more severe and will affect you much earlier than sea-level rise.

(table continues next page)

Climate Risk Checklist for Policy Makers

Process	Comment
e. Have you overlooked some other factor that would "swamp" any climate signal?	For example, studies of crop yields under climate change often predict decreases of 10 percent to 20 percent by 2050, although there is still great uncertainty about whether some of this loss will be compensated by "carbon dioxide fertilization." But the overall impact model may have a built-in assumption that technological improvement will increase crop yields by 1 to 2 percent per year, as it has over past decades. If the latter estimate is wrong by only half a percent or so, and technological improvement is declining in many regions, this trend will swamp any climate change effects. Beware of "crackpot rigor"—that is, a very detailed analysis of the wrong problem.
f. Are you prepared to recommend changes in plans on the basis of the results?	
g. Would you risk your own money, livelihood, or life on your advice? You might be!	

Source: Compiled by Ian Noble and Jens Christensen.

Notes

1. Intergovernmental Authority on Development (IGAD) Climate Prediction and Applications Centre (ICPAC), http://www.icpac.net.
2. See the IRI website at http://portal.iri.columbia.edu/portal/server.pt?space=CommunityPage&control=SetCommunity&CommunityID=580.
3. See the U.K. Met Office website at http://www.metoffice.gov.uk/science/specialist/seasonal.
4. See Hallegatte, Lecocq, and de Perthius (2011); Lempert and Groves (2010); Lempert et al. (2004); Wilby and Dessai (2010); Wilby et al. (2009); and World Bank (2009) for a wider discussion of adaptation in the context of wider development goals.

References

Almazroui, Mansour. 2011. "Sensitivity of a Regional Climate Model on the Simulation of High Intensity Rainfall Events over the Arabian Peninsula and around Jeddah (Saudi Arabia)." *Theoretical and Applied Climatology* 104 (1–2): 261–76. doi:10.1007/s00704-010-0387-3.

Bengtsson, Lennart, and Kevin I. Hodges. 2006. "Storm Tracks and Climate Change." *Journal of Climate* 19 (15): 3518–43. doi:10.1175/JCLI3815.1.

Black, Emily. 2009. "The Impact of Climate Change on Daily Rainfall Statistics in Jordan and Israel." *Atmospheric Science Letters* 10 (3): 192–200. doi:10.1002/asl.233.

Bonsal, Barrie R., Xuebin Zhang, Lucie A. Vincent, and William D. Hogg. 2001. "Characteristics of Daily and Extreme Temperatures over Canada." *Journal of Climate* 14 (9): 1959–76.

Brown, Casey. 2011. "Decision-Scaling for Robust Planning and Policy under Climate Uncertainty." World Resources Report, World Resources Institute, Washington, DC. http://www.worldresourcesreport.org.

Chaponniere, Anne, and Vladimir Smakhtin. 2006. "A Review of Climate Change Scenarios and Preliminary Rainfall Trend Analysis in the Oum er Rbia Basin, Morocco." Working Paper 110 (Drought Series: Paper 8), International Water Management Institute, Colombo, Sri Lanka.

Christensen, Jens H., Fredrik Boberg, Ole B. Christensen, and Philippe Lucas-Picher. 2008. "On the Need for Bias Correction of Regional Climate Change Projections of Temperature and Rainfall." *Geophysical Research Letters* 35 (20): L20709. doi:10.1029/2008GL035694.

Christensen, Jens H., Timothy R. Carter, Markku Rummukainen, Georgios Amanatidis. 2007. "Evaluating the Performance and Utility of Regional Climate Models: The PRUDENCE Project." *Climatic Change* 81 (Suppl 1): 1–6. doi:10.1007/s10584-006-9211-6.

Christensen, Jens H., and Ole B. Christensen. 2003. "Severe Summer Flooding in Europe." *Nature* 421: 805–6.

Christensen, Jens H., Bruce Hewitson, Aristita Busuioc, Anthony Chen, Xuejie Gao, Isaac Held, Richard Jones, Rupa Kumar Kolli, Won-Tae Kwon, René Laprise, Victor Magaña Rueda, Linda Mearns, Claudio Guillermo Menéndez, Jouni Räisänen, Annette Rinke, Abdoulaye Sarr, and Penny Whetton. 2007. "Regional Climate Projections." In *Climate Change 2007: The Physical Science Basis—Contribution of Working Group I to the Fourth Assessment Report of the Intergovernmental Panel on Climate Change*, ed. Susan Solomon, Dahe Qin, Martin Manning, Zhenlin Chen, Melinda Marquis, Kristen B. Averyt, Melinda Tignor, and Henry L. Miller. Cambridge, U.K.: Cambridge University Press.

Christensen, Jens H., Erik Kjellström, Filippo Giorgi, Geert Lenderink, and Markku Rummukainen. 2010. "Weight Assignment in Regional Climate Models." *Climate Research* 44 (2–3): 179–94. doi:10.3354/cr00916.

Christensen, Ole B., and Jens H. Christensen. 2004. "Intensification of Extreme European Summer Rainfall in a Warmer Climate." *Global and Planetary Change* 44 (1–4): 107–17.

Coppola, Erika, Fred Kucharski, Filippo Giorgi, and Franco Molteni. 2005. "Bimodality of the North Atlantic Oscillation in Simulations with Greenhouse Gas Forcing." *Geophysical Research Letters* 32: L23709.

Cullen, Heidi M., Alexey Kaplan, Phillip A. Arkin, and Peter B. de Menocal. 2002. "Impact of the North Atlantic Oscillation on Middle Eastern Climate and Streamflow." *Climatic Change* 55: 315–38.

Dahl-Jensen, Dorthe, Jonathan Bamber, Carl Egede Bøggild, Erik Buch, Jens H. Christensen, Klaus Dethloff, Mark Fahnestock, Shawn Marshall, Minik Rosing, Konrad Steffen, Robert Thomas, Martin Truffer, Michiel van den Broeke, and Cornelis J. van der Veen. 2009. *The Greenland Ice Sheet in a Changing Climate: Snow, Water, Ice and Permafrost in the Arctic (SWIPA).* Oslo: Arctic Monitoring and Assessment Programme.

Dai, Aiguo. 2010. "Drought under Global Warming: A Review." *WIREs Climate Change* 2 (1): 45–65. doi:10.1002/wcc81.

Déqué, Michel, and Samuel Somot. 2010. "Weighted Frequency Distributions Express Modelling Uncertainties in the ENSEMBLES Regional Climate Experiments." *Climate Research* 44 (2–3): 195–209.

Diffenbaugh, Noah S., Jeremy S. Pal, Filippo Giorgi, and Xuejie Gao. 2007. "Heat Stress Intensification in the Mediterranean Climate Change Hotspot." *Geophysical Research Letters* 34: L11706. doi:10.1029/2007GL030000.

Evans, Jason P. 2009. "21st Century Climate Change in the Middle East." *Climatic Change* 92: 417–32. doi:10.1007/s10584-008-9438-5.

———. 2010. "Global Warming Impact on the Dominant Rainfall Processes in the Middle East." *Theoretical and Applied Climatology* 99 (3–4): 389–402. doi:10.1007/s00704-009-0151-8.

Evans, Jason P., and Roland Geerken. 2004. "Discrimination between Climate and Human-Induced Dryland Degradation." *Journal of Arid Environments* 57 (4): 535–54.

Freiwan, Muwaffaq, and Mikdat Kadioğlu. 2008a. "Climate Variability in Jordan." *International Journal of Climatology* 28 (1): 68–89.

———. 2008b. "Spatial and Temporal Analysis of Climatological Data in Jordan." *International Journal of Climatology* 28 (4): 521–35.

Frich, Povl, Lisa V. Alexander, Paul Della-Marta, Byron Gleason, Malcolm Haylock, Albert M. G. Klein Tank, and Thomas C. Peterson. 2002. "Observed Coherent Changes in Climatic Extremes during the Second Half of the Twentieth Century. *Climate Research* 19 (3): 193–212.

Gao, Xujie, and Filippo Giorgi. 2008. "Increased Aridity in the Mediterranean Region under Greenhouse Gas Forcing Estimated from High Resolution Simulations with a Regional Climate Model." *Global and Planetary Change* 62 (3–4): 195–209.

Gesch, Dean B., Kristin L. Verdin, and Susan K. Greenlee. 1999. "New Land Surface Digital Elevation Model Covers the Earth." *Eos, Transactions, American Geophysical Union* 80 (6): 69–70.

Giorgi, Filippo, Colin Jones, and Ghassem R. Asrar. 2009. "Addressing Climate Information Needs at the Regional Level: The CORDEX Framework." *WMO Bulletin* 58 (3): 175–83.

Grinsted, Aslak, John Moore, and Svetlana Jevrejeva. 2010. "Reconstructing Sea Level from Paleo and Projected Temperatures 200 to 2100 AD." *Climate Dynamics* 34: 461–72.

Hallegatte, Stéphane, Franck Lecocq, and Christian de Perthius. 2011. "Designing Climate Change Adaptation Policies: An Economic Framework." Policy Research Working Paper 5568, World Bank, Washington, DC.

Hawkins, Ed, and Rowan Sutton. 2009. "The Potential to Narrow Uncertainty in Regional Climate Predictions." *Bulletin of the American Meteorological Society* 90 (8): 1095–107.

———. 2010. "The Potential to Narrow Uncertainty in Projections of Regional Precipitation Change." *Climate Dynamics* 37 (1–2): 407–18. doi:10.1007/s00382-010-0810-6.

Hurrell, James W., Thomas Delworth, Gokhan Danabasoglu, Helge Drange, Ken Drinkwater, Stephen Griffies, Neil Holbrook, Benjamin Kirtman, Noel Keenlyside, Mojib Latif, Jochim Marotzke, James Murphy, Gerald Meehl, Tim Palmer, Holger Pohlmann, Tony Rosati, Richard Seager, Doug Smith, Rowan Sutton, Axel Timmermann, Kevin Trenberth, Joseph Tribbia, and Martin Visbeck. 2010. "Decadal Climate Prediction: Opportunities and Challenges." In *Proceedings of OceanObs'09: Sustained Ocean Observations and Information for Society (Vol. 2), Venice, Italy, 21–25 September 2009*, ed. Julie Hall, D. E. Harrison, and Detlef Stammer. ESA Publication WPP-306, European Space Agency, Paris. doi:10.5270/OceanObs09.cwp.45.

IOP Science. 2009. "Climate Change: Global Risks, Challenges, and Decisions." *IOP Conference Series: Earth and Environmental Science* 6. http://www.iop.org/EJ/volume/1755-1315/6.

IPCC (Intergovernmental Panel on Climate Change). 2007. "Summary for Policymakers." In *Climate Change 2007: The Physical Science Basis—Contribution of Working Group I to the Fourth Assessment Report of the Intergovernmental Panel on Climate Change*, ed. Susan Solomon, Dahe Qin, Martin Manning, Zhenlin Chen, Melinda Marquis, Kristen B. Averyt, Melinda Tignor, and Henry L. Miller. Cambridge, U.K.: Cambridge University Press.

———. 2012. *Managing the Risks of Extreme Events and Disasters to Advance Climate Change Adaptation: Special Report of the Intergovernmental Panel on Climate Change*. Cambridge, U.K.: Cambridge University Press.

Kim, Do-Woo, and Hi-Ryong Byun. 2009. "Future Pattern of Asian Drought under Global Warming Scenario." *Theoretical and Applied Climatology* 98 (1–2): 138–50.

Kitoh, Akio, Akiyo Yatagai, and Pinhas Alpert. 2008. "First Super-High Resolution Model Projection That the Ancient 'Fertile Crescent' Will Disappear in This Century." *Hydrological Research Letters* 2: 1–4.

Kjellström, Erik, and Filippo Giorgi. 2010. "Introduction." *Climate Research* 44 (2–3): 117–19.

Knippertz, Peter, Michael Christoph, and Peter Speth. 2003. "Long-Term Precipitation Variability in Morocco and the Link to the Large-Scale Circulation in Recent and Future Climates." *Meteorology and Atmospheric Physics* 83: 67–88.

Krichak, Simon O., Pinhas Alpert, and Pavel Kunin. 2010. "Numerical Simulation of Seasonal Distribution of Rainfall over the Eastern Mediterranean with a RCM." *Climate Dynamics* 34 (1): 47–59.

Kummerow, Christian, Joanne Simpson, Otto Thiele, William Barnes, A. T. C. Chang, Erich Stocker, Robert F. Adler, Arthur Hou, Ramesh Kakar, Frank Wentz, Peter Ashcroft, Toshiaki Kozu, Ye Hong, Kenichi Okamoto, Toshio Iguchi, Hiroshi Kuroiwa, Eastwood Im, Ziad Haddad, George Huffman, Brad Ferrier, William S. Olson, Edward Zipser, Eric A. Smith, Thomas T. Wilheit, Gerald North, T. N. Krishnamurti, and Kenji Nakamura. 2000. "The Status of the Tropical Rainfall Measuring Mission (TRMM) after Two Years in Orbit." *Journal of Applied Meteorology* 39 (12): 1965–82.

Latif, Mojib, Thomas Delworth, Dietmar Dommenget, Helge Drange, Wilco Hazeleger, James W. Hurrell, Noel S. Keenlyside, Gerald Meehl, and Rowan

Sutton. 2010. "Dynamics of Decadal Climate Variability and Implications for Its Prediction." In *Proceedings of OceanObs'09: Sustained Ocean Observations and Information for Society (Vol. 2), Venice, Italy, 21–25 September 2009*, ed. Julie Hall, D. E. Harrison, and Detlef Stammer. ESA Publication WPP-306, European Space Agency, Paris. doi:10.5270/OceanObs09.cwp.53.

Lempert, Robert J., and David G. Groves. 2010. "Identifying and Evaluating Robust Adaptive Policy Responses to Climate Change for Water Agencies in the American West." *Technological Forecasting and Social Change* 77 (6): 960–74.

Lempert, Robert J., Nebojsa Nakicenovic, Daniel Sarewitz, and Michael Schlesinger. 2004. "Characterizing Climate-Change Uncertainties for Decision-Makers: An Editorial Essay." *Climatic Change* 65 (1–2): 1–9. doi:10.1023/B:CLIM.0000037561.75281.b3.

Lionello, Piero, and Filippo Georgi. 2007. "Winter Rainfall and Cyclones in the Mediterranean Region: Future Climate Scenarios in a Regional Simulation." *Advances in Geosciences* 12: 153–58.

Mahwed, K. 2008. "Vulnerability Assessment of the Climate Sector: Enabling Activities for Preparation of Syria's Initial Communication to the UNFCCC." Project 00045323, United Nations Development Programme, Damascus.

Mearns, Linda O., Michael Hulme, Tim R. Carter, Rik Leemans, Murari Lal, Penny Whetton, Lauren Hay, Roger N. Jones, Richard Katz, Timothy Kittel, J. Smith, Rob Wilby, Luis J. Mata, and John Zillman. 2001. "Climate Scenario Development." In *Climate Change 2001: The Scientific Basis—Contribution of Working Group I to the Third Assessment Report of the Intergovernmental Panel on Climate Change*, ed. John T. Houghton, Yihui Ding, David J. Griggs, Maria Noguer, Paul J. van der Linden, Xiosu Dai, Kathy Maskell, and Cathy A. Johnson. Cambridge, U.K.: Cambridge University Press.

Meehl, Gerald A., Lisa Goddard, James Murphy, Ronald J. Stouffer, George Boer, Gokhan Danabasoglu, Keith Dixon, Marco A. Giorgetta, Arthur M. Greene, Ed Hawkins, Gabriele Hegerl, David Karoly, Noel Keenlyside, Masahide Kimoto, Ben Kirtman, Antonio Navarra, Roger Pulwarty, Doug Smith, Detlef Stammer, and Timothy Stockdale. 2009. "Decadal Prediction: Can It Be Skillful?" *Bulletin of the American Meteorological Society* 90 (10): 1467–85. doi:15 10.1175/2009BAMS2778.1.

Meehl, Gerald A., Thomas F. Stocker, William D. Collins, Pierre Friedlingstein, Amadou T. Gaye, Jonathan M. Gregory, Akio Kitoh, Reto Knutti, James M. Murphy, Akira Noda, Sarah C. B. Raper, Ian G. Watterson, Andrew J. Weaver, and Zong-Ci Zhao. 2007. "Global Climate Projections." In *Climate Change 2007: The Physical Science Basis—Contribution of Working Group I to the Fourth Assessment Report of the Intergovernmental Panel on Climate Change*, ed. Susan Solomon, Dahe Qin, Martin Manning, Zhenlin Chen, Melinda Marquis, Kristen B. Averyt, Melinda Tignor, and Henry L. Miller. Cambridge, U.K.: Cambridge University Press.

Moss, Richard, Mustafa Babiker, Sander Brinkman, Eduardo Calvo, Tim Carter, Jae Edmonds, Ismail Elgizouli, Seita Emori, Lin Erda, Kathy Hibbard, Roger Jones, Mikiko Kainuma, Jessica Kelleher, Jean Francois Lamarque, Martin Manning, Ben Matthews, Jerry Meehl, Leo Meyer, John Mitchell, Nebojsa Nakicenovic, Brian O'Neill, Ramon Pichs, Keywan Riahi, Steven Rose, Paul Runci, Ron Stouffer, Detlef van Vuuren, John Weyant, Tom Wilbanks, Jean

Pascal van Ypersele, and Monika Zurek. 2008. *Towards New Scenarios for Analysis of Emissions, Climate Change, Impacts, and Response Strategies*. Geneva: IPCC.

Moss, Richard H., Jae A. Edmonds, Kathy A. Hibbard, Martin R. Manning, Steven K. Rose, Detlef P. van Vuuren, Tim R. Carter, Seita Emori, Mikiko Kainuma, Tom Kram, Gerald A. Meehl, John F. Mitchell, Nabojsa Nakicenovic, Keywan Riahi, Steven J. Smith, Ronald J. Stouffer, A. M. Thomson, John P. Weyant, and Thomas J. Wilbanks. 2010. "The Next Generation of Scenarios for Climate Change Research and Assessment." *Nature* 463: 747–56.

Mote, Thomas L. 2007. "Greenland Surface Melt Trends 1973–2007: Evidence of a Large Increase in 2007." *Geophysical Research Letters* 34: L22507. doi:10.1029/2007GL031976.

Murphy, James, Vladimir Kattsov, Noel Keenlyside, Masahide Kimoto, Gerald Meehl, Vikram Mehta, Holger Pohlmann, Adam Scaife, and Doug Smith. 2010. "Towards Prediction of Decadal Climate Variability and Change." *Procedia Environmental Sciences* 1: 287–304. doi:10.1016/j.proenv.2010.09.018.

Nakicenovic, Nabojsa, and Robert Swart, eds. 2000. *Special Report on Emissions Scenarios: A Special Report of Working Group III of the Intergovernmental Panel on Climate Change*. Cambridge, U.K.: Cambridge University Press.

Nasrallah, Hassan A., and Robert C. Balling. 1993. "Spatial and Temporal Analysis of Middle Eastern Temperature Changes." *Climatic Change* 25 (2): 153–61.

Nerem, R. Steven, Don P. Chambers, C. Choe, and Gary T. Mitchum. 2010. "Estimating Mean Sea Level Change from the TOPEX and Jason Altimeter Missions." *Marine Geodesy* 33 (1): 435–46.

Olesen, Jørgen E., Timothy R. Carter, Carlos H. Díaz-Ambrona, Stefan Fronzek, Tove Heidmann, Thomas Hickler, Tom Holt, María Inés Mínguez, Pablo Morales, Jean Palutikof, Miguel Quemada, Margarita Ruiz-Ramos, Gitte Rubæk, Federico Sau, Benjamin Smith, and Martin Sykes. 2007. "Uncertainties in Projected Impacts of Climate Change on European Agriculture and Ecosystems Based on Scenarios from Regional Climate Models." *Climatic Change* 81 (Suppl. 1): 123–43.

Önol, Bariş, and Fredrick H. M. Semazzi. 2009. "Regionalization of Climate Change Simulations over the Eastern Mediterranean." *Journal of Climate* 22 (8): 1944–61.

Prudhomme, Christel, Rob L. Wilby, Sue Crooks, Alison L. Kay, and Nick S. Reynard. 2010. "Scenario-Neutral Approach to Climate Change Impact Studies: Application to Flood Risk." *Journal of Hydrology* 390 (3–4): 198–209.

Racskó, Péter, László Szeidl, and Mikhail Semenov. 1991. "A Serial Approach to Local Stochastic Weather Models." *Ecological Modeling* 57 (1–2): 27–41.

Raible, Christoph C., Baruch Ziv, Hadas Saaroni, and Martin Wild. 2010. "Winter Synoptic-Scale Variability over the Mediterranean Basin under Future Climate Conditions as Simulated by the ECHAM5." *Climate Dynamics* 35 (2): 473–88.

Rauthe, Monika, Andreas Hense, and Heiko Paeth. 2004. "A Model Intercomparison Study of Climate Change Signals in Extratropical Circulation." *International Journal of Climatology* 24 (5): 643–62.

Reifen, Catherine, and Ralf Toumi. 2009. "Climate Projections: Past Performance No Guarantee of Future Skill?" *Geophysical Research Letters* 36: L13704. doi:10.1029/2009GL038082.

Rummukainen, Markku. 2010. "State-of-the-Art with Regional Climate Models." *Wiley Interdisciplinary Reviews: Climate Change* 1:82–96. doi:10.1002/wcc.8.

Schneider, Udo, Tobias Fuchs, Anja Meyer-Christoffer, and Bruno Rudolf. 2010. "Global Precipitation Analysis Products of the GPCC." Global Precipitation Climatology Centre, Deutscher Wetterdienst, Offenbach, Germany. http://climatedataguide.ucar.edu/guidance/gpcc-global-precipitation-climatology-centre.

Semenov, Mikhail A., Roger J. Brooks, Elaine M. Barrow, and Clarence W. Richardson. 1998. "Comparison of the WGEN and LARS-WG Stochastic Weather Generators for Diverse Climates." *Climate Research* 10 (2): 95–107.

Shahin, Mamdouh. 2007. *Water Resources and Hydrometeorology of the Arab Region*. Water Science and Technology Library, vol. 59. Dordrecht, Netherlands: Springer.

Simpson, Joanne, Robert F. Adler, and Gerald R. North. 1988. "A Proposed Tropical Rainfall Measuring Mission (TRMM) Satellite." *Bulletin of the American Meteorological Society* 69 (3): 278–95.

Smadi, Mahmoud M., and Ahmed Zghoul. 2006. "A Sudden Change in Rainfall Characteristics in Amman, Jordan, during the Mid-1950s." *American Journal of Environmental Sciences* 2 (3): 84–91.

Smith, Doug M., Stephen Cusack, Andrew W. Colman, Chris K. Folland, Glen R. Harris, and James M. Murphy. 2007. "Improved Surface Temperature Prediction for the Coming Decade from a Global Climate Model." *Science* 317 (5839): 796–99. doi:10.1126/science.1139540.

Steadman, Robert G. 1979. "The Assessment of Sultriness: Part I—Temperature-Humidity Index Based on Human Physiology and Clothing Science." *Journal of Applied Meteorology* 18 (7): 861–73.

Taylor, Karl E., Ronald J. Stouffer, and Gerald A. Meehl. 2011. "A Summary of the CMIP5 Experiment Design." Coupled Model Intercomparison Project, Program for Climate Model Diagnosis and Intercomparison, Livermore, CA. http://cmip-pcmdi.llnl.gov/cmip5/docs/Taylor_CMIP5_design.pdf.

Tebaldi, Claudia, Katharine Hayhoe, Julie M. Arblaster, and Gerald A. Meehl. 2006. "Going to the Extremes: An Intercomparison of Model-Simulated Historical and Future Changes in Extreme Events." *Climatic Change* 79: 185–211.

Tolba, Mostafa K., and Najib W. Saab, eds. 2009. *Arab Environment Climate Change: Impact of Climate Change on Arab Countries*. Beirut: Arab Forum for Environment and Development.

Tryhorn, Lee, Amanda Lynch, Rebecca Abramson, and Kevin Parkyn. 2008. "On the Meteorological Mechanisms Driving Postfire Flash Floods: A Case Study." *Monthly Weather Review* 136 (5): 1778–91.

Tschoegl, Liz, with Regina Below and Debarati Guhar-Sapir. 2006. "An Analytical Review of Selected Data Sets on Natural Disasters and Impacts." Prepared for the United Nation Development Programme–Centre for Research on the Epidemiology of Disasters Workshop on Improving Compilation of Reliable Data on Disaster Occurrence and Impact, Centre for Research on the Epidemiology of Disasters, Bangkok. April 2–4. http://www.em-dat.net/documents/Publication/TschoeglDataSetsReview.pdf.

van den Brink, Hendrik Willem, Günther P. Können, J. D. Opsteegh, Geert Jan van Oldenborgh, and Gerrit Burgers. 2005. "Estimating Return Periods of

Extreme Events from ECMWF Seasonal Forecast Ensembles." *International Journal of Climatology* 25 (10): 1345–54.

van Vuuren, Detlef, Keywan Riahi, Richard H. Moss, Jae Edmonds, Allison Thomson, Nabojsa Nakicenovic, Tom Kram, Frans Berkhout, Robert Swart, Anthony Janetos, Steve Rose, and Nigel Arnell. 2010. "Developing New Scenarios as a Common Thread for Future Climate Research." Intergovernmental Panel on Climate Change, Geneva. http://www.ipcc-wg3.de/meetings/expert-meetings-and-workshops/files/Vuuren-et-al-2010-Developing-New-Scenarios-2010-10-20.pdf/view?searchterm=New%20Scenarios.

Villholth, Karen G., Christian Tøttrup, Martin Stendel, and Ashton Maherry. 2012. "Integrated Mapping of Groundwater Drought Vulnerability in the SADC Region of Africa." Submitted to *Global Environmental Change*.

Wilby, Rob L., 2007a. "Climate Change Scenarios for Morocco." Background paper, World Bank, Washington, DC.

———. 2007b. "Decadal Climate Forecasting Techniques for Adaptation and Development Planning: A Briefing Document on Available Methods, Constraints, Risks, and Opportunities." Background paper, U.K. Department for International Development, London.

———. 2008. "Climate Change Scenarios for the Republic of Yemen." Background paper, World Bank, Washington, DC.

———. 2010. "Climate Change Projections and Downscaling for Jordan, Lebanon, and Syria." Draft synthesis report, World Bank, Washington, DC.

Wilby, Rob L., and Suraje Dessai. 2010. "Robust Adaptation to Climate Change." *Weather* 65 (7): 180–85.

Wilby Rob L., Jessica Troni, Yvan Biot, Leonard Tedd, Bruce Hewitson, Doug M. Smith, and Rowan T. Sutton. 2009. "A Review of Climate Risk Information for Adaptation and Development Planning." *International Journal of Climatology* 29 (9): 1193–215.

Willett, Katharine M., and Steven Sherwood. 2012. "Exceedance of Heat Index Thresholds for 15 Regions under a Warming Climate Using the Wet-Bulb Globe Temperature." *International Journal of Climatology* 32 (2): 161–77. doi:10.1002/joc.2257.

WMO (World Meteorological Organization) Secretariat. 2009. "Global Framework for Climate Services: Brief Note." Prepared for the World Climate Conference 3, Geneva, August 31–September 4. http://www.wmo.int/hlt-gfcs/documents/WCC3GFCSbrief_note_en.pdf.

World Bank. 2009. *World Development Report 2010: Development and Climate Change*. Washington, DC: World Bank.

Zhang, Xuebin, Enric Aguilar, Serhat Sensoy, Hamlet Melkonyan, Umayra Tagiyeva, Nader Ahmed, Nato Kutaladze, Fatemeh Rahimzadeh, Afsaneh Taghipour, T. H. Hantosh, Pinhas Albert, Mohammed Semawi, Mohammad Karam Ali, Mansoor Halal Said Al-Shabibi, Zaid Al-Oulan, Taha Zatari, Imad Al Dean Khelet, Saleh Hamoud, Ramazan Sagir, Mesut Demircan, Mehmet Eken, Mustafa Adiguzel, Lisa Alexander, Thomas C. Peterson, and Trevor Wallis. 2005. "Trends in Middle East Climate Extreme Indices from 1950 to 2003." *Journal of Geophysical Research* 110: D22104. doi:10.1029/2005JD006181.

Disaster Risk Management Increases Climate Resilience

Climate Disasters Are Increasing in Frequency

Worldwide, climate-related disasters are taking an increasing toll. From 1981 to 2010, climate disasters killed 1.4 million people and affected more than 5.5 billion people.[1] As described in chapter 2, recent observations and modeling show that climate change is leading to greater intensity and frequency of many weather phenomena, such as storms, floods,[2] and droughts. Climate change is likely to increase both rapid-onset disasters (such as storms or floods) and slow-onset disasters (such as drought and sea-level rise, leading to more coastal floods and salinization).[3] Climate change also has secondary impacts resulting in disasters such as dust storms, landslides, rockslides caused by heavy rain, and forest fires during droughts (UNISDR and LAS 2011).

Within the Arab region, the interplay of natural hazards, together with the impacts of climate change, water scarcity, and food insecurity, has emerged as a serious challenge for policy and planning for all states. Over the past 30 years, climate disasters affected 50 million people in the Arab region, with a reported cost of US$11.5 billion, although this estimate is clearly low because the costs of damages are reported for only 17 percent of disasters and rarely capture the suffering that follows loss of lives and livelihoods.[4]

Both disaster risk management (DRM) and climate change adaptation (CCA) have a common goal of risk reduction, but DRM is concerned with ongoing hazards, whereas CCA is principally concerned with emerging climate change challenges (McGray, Hammill, and Bradley 2007; UNISDR 2009). In terms of institutional structures and awareness, there is a disconnect between the DRM and CCA communities. Within most national and local governments, the two communities largely operate in isolation. This isolation can be partly attributed to thematic investments by donors and partly to existing in-house silos within governments (CCA usually falls under the mandate of the ministry of environment, and DRM under that of the ministry of interior, civil defense). Nevertheless, policy makers are increasingly recognizing the links between DRM and CAA through the reduction of hazard-specific vulnerabilities. Measures to reduce vulnerabilities include capacity building, improved warning and forecasting methods, and land-use planning.

This section outlines the main climate-related hazards, risks, and vulnerabilities in the Arab countries; the political commitment to DRM (national and regional initiatives); and key gaps and constraints. It concludes with a set of recommendations for better mainstreaming of DRM in national policies.

The Hazard Risk Profile of the Arab States Is Changing

Droughts and floods account for 98 percent of all people affected by climate-related disasters (figure S1.1). There is no clear trend in either their frequency or impact because the signal is dominated by occasional very large events. However, the data show a steady increase in the number of flash floods, with more than 500,000 people affected in the 2000s compared with only 100,000 in the 1990s. The region's high rate of urbanization, particularly in coastal areas, often exacerbates the effects of the floods, droughts, and landslides. Global climate change is projected to result in a greater intensity in rainfall events, leading to increased intensity and frequency of both droughts and floods and exposing up to 25 million urban dwellers to floods (Abu Swaireh 2009; Solomon et al. 2007). These same climate changes could lead to a 30 to 50 percent drop in water availability, thereby exacerbating existing severe water scarcity (World Bank 2007). This problem could lead to increased internal and external migrations, which would be compounded by sea-level rise. According to Erian, Katlan, and Babah (2010), 28 percent of the residents of Arab countries already live in areas vulnerable to drought.

In many Arab countries, the bulk of the population, physical assets, and government and

FIGURE S1.1

The Impact of Climate-Related Disasters across the Arab Region

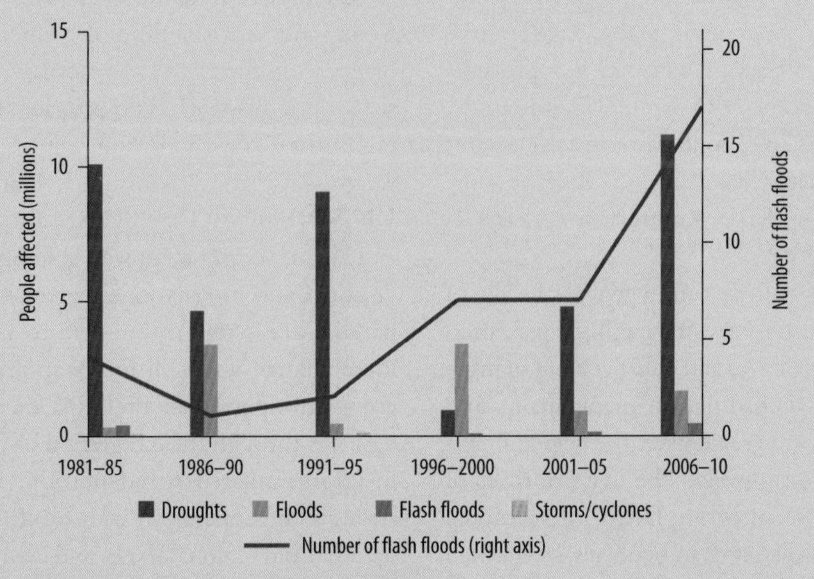

Source: Authors' compilation, based on EM-DAT 2010.

administrative centers are located close to coasts, thereby exposing a large portion of the population and major strategic and economic assets to the full range of hydrometeorological hazards. The *Global Assessment Report on Disaster Risk Reduction 2011* finds that although flood mortality risk has decreased globally since 2000, it is still increasing in Arab countries, indicating that growing exposure continues to outpace reductions in vulnerability (UNISDR 2011). The percentage of gross domestic product (GDP) exposed to floods has tripled in the four decades from 1970 to 2009 (UNISDR 2011, 32).

Arab States Are Responding to Mitigate Climate Threats

Although the region is affected by periodic earthquakes and droughts, DRM has not been a priority for the region's governments until recently (UNISDR 2011). At the regional level, the League of Arab States (LAS), the Council of Arab Ministers Responsible for the Environment, the United Nations International Strategy for Disaster Reduction (UNISDR), the Regional Office for the Arab States, and the Arab Economic and Social Council have approved a number of recent DRM and CCA initiatives, including the 2007 Arab Ministerial Declaration on Climate Change and the Arab Strategy for Disaster Risk Reduction (ASDRR) (UNISDR 2010).

ASDRR is a 10-year strategy with the aim of reducing disaster losses through identification of strategic priorities and enhancement of institutional and coordination mechanisms and monitoring arrangements at the regional, national, and local levels. The key priorities of ASDRR are to integrate DRM into national development planning and policies; strengthen commitment for comprehensive disaster risk reduction (DRR) across sectors; develop capacities to identify, assess, and monitor disaster risks; build resilience through knowledge, advocacy, research, and training; improve accountability for DRM at the subnational and local levels; and integrate DRR into emergency response, preparedness, and recovery. To achieve these goals, ASDRR aims at entrusting a ministry with strong political power with the DRM mandate—gradual decentralization according to resources and capacities. Local initiatives will be prioritized on the basis of their effectiveness in reducing risks to organizations such as grassroots women's organizations.

Meanwhile, individual Arab economies are making progress on the Hyogo Framework for Action (HFA). The Arab Republic of Egypt, Jordan, Morocco, the Syrian Arab Republic, and the Republic of Yemen are making advances in systematically reporting disaster losses for 2010. Jordan, Syria, and the Republic of Yemen have recently published national disaster inventories, and other countries are expected to soon follow. Nine Arab economies have completed their HFA progress reports for 2011: Algeria, Bahrain, the Comoros, Egypt, Lebanon, Morocco, the West Bank and Gaza, Syria, and the Republic of Yemen. Overall progress on the five HFA priorities show consistent progress with HFA priority 1; relatively high progress with HFA priority 2; and some progress with HFA priorities 3, 4, and 5 (LAS, UNISDR, and ROAS 2011; see also box S1.1).

LAS, in coordination with a number of Arab regional and international partners, has prepared a draft Arab action plan to address climate change issues in the Arab region. The cross-cutting program on DRM aims to follow up HFA through the integration of DRR in all programs related to adaptation, build and strengthen cooperation with UNISDR at the national and regional levels, and identify mechanisms and capacities to reduce disaster risk in

BOX S1.1

Hyogo Priority Actions

1. Ensure that disaster risk reduction is a national and a local priority with a strong institutional basis for implementation.
2. Identify, assess, and monitor disaster risks and enhance early warning.
3. Use knowledge, innovation, and education to build a culture of safety and resilience at all levels.
4. Reduce the underlying risk factors.
5. Strengthen disaster preparedness for effective response at all levels.

Source: Authors' compilation.

the planning and implementation of adaptation programs.

Gaps and Constraints Exist in Arab DRM Policies

Although the Arab states are moving in the right direction in the effort to integrate DRM in their national and regional policies, some specific gaps and constraints, observed in the progress reviews of HFA implementation in the Arab region for 2009 and 2011 (UNISDR and LAS 2009, 2011), are noteworthy:

- Coordination within the governments' relevant ministries and between different stakeholders at the national and local levels is weak, and institutional capacity to manage a cross-cutting sector such as DRM is low. This problem can be of particular importance when there is lack of coordination between institutions responsible for disaster preparedness and those responsible for hazard monitoring and early warning to communities.
- No consistent methodologies are used to conduct risk assessments, leading to lack of comparability within sectors and countries with similar risks.
- No resources are dedicated to implementing DRM and DRR actions at various administrative levels. In many cases, emergency plans exist, but resources for preparedness are inadequate, particularly at local levels.
- Comprehensive disaster data and information systems lack accessibility. Although information on disaster events and loss exist in many countries, there is a lack of adequate profiling of socioeconomic vulnerabilities and risks to populations.
- There is an absence of national strategies to integrate DRR in school curricula and public awareness activities.
- Links between DRM and CCA are unclear, leading to division of resources (financial and human) and duplicity.
- Recovery and reconstruction projects are stand-alone initiatives with time limits, re-

sulting in DRR being absent from rebuilding, regulation, and production and planning systems.

Recommendations to Build Resilience in the Arab Countries

The recommendations in this section are based on the trends of climatic hazards in the Arab region, the existing institutional capacity, the ongoing engagements in DRM and DRR, and the gaps identified in the previous section. The following recommendations, which may be used by policy makers at the national or international level, are derived from UNISDR and LAS (2009, 2011):

- Strengthen commitment at the national level for comprehensive DRM and DRR across all sectors through adequate and dedicated financing toward risk reduction activities. Another avenue for sustainable DRR is its inclusion in national development plans and legal frameworks.
- Develop capacities at the national, local, and community levels to identify, assess, and monitor disaster risks through multihazard risk assessments.
- Build resilience through knowledge, advocacy, research, and training by making information on risk accessible to all stakeholders. Educational materials and curricula, as well as public awareness and advocacy campaigns, should be used in this effort.
- Develop regional, national, and local early warning systems and networks and proper dissemination mechanisms.
- Identify institutional and administrative roles at all levels of government in the various stages of a climatic impact—before, during, and after the event—for timely information exchange, better coordination, and reduced overlapping of initiatives with stretched resources.

Box S1.2 provides an example of a successful initiative.

BOX S1.2

Drought Emergency in the Greater Horn of Africa: Early Response in Djibouti

Djibouti is highly vulnerable to natural hazards, particularly extended dry multiyear droughts that result in water scarcity for livestock, irrigation, and domestic uses. In addition, Djibouti is affected by frequent flash floods. Data from recent natural hazards suggest that such events have severely affected Djibouti's economic growth and sustainable development. According to the World Bank (2005) hotspots study, approximately 33 percent of Djibouti population lives in areas of high risk, and 35 percent of the economy is vulnerable to natural hazards. With less than 270 cubic meters of freshwater per year per capita in 2009, the country is classified as severely water poor (according to the World Health Organization definition of less than 1,000 cubic meters).

Djibouti's water crisis is exacerbated by rapid demographic growth and climate variability. Djibouti averages only 130 millime-

BOX S1.2 *Continued*

ters of rainfall a year (expected to fall to 100 by the end of the decade), whereas rainfall in the driest region of Obock averages only 50 to 100 millimeters a year. Moreover, Djibouti has no permanent rivers, streams, or freshwater lakes, and less than 5 percent of total rainfall replenishes the water table because of extreme evaporation.

Together, the Djibouti government, World Bank, United Nations Development Programme, and European Union, with support from the Global Facility for Disaster Reduction and Recovery (GFDRR), performed a postdisaster needs assessment (PDNA)[a] for the ongoing (2012) drought, which is in the fourth consecutive year of failed rainfall (in terms of quantity and regularity). Despite limited data, the PDNA had the following findings:

- The drought has affected more than 120,000 malnourished and food insecure, already vulnerable, rural poor. (Because of poor data, this estimate is very conservative.)
- Refugees entering Djibouti have increased considerably since May 2011, from 395 per month to 875 in August. When interviewed, 50 percent of the refugees explained that the drought was their primary reason for migration.
- The estimated economic losses from the drought are 3.9 percent of GDP over the period 2008–11.
- The identified need for drought and other hazard mitigation interventions for the next five years is US$196 million (about 4 percent of GDP).
- Over the same period, the greatest damage and losses were found in the agriculture livestock, water, and sanitation sectors. Costs amounted to US$96 million.
- Agriculture production and livestock losses attributable to the drought led to severe food insecurity in rural areas. They caused a 20 percent loss of kilocalories consumed per household and a 50 percent decrease in the consumption of goods and services (such as education, health and medicine, and kerosene).
- With 80 percent less rainfall since 2007, the aquifers, which are the only source of water for the capital, Ville de Djibouti, suffered a reduction in recharge equivalent to four years of water supply, resulting in an overall drawdown of the water table and severe increase of its salinity by 40 percent.
- During the four years of drought, 100 percent of the traditional wells and 80 percent of the community wells in Djibouti have been temporarily or permanently out of order because of water shortage or poor water quality. The result has been increased salinity and other types of contaminations of the aquifers.
- Drought social impacts include (a) increased vulnerability of communities caused by loss of their means of subsistence, (b) increased financial burden on host families, and (c) deteriorating health and basic living conditions—especially for pregnant women.
- Djibouti meteorological data show that multiyear droughts have historically been

(Box continues next page)

BOX S1.2 *Continued*

followed by floods, and given the drought of the previous years, 2012 could be characterized by intense floods. Thus, integrated risk management is a priority.

- Most of the impacts derived from extreme weather events (drought and floods) should be dealt with according to a reinforced vision on land-use planning. Overexploitation of land, water, and other natural resources must be prevented.
- Given the likelihood of continued water deficit events, quick exploration, assessment, and implementation of the feasibility of sustainable water-harvesting and desalination infrastructures are recommended.

As a result of the initial assessment, a number of priority drought mitigation activities have been identified, resulting in the mobilization of IDA Crisis Response Window funding in the amount of US$13.3 million. The PDNA identified mitigation priorities are in line with the government's seven-pillar mitigations priorities.[b] These priorities seek to maximize efficiency gains from existing World Bank operations and technical assistance, while promoting strategies that support sustainable, economically strong livelihoods and potentially negate the need for routine emergency appeals. These funds greatly complement the ongoing World Bank disaster management efforts (financed by GFDRR), which aim to establish a national risk assessment and communication system to better inform decision makers and start building a national culture of disaster mitigation.

a. A PDNA is a government-led exercise that provides a coordinated and credible basis for recovery and reconstruction planning, financing plans, and strategies. By incorporating risk reduction measures, it provides systemic links to sustainable development, thereby serving as a platform for national and international actors to assist governments and populations affected by a disaster.

b. The seven priorities are as follows: (a) strengthen price controls in the national market, (b) establish strategic foodstock and regulatory mechanisms, (c) expand social and productive safety, (d) establish conditional cash transfers, (e) create a more sustainable and drought-resilient agriculture, (f) strengthen water management and retention, and (g) strengthen existing disaster risk management mechanisms.

Source: Provided by Andrea Zanon.

Notes

1. Data are from EM-DAT, the International Disaster Database of the Centre for Research on the Epidemiology of Disasters at the Université Catholique de Louvain in Brussels. Data include droughts, extreme temperatures, floods, storms, and wildfires. EM-DAT can be accessed at http://www.emdat.be.
2. In the context of this section, floods and flooding include flash floods unless otherwise noted.
3. Although climate change may increase the frequency of an event, only rarely can a particular event be ascribed to climate change.
4. Data are from EM-DAT.

References

Abu Swaireh, Luna. 2009. "Disaster Risk Reduction Global and Regional Context." Presented at the Workshop on Climate Change and Disaster Risk

Reduction in the Arab Region: Challenges and Future Actions, organized by the United Nations International Strategy for Disaster Reduction, World Bank, Global Facility for Disaster Reduction and Recovery, League of Arab States, and Arab Academy for Science, Technology, and Maritime Transport, Cairo, November, 21–23.

EM-DAT. 2010. International Disaster Database. Centre for Research on the Epidemiology of Disasters. http://www.emdat.be/.

Erian, Wadid, Bassem Katlan, and Ouldbdey Babah. 2010. "Drought Vulnerability in the Arab Region: Special Case Study—Syria." Background paper prepared for *Global Assessment Report on Disaster Risk Reduction 2011: Revealing Risk, Redefining Development*, United Nations International Strategy for Disaster Reduction, Geneva.

LAS (League of Arab States), UNISDR (United Nations International Strategy for Disaster Reduction), and ROAS (Regional Office for the Arab States). 2011. "Progress Review of the Implementation of HFA in the Arab Region." Presented at the Third Global Platform for Disaster Risk Reduction, Geneva, May 9.

McGray, Heather, Anne Hammill, and Rob Bradley. 2007. "Weathering the Storm: Options for Framing Adaptation and Development." World Resources Institute, Washington, DC. http://pdf.wri.org/weathering_the_storm.pdf.

Solomon, Susan, Dahe Qin, Martin Manning, Zhenlin Chen, Melinda Marquis, Kristen B. Averyt, Melinda Tignor, and Henry L. Miller. 2007. *Climate Change 2007: The Physical Science Basis—Contribution of Working Group I to the Fourth Assessment Report of the Intergovernmental Panel on Climate Change*. Cambridge, U.K.: Cambridge University Press.

UNISDR (United Nations International Strategy for Disaster Reduction). 2009. *Global Assessment Report on Disaster Risk Reduction 2009: Risk and Poverty in a Changing Climate*. Geneva: UNISDR.

———. 2010. "The Arab Strategy for Disaster Risk Reduction 2020." UNISDR, Geneva. http://www.unisdr.org/files/17934_asdrrfinalenglish-january2011.pdf.

———. 2011. *Global Assessment Report on Disaster Risk Reduction 2011: Revealing Risk, Redefining Development*. Geneva: UNISDR.

UNISDR (United Nations International Strategy for Disaster Reduction) and LAS (League of Arab States). 2009. "Progress in Reducing Disaster Risk and Implementing Hyogo Framework for Action in the Arab Region." UNISDR, Geneva.

———. 2011. "Progress Review of the Implementation of Hyogo Framework for Action in the Arab Region 2011." UNISDR, Geneva.

World Bank. 2005. *Natural Disaster Hotspots: A Global Risk Analysis*. Washington, DC: World Bank.

———. 2007. *Making the Most of Scarcity: Accountability for Better Water Management Results in the Middle East and North Africa*. Washington, DC: World Bank.

CHAPTER 3

Climate Change Contributes to Water Scarcity

Water has always been a central issue in the Arab region. In fact, many anthropologists believe that human civilization first emerged in this part of the world as an adaptation to the region's desiccation that started at the onset of the Holocene a few thousand years ago. Faced with long rainless summers and short rainy winters, early inhabitants sought to settle near perennial springs (for example, in Damascus and Jericho) where they could secure a steady supply of food and shelter by domesticating plants and animals—which heralded the agricultural revolution. As small settlements amalgamated into larger towns, new empires sought to regulate and secure access to water for their subjects. Across the region, remnants of water infrastructure are a testimony to human ingenuity and the capacity to adapt to harsh natural conditions characterized by severe droughts and marked seasonality.

The bulk of the Arab region lies in the horse latitudes, a region characterized by its aridity as global climate circulations drive moisture to the low and high latitudes, causing the formation of the vast Sahara and Arabian Deserts. Before their desiccation several thousand years ago, these deserts received substantial precipitation that percolated into deep layers to form vast fossil aquifers. The region receives substantial runoff from neighboring areas. The Taurus and Zagros Mountains to the north and east are the main headwaters of the Euphrates and Tigris Rivers. The Arab Republic of Egypt relies virtually exclusively on runoff from the Nile's headwaters in the Ethiopian and Equatorial highlands several thousand kilometers to its south.

Narrow strips of coastal plains in North Africa, the eastern Mediterranean, and the southwestern corner of the Arabian Peninsula receive

Photograph by Dorte Verner

MAP 3.1

Middle East and North Africa Aridity Zoning: Precipitation Divided by Reference Evapotranspiration

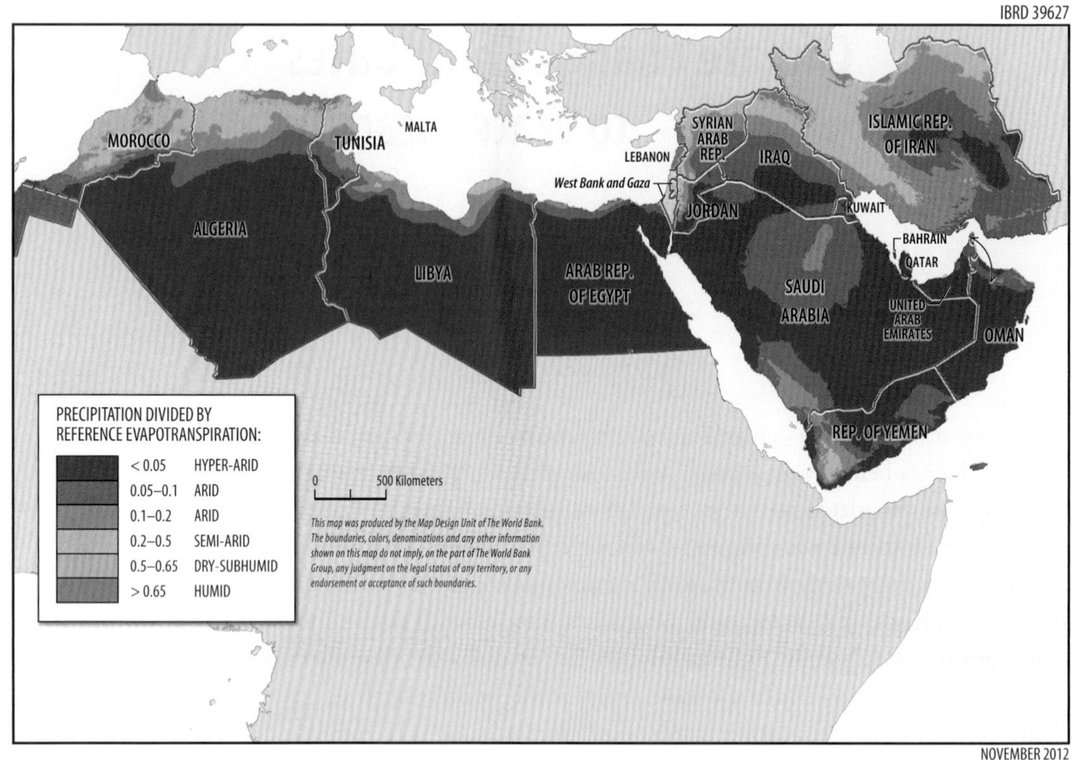

Source: Bucknall 2007.

significant runoff from mountain ranges that separate them from the interior deserts (see map 3.1). Deep in the Sahara and Arabian Deserts, several oases create microclimates where limited agriculture can be practiced. High evapotranspiration rates greatly reduce the amount of water that turns into surface runoff or percolates through the soil to recharge the aquifers. For example, in Jordan, an estimated 90 percent or more of the rain evaporates, leaving a fraction to recharge aquifers and feed surface runoff (ESCWA 2005).

Climate Change Will Have Diverse Impacts on Hydrometeorological Conditions

Chapter 2 describes the use of global circulation models (GCMs) in making climate projections and applying them to the Arab region, but GCMs are suitable to assess only the general characteristics of potential changes; because of their coarse spatial resolution, GCMs are unable to capture effectively many smaller-scale processes, such as cloud formation and the effect of sharp topographic variations. The performance of GCMs can be improved by downscaling their outputs to better represent local conditions (see chapter 2).

The Intergovernmental Panel on Climate Change identifies the Middle East and North Africa region as the region most severely affected by climate change, particularly because the effects will accentuate already-severe water scarcity (Parry et al. 2007). Most GCMs project that much of the Arab region will undergo significant reductions in precipitation levels and increases in temperatures that will increase evapotranspiration rates. The net effect will be a severe reduction in river runoff and soil moisture. Simulations using a middle path greenhouse gas emissions scenario (SRES A1B) (Bates et al. 2008) show that 80 percent of the GCMs agree on the direction of the change over the most densely inhabited northern part of the region, although the simulations agree less over the Arabian Peninsula (map 3.2; see also chapter 2).

Climate change projections clearly show stark differences in the effects across the region. Although runoff in North Africa and the eastern Mediterranean (including the headwaters of the Euphrates and Tigris) is expected to drop by up to 50 percent, southern Saudi Arabia and East Africa (including the headwaters of the Nile) will experience increases in runoff by up to 50 percent. Consequently, climate change is projected to reduce water supplies in the northern and western parts of the Arab region and to increase those of Egypt and the southern part of the Arab world.

A World Bank study assessed the impact of climate change on water resources in the Arab counties[1] and identified options to manage these resources under future conditions of higher water demands (World Bank 2012). The study involved first assessing potential spatiotemporal distributions of surface and groundwater resources in the region over the next four decades based on output from nine GCMs for the A1B SRES scenario; the study then downscaled a 10-kilometer by 10-kilometer grid covering the Arab region and the headwater areas of the Tigris and Euphrates and Nile Rivers.

A distributed hydrological model, PCR-GLOBWB, processed the downscaled GCM output and reference data to simulate runoff, groundwater, and soil moisture. Outputs from the hydrological model were then run through a water resources planning model, WEAP, to determine corresponding scenarios for municipal, industrial, and agricultural water demands. Marginal cost analysis was then used to assess the economic efficiency of alternative adaptation options, described in a later section.

Results from this study show that the majority of Arab countries are already experiencing deficits in internal and external renewable water resources. By midcentury all Arab countries will face serious water deficits

MAP 3.2

Annual Mean Changes in Hydrometeorological Variables for the Period 2080–99 Relative to 1980–99

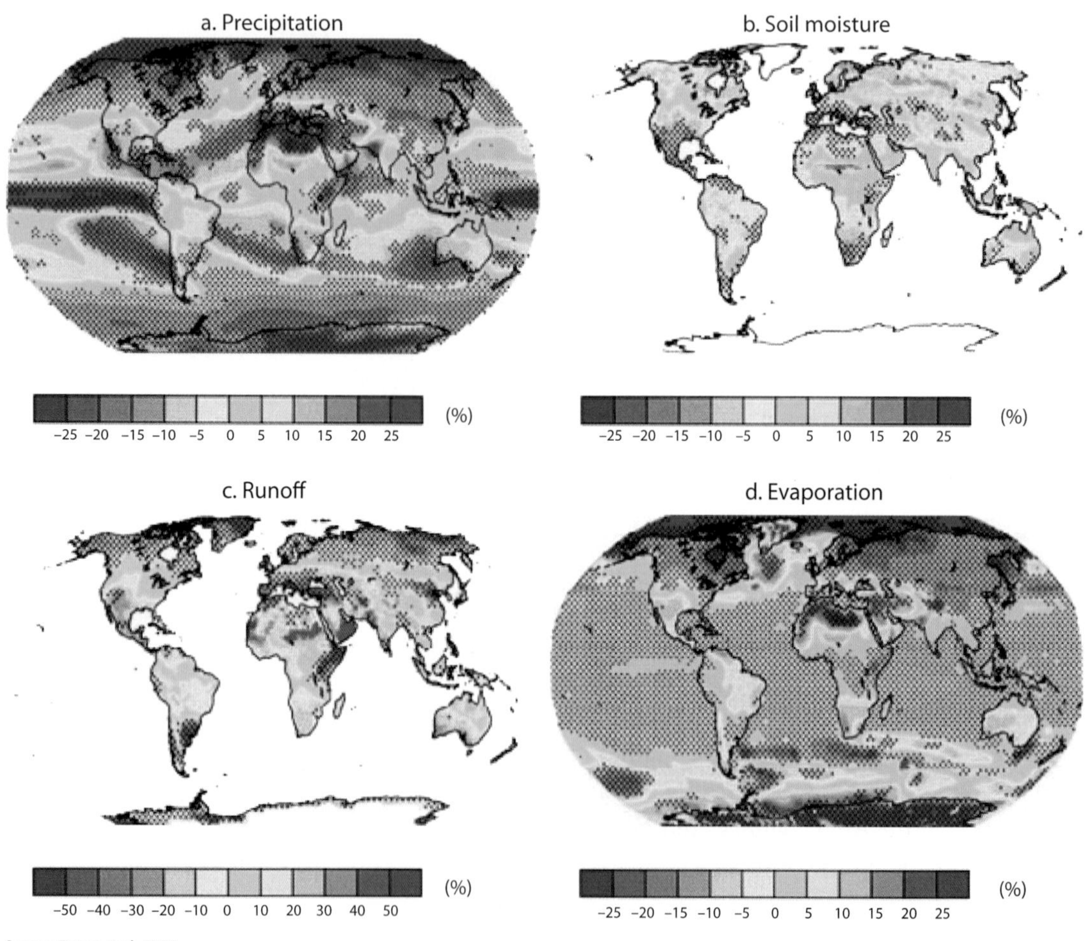

Source: Bates et al. 2008.

Note: Based on simulation results from 15 GCMs for the greenhouse gas emissions scenario A1B. Stippled areas indicate those where at least 80 percent of the GCMs agree on the direction of change.

as demand and supply continue to diverge. Total regional renewable water shortage will be about 200 cubic kilometers per year in 2040–50 based on the average climate change projection.[2] The demand is expected to rise by about 25 percent in 2020–30 and up to 60 percent in 2040–50, whereas renewable supply will drop by more than 10 percent over the same period in the region. As a result, unmet demand for the entire Middle East and North Africa region, expressed as a percentage of total demand, is expected to increase from 16 percent currently to 37 percent in 2020–30 and 51 percent in 2040–50 (figure 3.1).

Water shortages for the individual countries will vary substantially (table 3.1 and figure 3.2). Many countries that are not currently facing any

FIGURE 3.1

Renewable Water Resources versus Total Water Demand through 2050

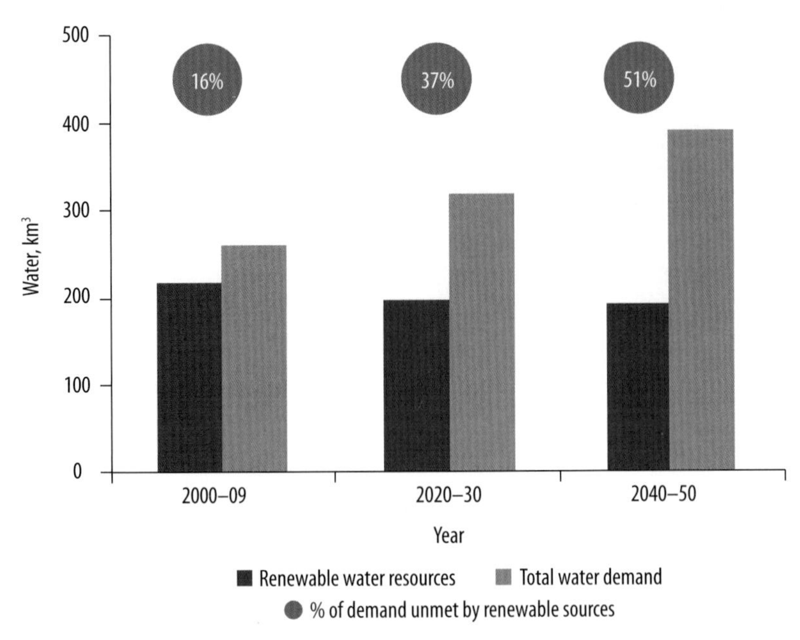

Source: World Bank 2012.

shortages will be confronted with huge deficits in the near and distant future. The situation will be particularly troublesome for Jordan, the Gaza Strip, the West Bank, and the Republic of Yemen, which do not have the funds to procure additional expensive water supplies.

Evans (2009) analyzed the impact of climate change on an area covering the Islamic Republic of Iran, the Mashreq, the northern Arabian Peninsula, and Turkey using simulation results from 18 GCMs under the SRES A2 emissions scenarios, an area that represents a high emission pathway. His analysis shows that most of the Arab region will become warmer and will undergo a significant reduction in precipitation. The 200-millimeter isohyet—a threshold for viable rainfed agriculture—will move northward as the climate warms. By midcentury, 8,500 square kilometers of rainfed agricultural land will be lost. By the end of the century, the 200-millimeter isohyet is projected to move northward by about 75 kilometers, resulting in the loss of 170,000 square kilometers of rainfed agricultural land over an area covering the Islamic Republic of Iran, Iraq, Lebanon, the Syrian Arab Republic, and the West Bank and Gaza. Evans also concluded that the dry season will grow longer by about two months, reducing the rangelands in Iraq and Syria and necessitating the reduction of herd sizes or increasing water requirements and imports of feed.

TABLE 3.1

Current and Projected Water Demand and Supply for Arab Countries
(million cubic meters)

		2000–09						
		Demand				Supply		
Region	Country Name	Agricultural	Municipal	Industrial	Total	Surface	Groundwater	Total
Maghreb	Algeria	3,955	1,523	878	6,356	4,622	1,733	6,355
	Libya	3,287	691	147	4,125	1,612	2,512	4,124
	Mauritania	—	—	—	—	—	—	—
	Morocco	13,942	1,403	395	15,740	10,440	3,208	13,648
	Tunisia	1,938	417	417	2,772	2,210	562	2,772
Nile Valley	Egypt, Arab Rep.	45,371	6,003	4,462	55,836	47,470	5,509	52,979
	Sudan	—	—	—	—	—	—	—
Mashreq	Iraq	34,084	4,942	4,942	43,968	29,591	7,105	36,696
	Jordan	789	286	38	1,113	193	67	260
	Lebanon	677	376	149	1,202	830	231	1,061
	Syrian Arab Republic	13,202	1,490	1,490	16,182	14,835	474	15,309
	West Bank and Gaza	323	122	122	567	78	92	170
Gulf	Bahrain	20	184	21	225	14	16	30
	Kuwait	44	442	23	509	306	203	509
	Oman	596	148	148	892	723	168	891
	Qatar	144	173	173	490	137	118	255
	Saudi Arabia	17,788	1,972	1,972	21,732	7,335	3,825	11,160
	United Arab Emirates	2,691	610	68	3,369	169	164	333
	Yemen, Rep.	5,137	341	341	5,819	3,858	699	4,557
South	Comoros	—	—	—	—	—	—	—
	Djibouti	9	18	1	28	28	—	28
	Somalia	—	—	—	—	—	—	—

Source: World Bank 2012.
Note: — = not available. For Djibouti, model outputs are not completely reliable because of poor-quality input data.

The Arab Region Is Water Scarce

To manage growing deficits in water, many countries have unsustainably tapped freshwater aquifers and seriously depleted strategic fossil water stocks. Water resource development in upstream countries has significantly reduced river runoff in downstream Arab countries. Rapid population growth, urbanization, and industrialization have contributed to pollution of vital water resources, including strategic aquifers. The risk of flooding has also increased from the higher frequency and intensity of extreme events, poor urban planning, and inadequate preparedness. Water governance is a major concern for many countries

2020–30							2040–50						
Demand				Supply			Demand				Supply		
Agricultural	Municipal	Industrial	Total	Surface	Groundwater	Total	Agricultural	Municipal	Industrial	Total	Surface	Groundwater	Total
4,621	2,944	1,221	8,786	4,939	3,711	8,650	5,059	5,814	1,463	12,336	4,786	3,431	8,217
3,597	1,163	214	4,974	1,448	2,062	3,510	3,917	1,799	265	5,981	1,091	1,184	2,275
—	—	—	—	—	—	—	—	—	—	—	—	—	—
16,115	2,691	551	19,357	7,581	2,347	9,928	18,173	5,386	665	24,224	6,678	1,855	8,533
2,304	841	841	3,986	1,845	2,054	3,899	2,648	1,634	1,634	5,916	1,829	1,125	2,954
53,478	10,284	6,646	70,408	41,573	5,485	47,058	61,712	17,525	8,443	87,680	48,703	5,640	54,343
—	—	—	—	—	—	—	—	—	—	—	—	—	—
40,521	8,304	8,304	57,129	25,471	4,175	29,646	47,901	11,606	11,606	71,113	23,802	3,561	27,363
871	600	56	1,527	146	19	165	975	1,233	68	2,276	158	17	175
781	557	187	1,525	907	122	1,029	893	776	200	1,869	859	108	967
14,358	2,544	2,544	19,446	13,061	1,605	14,666	15,973	4,222	4,222	24,417	12,438	2,237	14,675
370	291	291	952	86	10	96	413	587	587	1,587	87	9	96
19	271	30	320	8	1	9	17	337	36	390	4	1	5
42	789	36	867	282	254	536	38	1,133	44	1,215	224	177	401
571	487	487	1,545	639	209	848	521	1,143	1,143	2,807	498	56	554
139	231	231	601	120	50	170	127	257	257	641	105	43	148
16,450	5,108	5,108	26,666	5,974	2,065	8,039	15,062	10,098	10,098	35,258	4,797	1,314	6,111
2,517	886	91	3,494	166	69	235	2,279	1,014	96	3,389	122	47	169
5,623	1,270	1,270	8,163	3,700	751	4,451	6,081	6,492	6,492	19,065	3,784	612	4,396
—	—	—	—	—	—	—	—	—	—	—	—	—	—
11	34	1	46	45	—	45	11	72	1	84	62	—	82
—	—	—	—	—	—	—	—	—	—	—	—	—	—

because of a lack of accountability and weak institutional capacity. The scale and nature of these challenges and how they will be influenced by climate change vary considerably across the region, but they clearly point to the need for action. Climate adaptation is closely linked to a general need for good management of resources for a growing population under conditions of high water variability.

Water Resources Are Scarce, Highly Variable, and Unevenly Distributed

Water resources management aims to secure supplies to meet demands by matching supply to demand, not only in quantity and quality but also in location and timing. Water demand in the Arab region is already sur-

FIGURE 3.2

Current and Projected Water Demand and Supply for Selected Arab Countries

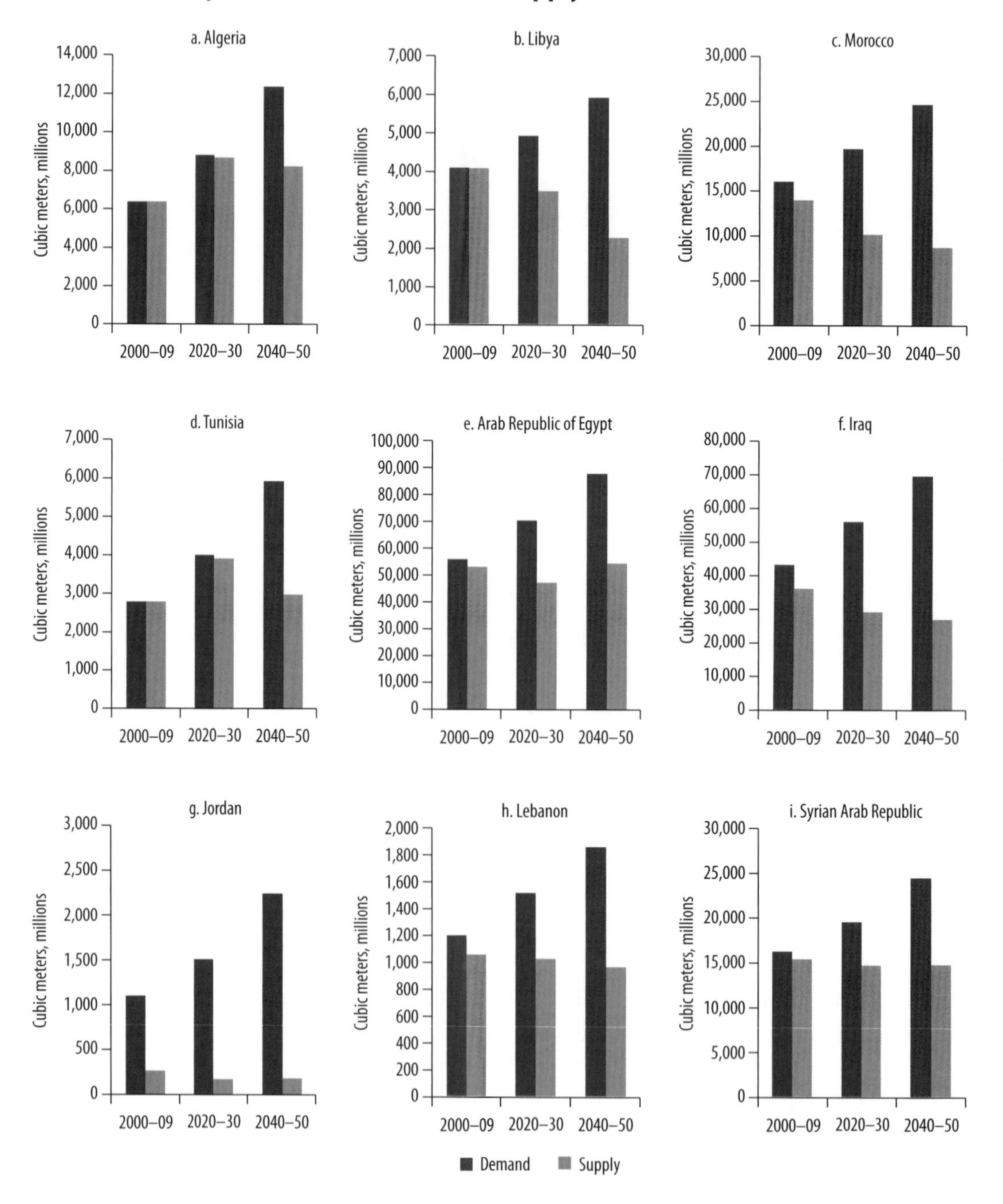

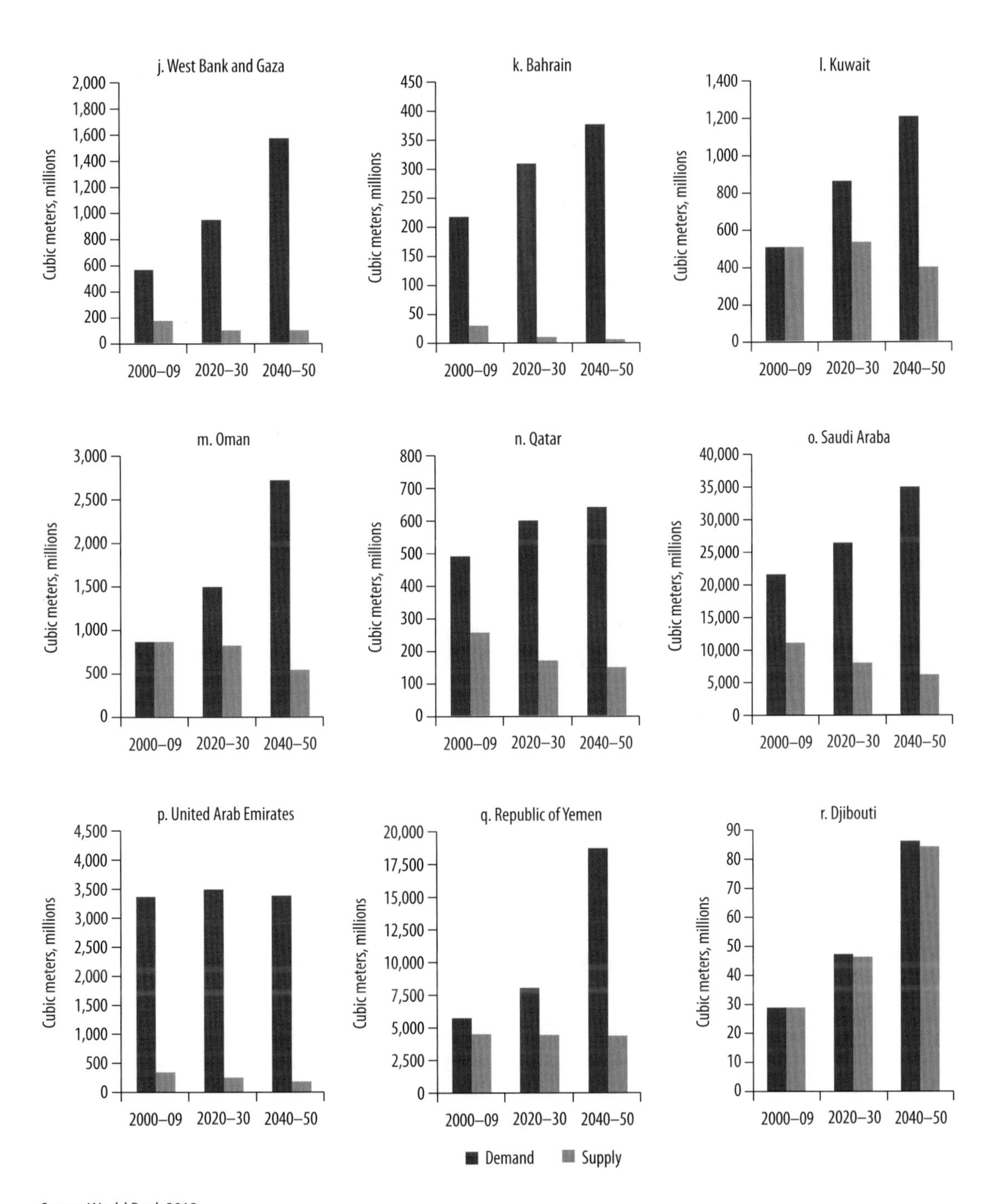

Source: World Bank 2012.

FIGURE 3.3

Variability and Level of Precipitation in Arab Countries versus Other Countries Worldwide

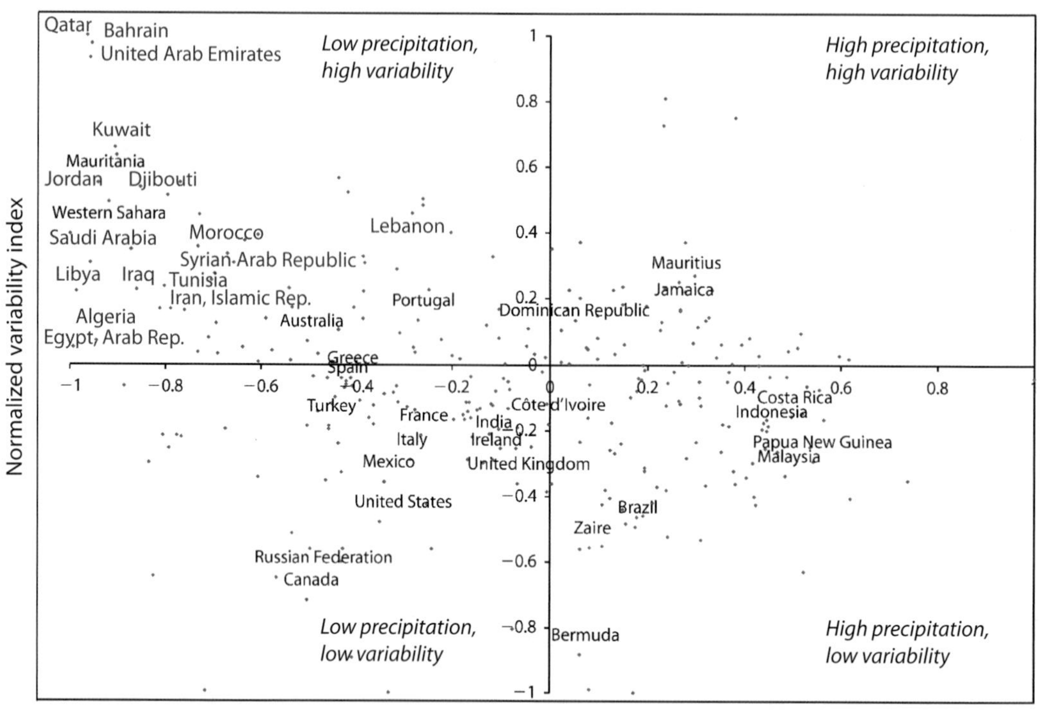

Source: Bucknall 2007.

passing supply and rising rapidly. Demand is generally concentrated in large urban areas, but irrigation for agriculture increases demand during the drier seasons. In most areas, water resources are scarce, highly variable, unevenly distributed, and seasonally out of sync with demand.

The Arab region is characterized by uneven topographical and climatic conditions; the bulk of the region is very arid, but arid areas are flanked by more humid mountainous and coastal plains. Precipitation levels are low and highly variable in time and location. Globally, Arab countries have the least favorable combination of the lowest levels of precipitation and the highest level of variability (figure 3.3).

The ramifications of these conditions vary across the region. In countries where water resources are derived from relatively higher levels of precipitation—mostly in North Africa and the eastern Mediterranean—water supplies are sizable yet highly variable and susceptible to frequent droughts. Although precipitation levels in these countries were historically adequate to support demand, they are facing serious challenges in

FIGURE 3.4

Yearly Inflows to Lake Qaraoun, Lebanon, 1962–2004

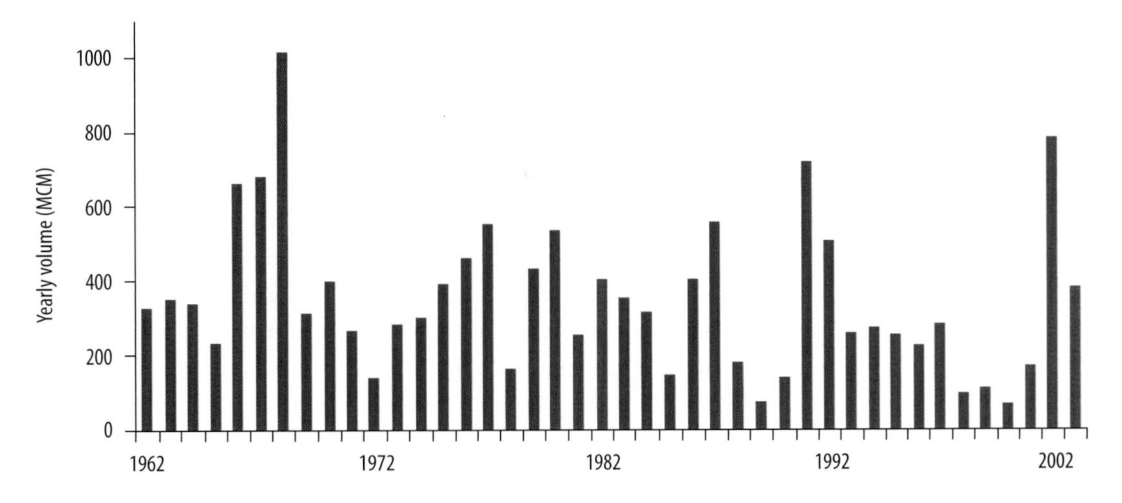

Source: Assaf and Saadeh 2008.

meeting current demand given their high variability. For example, yearly inflows to Qaraoun Lake from the Litani River—the most significant water resource in Lebanon—is extremely variable, with maximum flow more than an order of magnitude higher than minimum flow (figure 3.4).

The large fluctuations in runoff across North Africa and the eastern Mediterranean are strongly linked to the North Atlantic Oscillation (NAO) global teleconnection pattern. A stronger NAO anomaly shifts the moisture-bearing westerly winds northward, depriving the region of a substantial amount of rainfall. The devastating droughts of the mid-1980s to 1990s in the region were attributed to this phenomenon. During that period, dams in Morocco did not fill beyond half their maximum capacity (Bucknall 2007).

Water resources in the riparian countries of Egypt, Iraq, and Syria are mainly derived from very large catchments in the more humid regions to the south and north of the Arab region that are characterized by higher and more consistent precipitation. For example, Turkey, location of the main headwaters of the Euphrates and Tigris Rivers, has much higher precipitation and less variability than neighboring Arab countries (figure 3.3). In the Nile River basin, the Nile is fed by the monsoon-dominated Ethiopian highlands and equatorial Lake Victoria. The discharge from the Ethiopian highlands peaks at a different period—from July to September—than runoff from Lake Victoria, which has two peak periods: a long one from March to May and a less intense one from October to December (Conway 2005). These staggered seasons have largely stabilized runoff patterns in the region. Before the construction of the Aswan High Dam, however, Egypt was exposed to several devastating floods and droughts. The dam has drastically reduced multiyear fluctuations; however, it was drawn down to alarmingly low levels as a record-breaking, severe drought extended from 1978 to 1987. The drought was mainly attributed to a drastic reduction in precipitation over the Ethiopian highlands associated with an El Niño event (Conway 2005).

Seasonal and multiyear variability have been managed on the Euphrates, Nile, and Tigris Rivers through extensive development of storage and conveyance infrastructure. However, more pressing issues are related to sharing water resources and management of multidecadal droughts. Climate change is projected to have different, and almost opposite, impacts on the Nile and Euphrates-Tigris basins. The former is mainly influenced by the monsoon system, which will gain strength in a warmer world. Precipitation over the latter is highly influenced by the NAO, which will lead to drier conditions as a result of climate change (Cullen and deMenocal 2000).

In the already extremely arid regions in the Gulf countries and Libya, precipitation levels are very low and extremely variable. Extreme water scarcity in these regions has, until modern times, suppressed growth in population and limited human activities to pastoralism and subsistence agriculture in oases and coastal regions with access to springs. But the discovery of oil resulted in a dramatic increase in population and living standards, increasing demand for water beyond the capacity of renewable resources. This sharp water imbalance is filled through desalination in most countries and through excessive reliance on fossil water, such as Libya and Saudi Arabia. Although climate change is not expected to significantly alter water balance in these countries, it is projected to increase the intensity and frequency of extreme rainfall outbursts that may result in extensive damage and loss of life similar to those experienced recently in Jeddah, Saudi Arabia (Assaf 2010).

The southern part of the Arabian Peninsula is more humid than the northern and central regions. In the Republic of Yemen, relatively more abundant natural water supplies on the order of 2.1 cubic kilometers per year (table 3.1) have, however, been outstripped by a relatively large and rapidly growing population—24 million people in 2010 (see chapter 1). In comparison, the Republic of Yemen's eastern neighbor Oman, with 1.4 cubic kilometers per year of renewable water resources, has a more favorable water balance, given its much smaller population of only 2.8 million. Being in the domain of the monsoon system, the southern part of the Arabian Peninsula is expected to receive more precipitation—albeit in the form of extreme events, such as those that hit Oman recently—as global climate continues to warm.

The human suffering, hunger, and potential famine that may result from extreme drought conditions will be felt mostly in the least developed Arab countries (the Comoros, Djibouti, Mauritania, Somalia, Sudan, and the Republic of Yemen), where most of the economically active population are engaged in agriculture (see chapter 1). Many are dependent on pastoralism and subsistence rainfed farming, making them highly vulnerable to rainfall variability. The recent and ongoing drought in eastern Africa has taken a large toll on rural populations that incurred heavy income losses and are suffering from chronic hunger that could develop into wide-scale famine.

Droughts have also had heavy impacts on other Arab countries, particularly Algeria and Syria, where rainfed agriculture is prevalent. In Syria, the wheat-producing northeast was ravaged in 2006 by a four-year drought that dried up the Khabur River (Erian, Katlan, and Babah 2010). Although farmers initially adapted by tapping shallow aquifers, the continuation of the drought has led them to draw down groundwater levels significantly. Hundreds of thousands of farmers abandoned their villages to look for livelihoods in cities and in neighboring Arab countries.

As climate continues to change, precipitation—and consequently droughts and floods—is expected to change in frequency, intensity, and distribution. This shift will challenge the hypothesis of stationarity of statistical means, which water planners and managers have traditionally applied to the design and operation of water resource systems. A shift in hydrological conditions in North Africa has left many water projects with overdesigned infrastructure (Bucknall 2007). A recent policy document has identified hydrological nonstationarity as a great challenge to water resource planners in the United States (Brekke et al. 2009).

Population Growth and Urbanization Will Add Pressures to Water Resources

Among the pressing challenges to water resources in the Arab region is the rapid growth in population and improvement in living standards. Although population growth rates have subsided over the past two decades (Dyer 2008), population is still expanding at some of the highest global rates in some countries—particularly the least developed ones—driving renewable water resources per capita in most Arab countries well below the absolute water-scarcity level of 500 cubic meters per capita (see table 3.1). This situation has been compounded by high domestic per capita water consumption, which in some Arab countries dwarfed rates in the developed world. Rising living standards are expected to further drive this trend (figure 3.5).

FIGURE 3.5

Estimated Increase in Domestic Per Capita Consumption in Selected Arab Countries

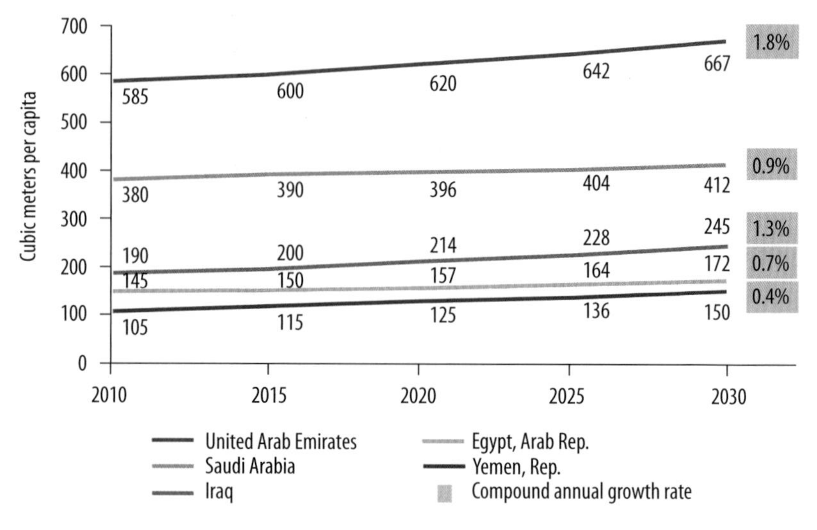

Source: Chatila 2010.

A major concern in water management around the region is rapid urban sprawl, particularly in areas away from water-supply sources (Brauch 2003). This growth is a consequence of an ongoing urbanization process as people from rural areas continue to abandon farming in search of a better life in cities, in part because of the difficulty of maintaining viable agriculture as water resources become scarcer. The decline in water resources projected under climate change is expected to accelerate this process, particularly since adaptation measures will probably lead to a further reduction in agricultural activities.

The challenge presented by this ongoing redistribution of population is to secure water supplies and to provide water services. In coastal cities in Lebanon, for example—particularly in the capital Beirut where half the population lives—water shortages are frequent because local supplies are incapable of meeting rising demand. Lacking access to adequate water services, people often illegally tap shallow aquifers, resulting in serious seawater intrusion. In an attempt to reduce pressure on heavily populated Cairo, the government has encouraged urban development in desert areas, which has presented serious challenges in procuring water supplies over large distances. In Jordan, the population is increasingly concentrated in the highlands, several hundred meters above most prospective water resources.

Urban sprawl in several Arab cities has brought increasing numbers of people and economic assets into harm's way from incidents of extreme flooding that have increased in frequency and intensity.

Already-High Agricultural Water Requirements Will Increase Further

High evapotranspiration and soil infiltration rates in the arid Arab region reduce soil moisture—green water—and consequently increase irrigation requirements that typically surpass 80 percent of total water withdrawals in most Arab countries (chapter 2). With climate change, the predicted increases in evapotranspiration rates will lead to higher irrigation requirements.[3] For pastoralism, which currently does not normally use irrigated sources of fodder, the lengthening of dry periods will reduce available rangeland and consequently increase the need for irrigated fodder to maintain the same level of livestock (Evans 2009).

In the long term, difficult questions will be raised regarding the allocation of water to agriculture, particularly in areas of marginal returns, with likely increased competition from high-value uses in the industrial and urban sectors. Any solutions to managing water will require the inclusion of agriculture within an integrated national socioeconomic development strategy that involves all sectors. This approach will be particularly important given that the agriculture sector is the largest employer in many Arab countries and contributes significantly, yet decreasingly, to meeting food requirements (see chapter 4 for more details). One way of looking at trade-offs between social security and food security considerations and water use for agriculture is to consider the economic returns of water used for irrigation for different crops, which differ significantly among Arab countries (figure 3.6). Countries can optimize their return on water by

FIGURE 3.6

Agricultural Value Added GDP per Cubic Kilometer of Water Used in Agriculture

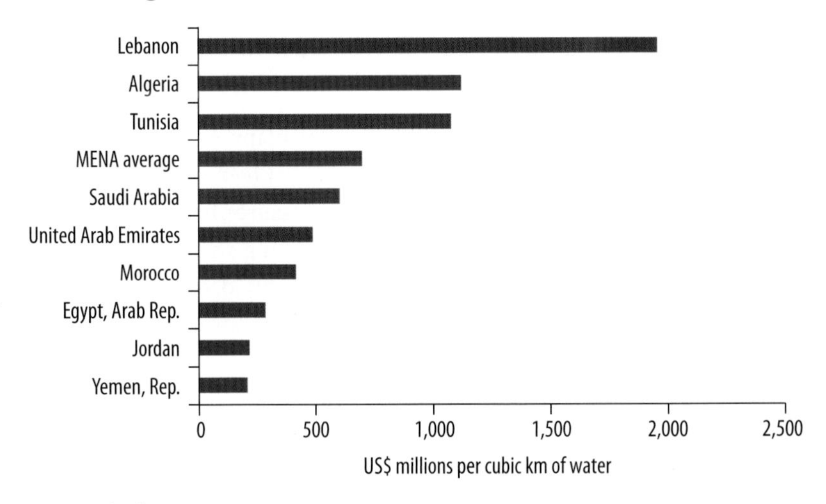

Source: Bucknall 2007.

FIGURE 3.7

Water Productivity in Irrigated Crop Fields

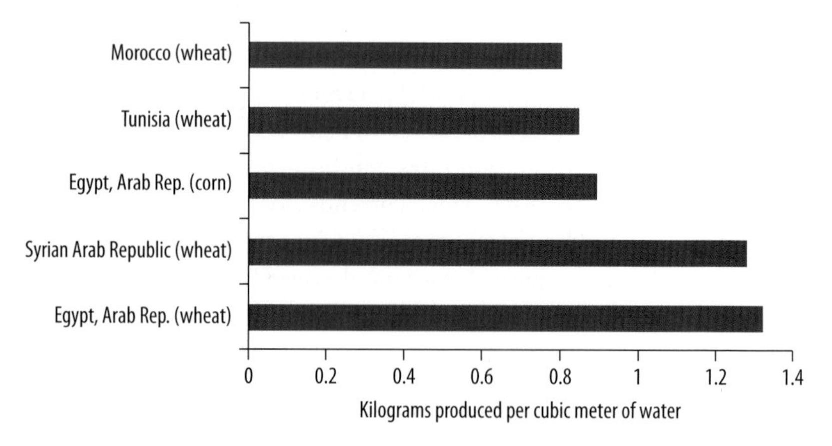

Source: Renault 2002.

choosing different crop mixes, which will lead to different returns on the agricultural water used. Although vegetables have a high economic return per cubic meter of water, the return of cereals is significantly lower. The cost of producing crops will continue to rise in many Arab countries, as fossil groundwater resources are depleted and groundwater levels sink. Wells now require deep drilling, and the cost per cubic meter is increasing.

The environmental and social conditions of a given country, as well as water availability (from irrigation and rainfed agriculture), will determine the most viable crop mix. Returns on water for specific crops can differ among countries, as illustrated by wheat and corn for selected Arab countries (figure 3.7).

Strategic Groundwater Reserves Are Being Depleted at an Alarming Rate

In an attempt to meet rising demand for water, many Arab countries resorted to mining their groundwater reserves. Over several decades, roughly from the 1960s to the 1990s, these measures have drawn down levels in many aquifers by tens of meters rendering them economically unusable; in several cases, aquifers were irreversibly damaged by salination from rising underlying saline waters or by seawater intrusion in coastal areas. This period also witnessed attempts by several Arab countries to achieve food sufficiency at the expense of depleting vast nonrenewable fossil aquifers. In Saudi Arabia, an estimated 50 percent or more of fossil water was used to produce wheat that could have been bought in the global market at much lower costs. The opportunity cost of these lost water resources is enormous; Riyadh, the country's capital and largest city, is mostly supplied with desalinated water from the Gulf coast that is pumped over 450 kilometers at a cost of about US$1.50 per cubic meter (Allan 2007).

Some socioeconomic development policies have been detrimental to strategic aquifers. For example, the policies to settle nomads in the northern Badia region of Jordan gave unlimited access to underlying renewable aquifers, which get recharged from winter precipitation. Over two decades, water tables declined by several meters and water became too saline for use in agriculture. Poor groundwater licensing, water pricing, and energy subsidies encouraged farmers to mine aquifers unsustainably. The net effect of those policies was that farmers did not take into account the social opportunity cost of water and used high-quality water to grow low-value crops, whereas domestic users nearby in Amman were willing to pay very high prices for water (Chebaane et al. 2004). Although the opportunity cost of water stored in aquifers is relatively well understood, a less obvious and equally important value is the opportunity cost of storage, which is exemplified in the current estimate of the cost required to develop a strategic reserve in the United Arab Emirates for desalinated water. Along the Gulf, water storage is very low, ranging from a one- to five-day supply at best (Dawoud 2009). Low storage volumes place these countries at the mercy of interruptions in desalination even for very short periods.

A Heavy Dependence on Shared Water Resources Makes Countries More Vulnerable

A widely recognized fact is that water resources should ideally be managed at the watershed level. Integrated watershed or aquifer management facilitates optimal and balanced allocation of water resources among a watershed's inhabitants and ecosystems. However, such an approach faces major obstacles if these natural basins are shared among different countries, or even among different administrative divisions within the same country. First, national socioeconomic development objectives could conflict with those of integrated watershed and aquifer management, since countries seek to use natural resources within their national boundaries for the benefit of their citizens. This approach may include not only using water resources within a watershed or aquifer but also transferring that water to other parts of the country. Second, technological advances have made it possible to develop large water storage and conveyance infrastructure and to use deep aquifers that were inaccessible in the past. Third, the high variability and uneven distribution of water make its value—and the potential for conflict—vary over time and space.

In the aftermath of World War I, newly formed political boundaries crossed natural water basins and aquifers. Following independence, Arab countries sought to develop their water resources to expand their agriculture and meet rising domestic and industrial demands. This endeavor has brought several countries into competition—and potential conflict—with one another over shared water resources. The significance of these issues to water resources management varies across the region and depends on the level of dependence on shared water resources, the upstream or downstream position of the country, its economic and military stature, and the political relationships among sharing countries.

Turkey has the most favorable position in the Euphrates-Tigris basin as an upstream country; despite repeated protests from downstream Syria and Iraq, Turkey was able to dam the Euphrates basin extensively and to pursue development in the Tigris basin that jeopardized runoffs to Syria and Iraq. Tension also arose between Syria and Iraq over filling a major reservoir in Syria. A less conspicuous tension is broiling over the Islamic Republic of Iran's recent diversion of major tributaries to the Tigris, which has significantly reduced runoff to the marshes in southern Iraq.

Egypt has maintained dominance over the Nile basin. Despite being at the extreme downstream end of the watershed, Egypt's policies influence water resource development in upstream countries. Egypt has been part of the Nile Basin Initiative to facilitate collaboration in managing the Nile basin. A new agreement, the Cooperative Framework Agreement, which was recently signed by most riparian countries, calls for replacing the initiative with a basin commission that manages water resources in the Nile basin on "behalf of all the Nile Basin states" (Stephan 2010). Egypt and Sudan are concerned that the Cooperative Framework Agreement would effectively reduce their current water allocations, and they are currently not part of the agreement.

Deaths and Damage from Extreme Flooding Are Expected to Rise

Climate change is expected to increase the frequency and intensity of flooding. A flooding disaster is a construct of the physical flooding of a massive and fast-moving body of water and an affected area that contains people, infrastructure, and other vital economic assets (Assaf 2011). An intense rainfall in an open desert is hardly an issue, whereas a much less intense rainfall in a crowded, highly built, and poorly drained area is of great concern since it may lead to torrents that sweep people away to their deaths.

The 2009 flooding in Jeddah, Saudi Arabia, which killed over 150 people and caused great economic losses, was initiated by an intense rainfall that dumped 90 millimeters of rain in four hours over an area that normally receives 45 millimeters per year. Although the storm was unprecedented, the resultant torrents would have been reduced significantly had the area been equipped with an adequate drainage system. More significantly, the death and damage could have been reduced, or even eliminated, had development been avoided in the natural drainage area, known as a *wadi*. Many of the victims were migrant workers who lived in slums in the *wadi* area. The area also contains a major highway junction, causing considerable damage to cars and the deaths of some of their occupants. To make matters worse, the police and civil defense units were poorly prepared to handle large-scale disasters (Assaf 2010).

Water Quality Is Deteriorating

Deteriorating water quality is making a significant quantity of water unusable, even for applications that require substandard water quality. For example, domestic sewage, industrial waste, and agricultural return flows from Cairo are sent mostly untreated through the 70-kilometer Bahr El Baqar channel to discharge into the 1,000-square-kilometer Lake Manzala in the northeastern Nile Delta. The discharge from Bahr El Baqar is heavily loaded with a wide range of contaminants, including bacteria, heavy metals, and toxic organics. Local fisheries have suffered significantly because of the widespread public aversion to consuming the lake's fish, which in the past represented a third of the total fish harvest in Egypt (USAID 1997). The Upper Litani basin in Lebanon provides another stark example of how years of poor wastewater management has turned the river, mostly fed by freshwater springs, into a sewage channel during most of the year (Assaf and Saadeh 2008). The situation is also compounded by the uncontrolled use of fertilizers, which has increased contamination of underlying aquifers (Assaf 2009). Climate change would exacerbate these problems because higher temperatures will increase bacterial activity and lower freshwater supplies, increasing the pollution content of wastewater.

Overuse of aquifers has caused salinization throughout the region, especially in heavily populated coastal areas, including Beirut, Gaza, Latakia in Syria, and along the Gulf. Interior aquifers (for example, the Amman-Zarqa basin in Jordan) have also been affected by the problem since excessive abstraction draws up underlying saline waters. Salinization is very difficult to reverse because it requires large amounts of freshwater to bring down the freshwater-saline interface. Lacking any control measures, climate change is projected to intensify the salinization of aquifers as the increased supply-demand gap will encourage further abstraction of groundwater.

Managing Water Resources Is Critical in a Changing Climate

A warmer, drier, and more volatile climate in the Arab region will most likely exacerbate already-adverse water conditions; holistic water strategies that can respond to a multitude of complex, intertwined, and often-conflicting challenges need to be adopted. These strategies need to be flexible and adaptive to address the great uncertainties about conditions in the future. In this water-scarce region, water is often the most limiting factor in key socioeconomic sectors. Adaptation strategies must incorporate water issues in all sectors, including agriculture, urban development, trade, and tourism.

To facilitate developing these strategies, we propose several adaptation options organized under the umbrella of Integrated Water Resources Management (IWRM), as well as a socioeconomic development framework. IWRM seeks to balance water-supply development with demand management within a framework of environmental sustainability. The IWRM components are complemented by measures that address the impact of water resource availability on socioeconomic development.

A central theme in this approach is that no one-size-fits-all adaptation solutions apply to all Arab countries. Even at the national level, adaptation measures have to consider variations from one locality to another. An effective strategy is a portfolio of adaptation options from among a pool of measures tailored to suit each country's political, socioeconomic, and environmental conditions. Gulf countries, for example, will need to focus on enhancing their desalination capacities, reusing wastewater, and developing strategic reserves, while pursuing aggressive water-demand management programs. Arab countries dependent on shared water resources would have to place a high priority on reaching international water resource agreements with border countries.

To simplify this discussion, adaptation options are grouped into two main categories: (a) water management and (b) water-related development policies in nonwater sectors. The first category captures the two main IWRM branches of supply-and-demand management in addition to other relevant water issues, such as governance, disaster risk management, and cooperation in managing shared water resources.

Integrated Water Resources Management Will Be Critical

IWRM is based on four principles presented by the International Conference on Water and the Environment in Dublin in 1992, and later adopted by the United Nations Conference on Environment and Development in Rio de Janeiro in 1992 (Agarwal et al. 2000):

- Freshwater is a finite and vulnerable resource, essential to sustain life, development, and the environment.
- Water development and management should be based on a participatory approach, involving users, planners, and policy makers at all levels.
- Women play a central part in the provision, management, and safeguarding of water.
- Water has an economic value in all its competing uses and should be recognized as an economic good.

Subsequent efforts by the Global Water Partnership (GWP) focused on developing implementation frameworks for IWRM. They include balancing water-demand management with supply management, ecosystem protection, and social equity. It also emphasized the importance of water as an economic commodity that needs to be managed to reflect its scarcity and optimize its socioeconomic and environmental services.

Supply-Side Management Will Need to Be Optimized

Water resources management traditionally focused on developing water supplies. Although water resources management efforts are leaning toward better demand management and governance, water-supply development is necessary to ensure reliability of water resource systems and optimal use of resources.

Storage and conveyance

Currently, a number of Arab countries have invested significantly in water-supply infrastructure. The region now has the world's highest storage capacity per cubic meter (Bucknall 2007). These investments have greatly enhanced access to water supplies and have facilitated a dramatic expansion of agriculture. Before the construction of the Aswan High Dam, Egypt suffered major debilitating droughts, and a large part of its population was at the mercy of devastating seasonal floods. The Aswan High Dam offered Egyptians a way to control runoff from the Nile and to provide a stable water supply. Combined with good forecasting and operational systems for the entire Nile basin, the dam also enabled Egypt to effectively weather the extended drought of the mid-1980s (Conway 2005).

Several factors influence the effectiveness of storage strategies: size, cost, rate of loss, and externalities. Large reservoirs are needed to provide adequate and reliable water supplies for large communities and to secure irrigation for agriculture during the rainless growing season. Large reservoirs also have the advantage of economy of scale. However, they are costly to build and maintain, and they can result in massive social and environmental disturbances. For example, Egypt was hard-pressed financially and politically to secure funding for the Aswan High Dam. Although the dam successfully stopped damage from seasonal flooding, it was at the expense of forfeiting the nutrient-laden sediment the floodwater brought with it to replenish the Nile Delta (Syvitski 2008). The reduced sediment has also resulted in shrinking the delta since less sediment is deposited to replace that lost by erosion.

Integrated surface and groundwater storage strategy

Reservoirs in arid and semiarid regions sustain significant evaporation and seepage from flat terrain, permeable geological formations, and long, hot summers (Sivapragasam et al. 2009). The evaporation from Lake Nasser (a reservoir of the Aswan High Dam) is estimated to consume about 5 percent of the total Nile flow (Sadek, Shahin, and Stigter 1997). Lake Assad in Syria also loses substantial amounts of water from evaporation. In warmer climates, higher evaporation rates reduce the storage value of these reservoirs. Historically, underground storage and tunnels were used to reduce evaporative losses. Evaporation is effectively eliminated from water cisterns and the underground *aflaj* system. These practices can be reinstated to complement existing storage facilities.

A more promising implementation of underground storage is to use the vast natural aquifer storage capacity to store and improve the quality of water. Because they are not subject to evaporation, aquifers have a distinct advantage over surface reservoirs in semiarid regions. The Arab region also has few suitable sites for surface storage but has ample aquifer capacity. Aquifer storage can be used for excess winter runoff and treated wastewater. In Saudi Arabia, a large network of recharge dams dots the arid Arabian Desert. Al-Turbak (1991) indicated that these dams are highly effective in recharging shallow aquifers. Abu Dhabi has embarked on a massive US$5 billion program, based on the aquifer storage and recovery approach, to use local aquifers as strategic reserves for desalinated water. Currently, the United Arab Emirates has only a two-day desalinated water storage capacity, making the country extremely vulnerable to any disruption in its desalination plants. Other Gulf Cooperation Council (GCC) countries have similar limited storage capacity, with the highest not exceeding five days (Dawoud 2009).

In the face of projected increases in the frequency and intensity of droughts, Arab countries may develop long-term plans to manage their natural and constructed storage to offer reliable water supplies on a multiyear basis. Storage facilities can be managed to strike a value-driven balance between supply and demand through systems that involve forecasting and monitoring water inputs, outputs, and stock levels and protecting water quality. If properly managed, water storage—both surface and groundwater—can be a cost-effective way to mitigate seasonal and multiyear variability in weather conditions.

Management of groundwater resources

Fossil groundwater is particularly important, and its strategic importance will rise as climate change further shrinks water supplies in the Arab region. These resources are best reserved for domestic use and high-value industrial and agricultural activities. To retain the vital socioeconomic role of these resources, Arab countries may consider placing strict regulations on their use and developing programs to rehabilitate and recharge fossil aquifers.

Renewable groundwater resources are in theory best managed by maintaining a balance between supply and optimal allocation of water withdrawals. In practice, however, two main barriers stand in the face of proper management of groundwater resources. First, many of these resources stretch over several countries that in most cases have not entered into a resource-sharing agreement, which has encouraged overexploitation. Second, encouraged by past agricultural policies, many of these resources are already being used by farmers, and halting these activities without endangering established livelihoods is difficult. In many cases, farming has stopped only after water levels dropped below exploitable levels or water became too saline to be used in agriculture (Chebaane et al. 2004).

However, after several decades of improper management, many Arab countries—alarmed by the loss of valuable groundwater stocks—have implemented policies that restrict groundwater extraction and curtail agricultural activities based on groundwater. Jordan has restricted abstraction and stopped issuing licenses for drilling wells in the Amman-Zarqa basin after aquifers dropped several meters following years of excessive abstraction (Chebaane et al. 2004). Saudi Arabia has phased out wheat farming that uses fossil water, which reached a peak several years ago at the expense of valuable nonrenewable water reserves. Sowers and Weinthal (2010) indicate, however, that these restrictions are facing resistance from highly influential agricultural firms, and some have circumvented the regulations by switching to other crops.

Protection of water resources

The relentless and growing pollution of water resources in the region is depriving its people—sometimes irreversibly—of vital natural assets that are very costly to replace; laws and regulations are urgently needed to stem pollution. Although several Arab countries have strict laws and regulations for protecting water resources, few have implemented them effectively. A notable exception is Jordan, which has recently created a water and environmental protection program that includes a dedicated law enforcement force—the first of its kind in the region (Subah and Margane 2010).

Artificial recharge could be used to retard seawater intrusion by creating a barrier of freshwater at the seawater–freshwater interface. The coastal aquifers in Lebanon are currently in great danger of being overwhelmed by seawater intrusion from the excessive extraction of groundwater, especially in the drier periods of the year (Saadeh 2008). Before the urbanization of the coastal area, seawater was kept in check by the hydraulic pressure of inflow from the neighboring mountains. To restore this balance, several measures need to be taken, including controlling illegal pumping, recharging the aquifer with excess runoff in the winter, and treating wastewater throughout the year.

Wastewater treatment and reuse

Improperly managed, highly contaminated wastewater will very likely find its way to streams and aquifers, endangering public health, damaging vital ecosystems, and rendering valuable water resources unusable. Farmers occasionally access disposed untreated wastewater to try to manage through drier seasons or simply to avoid paying for water services. Unless proper action is taken, this maladaptation practice is expected to intensify in a warmer and drier climate.

High capital and operational costs are the main obstacles to setting up wastewater treatment systems. They can, however, be defrayed by reusing treated wastewater for agriculture and freeing up high-quality freshwater for domestic use. The nutrient-laden, treated wastewater has the added benefit of reducing the need for costly and environmentally unfriendly fertilizers. Given the choice, however, farmers prefer to use freshwater, fearing restrictions by importing countries on wastewater-grown produce and the public's aversion to such produce. Gulf countries, for example, imposed restrictions in the 1980s on importing Jordanian produce when the country expanded the use of treated wastewater in agriculture. Egypt generates a substantial amount of wastewater, which is mostly treated and reused outside the Nile Delta to support expanding desert reforestation schemes and cultivation of jatropha to produce biodiesel (AHT Group 2009).

Several measures are required to expand the use of treated wastewater in agriculture. Stringent public health regulations in the application of wastewater and the handling of produce are necessary to reduce risk and increase public confidence and acceptance. Treatment methods need to be fine-tuned and optimized for specific applications. Less stringent, and consequently less costly, quality requirements are adequate if irrigation methods and crop choices reduce the risk of exposure of workers and produce to treated wastewater. More stringent standards are required to treat wastewater at the tertiary level for recharging aquifers used for drinking water. Religious fatwas have cleared the way for using treated wastewater to grow food.

Using treated wastewater has to be well integrated into the overall water resources management strategy. In particular, regulation and pricing of freshwater for agriculture must be in tune with those of treated wastewater. In Tunisia, preferential pricing of treated wastewater over freshwater has encouraged wider use of treated wastewater. Jordan applies a combination of restriction and pricing to expand the use of treated wastewater—drawn by gravity from Amman to the Jordan valley from which freshwater is pumped back to Amman. Along with other measures, the use of treated wastewater allows the country to defer capital investment in expensive water-supply projects.

Desalination

For Arab countries with extreme water scarcity, desalination is the primary source of water supply. Historically, these countries—mainly concentrated in the Gulf region—were very thinly populated. The advent of oil and the consequent population boom have necessitated tapping into seawater to meet unabated increases in demand. The GCC countries are at the lead worldwide in using desalination technology. Today, nearly 50 percent of the world's total desalination production is concentrated in those countries (Bushnak 2010).

After decades of contemplating securing water supplies through piping schemes from other countries (for example, the Peace pipeline from Turkey and the Green pipeline between the Islamic Republic of Iran and Qatar), the GCC countries have adopted desalination as their long-term strategy to achieve water security. Desalination offers exclusive sovereignty over produced water resources. However, the technology is energy intensive and consequently has a large carbon footprint. Brine and heat from desalination plants have detrimental environmental impacts that can be costly to manage. In addition, the GCC countries have very limited storage capacity to maintain supplies during interruptions in plant operations. In the Gulf region, operation of desalination plants can be suspended for days during red tides.

Several options for enhancing the reliability of water supplies in the GCC countries include (a) developing surface-water facilities, (b) constructing a large network connecting desalination plants in GCC countries, and (c) using local aquifers for strategic storage of desalinated water. The first option was assessed to be too costly and results in water stagnation. The second is very costly and requires coordination among different countries. The third option is currently being considered by several GCC countries. As mentioned earlier, Abu Dhabi has embarked on developing strategic aquifer storage system that would provide several months of storage capacity (Dawoud 2009).

Because of desalination's high financial and environmental costs, large-scale desalination is a last-resort measure after exhausting more cost-effective and sustainable supply-side and demand-side options. Even in GCC countries, investment in additional desalination capacity can be deferred by adopting better demand management through effective pricing and by reducing unaccounted-for water in distribution networks. In coastal cities in other parts of the Arab world, desalination can be used to enhance water-supply systems and augment other water supplies. Desalination can also be a flexible and cost-effective water-supply solution in isolated areas, or tourist destinations, where retrieving water from distant water sources is prohibitively expensive. Brackish water, which is abundant in some Arab countries, is generally less costly to desalinate than seawater. However, treating effluents of desalination plants in interior areas can be quite costly.

Advancements in using solar energy in desalination is making this option increasingly more competitive in the long term compared with fossil fuels, particularly given that oil prices are projected to continue their upward trend. The Arab region has vast areas that are rated as prime sites for solar energy production. Several GCC countries are investing in renewable energy, particularly in desalination applications. Both Masdar in Abu Dhabi and the recently established King Abdullah City for Atomic and Renewable Energy have ambitious research and development programs in solar energy and desalination.

Demand-Side Management Will Become Increasingly Important

Most Arab countries have already developed most of their renewable water resources; managing demand offers an effective and realistic option since those resources will shrink further under climate change. An emerging set of management best practices has greatly enhanced water efficiency in several water-scarce environments. Chief among them is using market-based instruments to encourage efficiency and to ensure economic viability of water utilities.

Market-based instruments

In setting the "Arab Regional Strategy for Sustainable Consumption and Production," the Council of Arab Ministers Responsible for the Environment has called for adopting "policies, including market-based instruments, for water cost recovery" (CAMRE 2009). A study commissioned under the auspices of the joint Water Policy Reform Program by the Egyptian government and USAID-Egypt assessed 20 market-based instruments to enhance water resources management in Egypt (McCauley et al. 2002). Those instruments were assessed using several criteria, including economic efficiency, equity, and political, social, and cultural acceptability, as well as institutional capacity. Initial assessments were further refined through a consultation process involving several government agencies, particularly the Ministry of Water Resources and Irrigation.

Several market-based instruments were deemed suitable or warranted further assessment for implementation in Egypt. Groundwater extraction charges and subsidies of urban water meters and water-conserving equipment were considered most suitable for reducing water use. The study recommended further assessment of area-based and volumetric irrigation charges and increased urban and industrial water service tariffs. Increasing user wastewater treatment fees and subsidizing rural sanitation and pollution control equipment were deemed necessary to manage water quality. The study recommended encouraging voluntary agreements by the industry to control pollution and publicly disclose environmental information.

Regardless of how water service charges are ultimately set, the real full cost of water must be understood. According to Agarwal et al. (2000), the full cost of water is composed of a full economic cost and the environmental cost of forfeiting water's ecosystem services (figure 3.8). The full economic cost is widely confused with the full supply cost, which captures only the actual cost of providing water and services, including a profit margin. Commonly overlooked is the opportunity cost, which reflects

FIGURE 3.8

Components of the Full Cost of Water

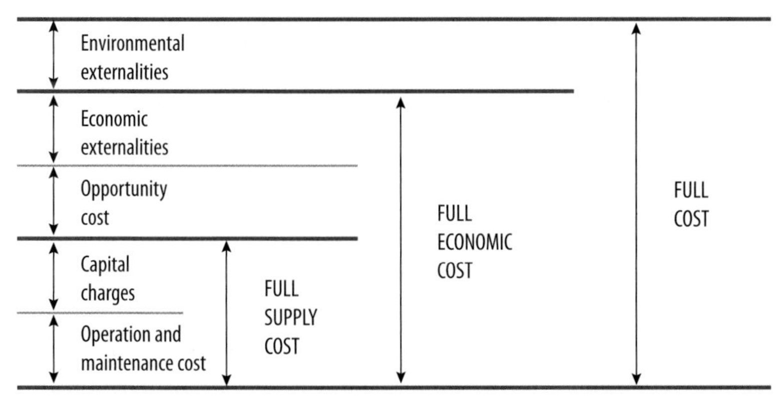

Source: Agarwal et al. 2000.

the additional benefit forfeited from not using water in higher-value applications.

Opportunity cost is quite significant under scarce water conditions. For example, water supplied with minimal or no cost for irrigation is valued considerably higher by domestic users. Another example is the cost of desalinating and pumping water to Riyadh from the Arabian Gulf is valued less than the nominal amount charged for using fossil water for agriculture. Another component of the full economic cost is lost benefits by indirect users, including, for example, a reduction in return flows from springs and streams.

The wastewater treatment cost, which dwarfs the supply cost in certain locations, is an important component of the full economic cost. The cost can be defrayed by reusing treated wastewater.

Basing water service charges on full cost recovery—although necessary for maintaining the economic sustainability of water utilities—is a highly contentious issue in the Arab region because many Arab countries adopt complex social welfare systems that are hinged on food and agricultural subsidies, which are in turn dependent on subsidized water services and irrigation. Reducing these subsidies—even only to recover the supply costs—may trigger social and economic upheavals that could have serious economic, social, and political ramifications. Such measures must be part of a larger development strategy that seeks to diversify the economy and support the agricultural communities and workforce to transition to more productive agriculture and less water-intensive industrial sectors (see chapter 4).

To protect the poor, water tariffs can be structured in a progressive tariff system to allow for below-cost rates on a threshold of water usage necessary to maintain good health and well-being. Additional units of water can be charged at a progressively higher rate to restrain excessive and wasteful uses. Many Arab countries have highly subsidized rates, including the GCC states where full supply costs are very high. Other countries, such as Jordan and Tunisia, have set more effective tariff schemes. Tunisia employs favorable differential rates for treated wastewater to encourage its use in agriculture. Well-structured and consistent demand-management measures in the Rabat-Casablanca area in Morocco drastically suppressed projected water demand, deferring planned major water supply projects (DGH Rabat 2002).

Effective water-use tariffs require adequate institutional and regulatory conditions. Users are willing to pay higher rates for better-quality water services. The overall impact would be positive if higher water-service charges were accompanied by improved services and transparent accounting, which requires a reliable metering system. But reliable water metering is a challenge for many Arab countries because when water service is intermittent, only the more expensive types of meters are accurate. The World Bank (2010) recommends that Arab cities maintain a 24–7 service.

Reducing unaccounted-for water in distribution networks

Unaccounted-for water (UfW) is defined as the difference between the amount of water delivered by the water utility and the amount actually billed. UfW includes losses in the network, illegal water use, and inaccurate metering. UfW can surpass 60 percent in poorly maintained distribution networks in some Arab cities, such as Beirut, Jericho, and Sanaa. Potential water savings from reducing UfW are substantial given that their rates can be as low as 5 percent in highly pressurized distribution networks, such as those in Japan and Singapore (Ueda and Benouahi 2009). UfW has huge opportunity costs that are equal to the cost of desalination and pumping in the GCC countries. In Libya, the cost is at least equal to the capital and operational costs of delivering water through the Great Manmade River system, which taps the country's main fossil aquifers in the south and delivers water hundreds of kilometers north to the coast. Rehabilitating old networks also has large public health benefits in that it reduces the risk of contaminated water entering the network through leaks, which is exacerbated by the relatively low pressure in many networks in the region.

Raising public awareness

Despite evident water scarcity across the Arab region, the relatively widespread perception is that groundwater is abundant and governments are withholding information on its availability to stop people from using it. Although information on water resources is still lacking in the Arab region, the perception that groundwater is abundant is false and counterproductive. People need better information about the water resources in the Arab region and how climate change is likely to affect them.

Several public awareness campaigns have been launched across the region, but more are needed. The public must be engaged in the process of water resources management. Engaging the public requires a transparent process involving the media, schools, and nongovernmental organizations.

Managing Water Resources Requires Multifaceted Approaches

Management of water resources and provision of water services involve a complex array of formal and informal organizations, different and often competing users, and other stakeholders under varying socioeconomic, environmental, and political conditions. Separate bodies often exist for water services (the urban domestic and industrial sectors) and water resources (the rural domestic and agriculture sectors). In an increasing number of Arab countries, the private sector has become an important factor in the provision of water services, encouraged under public-private partnerships.

Water governance

Negotiating the complexities of water management requires effective and efficient governance, with actors with the authority and responsibility to make and implement decisions across other ministries. The water governance system should establish the mandates, authorities, and responsibilities of different institutions and delineate relationships among them and other actors. Such governance systems are necessary for establishing policies and enacting laws and regulations to facilitate implementation of the IWRM principles of economic optimization, social equity, and environmental sustainability.

Rogers and Hall (2003) have identified several conditions to achieve good governance: inclusiveness, accountability, participation, transparency, predictability, and responsiveness. In water governance, these conditions would entail involving civil society in decision making, developing a fair and transparent water-rights system, properly monitoring and sharing information, and balancing the involvement of the public and private sectors in managing water resources and delivering services. Such principles formed the basis of traditional community-based water systems in the Arab world, such as *aflaj/foggara/qanat*. Important lessons can be learned from these ancient systems on establishing and enforcing rules of water allocation and quality control, and in participatory decision making.

Progress in reforming the water governance systems in Arab countries has been generally limited to date. Few governments have enacted specific water legislation, and even those that have often do not have the means or power to enforce the measures (Majzoub 2010). The World Bank identified poor accountability as the main obstacle to effective management of water resources because of the lack of political will to undertake necessary but unpopular water reforms (Bucknall 2007). The World Bank advocates more transparent approaches to engage the public in the process of making difficult reforms, such as deregulation of utilities, pricing, and restrictions on water use (Bucknall 2007).

The Global Water Partnership offers a wide range of options based on the IWRM approach to achieve better governance through its "Toolbox for IWRM" (GWP 2003c). Several Arab countries have applied some of those tools to enhance water governance. In Egypt, GWP Toolbox methods were used to support institutional strengthening of the Alexandria General Water Authority (GWP 2003a). In Jordan, GWP Toolbox instruments were used to facilitate the reforming of the public Jordan

Valley Authority from a service provider to a regulatory agency overseeing provision of water services by private entities (GWP 2003b).

Disaster risk management

Climate change is expected to exacerbate the risk of two types of precipitation-related hazards: flash floods and severe droughts. An increase in the frequency and intensity of flooding has already been observed across the region, particularly in areas that have recently been settled with virtually no consideration of flood risk. Extended droughts have not only devastated the least developed and most vulnerable countries, such as Somalia, but they have also severely affected localities within more developed countries, such as Syria. Addressing both types of hazards requires not only reducing exposure to these hazards but also developing capacity to cope with their aftermaths.

Reducing vulnerability to flooding requires coordinated efforts among different public agencies and the active participation of all stakeholders. Flooding risk—and its projected increase under climate change—should be considered in urban planning (see chapter 5 and spotlight 1 on disaster risk management). Preparedness for flooding events is an important component of flood risk management that includes raising public awareness and training police and civil defense emergency units complemented by a comprehensive and responsive flood-forecasting system.

Dealing with severe droughts requires developing adequate storage and conveyance and reducing the socioeconomic vulnerability of the affected population. Rural communities that depend on rainfed agriculture and pastoralists are particularly at risk. In the least developed countries, such as Somalia, socioeconomic activities depend heavily on rainfall patterns. Both farmers and pastoralists are vulnerable to extended droughts, including the risks of loss of income, malnutrition, and famine. Enhancing the resilience of these groups requires reducing their dependence on rainfed agriculture by diversifying the economy and creating alternatives to climate-dependent economic activities (see chapter 4). In the shorter term, authorities and nongovernmental organizations can support these societies during droughts through relief efforts, including relocation to other parts of the country.

Cooperative management of shared water resources

In the absence of cooperation on managing shared water resources, unilateral adaptation to the impact of drying conditions may undermine these scarce resources, cause socioeconomic and environmental harm to other parties, and increase tension and the potential for conflict among water-sharing countries.

Few ratified agreements on shared water resources exist in the Arab region, and none has led to an effective joint management of shared resources (ESCWA 2009; Stephan 2010). Syria and Lebanon reached an agreement on the Orontes River, which was signed in 1994 and ratified in 2001. The two countries also entered into agreement on sharing the Nahr al-Kabir al-Janoubi in 2002. Jordan and Syria first signed an agreement on the Yarmouk River in 1953, which was later superseded by another in 1987, eventually leading to the joint development of the Unity Dam.

A comprehensive agreement on the Euphrates-Tigris basin is still elusive, despite several decades of technical cooperation and bilateral agreements among the three riparian countries—Iraq, Syria, and Turkey. Egypt and Sudan entered into an agreement in 1992 on the sharing of the Nubian Sandstone Aquifer. The two countries also signed an agreement in 1957 to share water resources from the Nile, although that agreement has been contested by upstream countries, which have recently signed another agreement that has in turn been rejected by both Egypt and Sudan (discussed earlier).

Stephan (2010) proposes several actions to enhance more sound management of shared water resources in the region. In particular, she emphasizes initiating joint projects, involving regional and international organizations such as the UN Economic and Social Commission for Western Asia, ratifying international water laws and conventions, and reaching agreements on sharing these resources. Although many of the workshops and conferences attended by these countries discussed most of these recommendations extensively, they seem to have had little effect in driving further cooperation.

A large part of this inaction is attributed to power imbalances among sharing countries (Zeitoun and Warner 2006), the generally lukewarm if not hostile political relations among many of these countries, and to the perception by many countries that agreements will impose constraints on developing water resources within their boundaries. Turkey's current rapprochement with Arab countries is easing tension over disputes about the Euphrates-Tigris basin. But it is unclear if it will eventually lead to an overarching agreement on this basin, given that Turkey is still pursuing the development of its massive Southeastern Anatolia Project (Mutlu 2011).

Climate change and agricultural policies

A multipronged and multisectoral strategy will be needed by all countries in which agriculture currently uses the lion's share of water (box 3.1).

First, the agriculture sector needs to implement more water-efficient technologies and to use marginal water sources (see chapter 4) to protect freshwater reserves. Second, mechanisms should be established to reduce

BOX 3.1

Consuming versus Using Water: The Double-Edged Nature of Irrigation Efficiency

Water is distinctive in that its use at one point does not necessarily preclude its further use at another point. For example, a large part of the water used in showering and bathing may find its way back to the water resource system through drainage and treatment. Even leakage from a domestic water supply network can recharge local aquifers. Obviously, water does not usually return with the same quality with which it was delivered. In comparison, water lost through evaporation or drainage to saline water bodies is not retrievable and is considered "consumed." These issues are important when considering improving irrigation efficiency and allocating water among upstream and downstream users. Ward and Pulido-Velazquez (2008) have found that programs designed to improve irrigation efficiency in the Rio Grande basin shared between the United States and Mexico have increased crop yield at the expense of increasing evapotranspiration and reducing recharge to aquifers. They propose that irrigation efficiency initiatives are better conducted in an integrated manner that takes full account of the water balance throughout the whole watershed. These findings are particularly important in assessing the impact of irrigation practices in shared aquifers and basins. For example, improving irrigation efficiency on the Turkish side of the Euphrates could have negative consequences for groundwater levels in downstream Syria.

Source: Authors' compilation.

the subsidies associated with agricultural water use (subsidized electricity, water, and pumps) and to ensure that market mechanisms are used that reflect the resource value and supply costs. Such mechanisms will provide an incentive to increase efficiency and facilitate more effective allocation of water resources. Such measures are likely to lead to reallocation of water to sectors with greater economic returns.

Like in many countries of the Arab region, agricultural water use accounts for the majority of water consumption in Mexico, consuming 70 percent of the available freshwater. Subsidies of electricity tariffs contributed to this high agricultural water consumption, because they not only incentivize the overextraction of pumped groundwater but also discouraged the adoption of energy- and water-saving techniques. The National Ecological Institute of Mexico found that even in areas with extreme water scarcity, farmers use some of the least efficient irrigation technologies, such as open dirt canals. In response to overextraction rates of renewable groundwater of close to 200 percent, the Mexican government decided to remove the price distortion and is piloting the replacement of the price subsidy with direct cash transfers to farmers. Although the case of Mexico can serve as an example of an economic management tool, the context of scarcity and transboundary aquifers is distinct in the Arab region and not immediately comparable.

Some countries have already started to curtail agricultural activity by limiting water allocations and through more realistic pricing. These actions need to be coordinated, however, with human resources development strategies that aim at developing skills for farmers to move into more productive sectors (see chapter 4 for more details). Economic water cost curves are a decision support tool for supply-demand management

Several factors need to be considered in choosing the right mix of adaptation measures, which are country specific and depend on political, socioeconomic, and cultural choices. Marginal economic cost curves for water are intuitive tools used to identify the optimal mix of technical measures to close a given supply-demand gap. They aim to compare the availability and cost-effectiveness of options for alternative adaptation measures. The World Bank (2012) produced a regional water cost curve as well as a set of marginal water cost curves for most Arab countries.[4] The study considered nine adaptation options and assessed their costs in 2030 (table 3.2). These costs are assumed homogeneous among Arab countries, meaning cost curves will differ with regard to the volume potential of each measure to close the demand gap, rather than in the costs of the measure. The cost of each measure is indicative of the relative financial magnitude of each measure.

Overall the curves show that desalination—even with expected efficiency improvements—is vastly more expensive than storage expansion

TABLE 3.2

Cost of Water Adaptation Measures

Adaptation measure	Cost (US$/m³ water)
Improve agricultural practice	0.02
Expand reservoir capacity (small scale)	0.03
Reuse domestic and industrial water	0.03
Reuse irrigation water	0.04
Expand reservoir capacity (large scale)	0.05
Reduce irrigated areas	0.10
Desalinate using renewable energy	1.30
Desalinate using conventional energy	1.85
Reduce domestic and industrial demand	2.00

Source: World Bank 2012.

measures, which in turn are often much more expensive than measures to improve water productivity.

By overlaying the projected supply-demand gap (unmet demand) for the region for 2040–50 on the cost curve, figure 3.9 shows that the region has choices in how to close the gap. If the cheapest options are selected, the total annual costs in 2050 of bridging the unmet water gap of 199 cubic kilometers are about US$104 billion.

FIGURE 3.9

Regional Water Marginal Cost Curve for the Average Climate Projection for Arab Countries

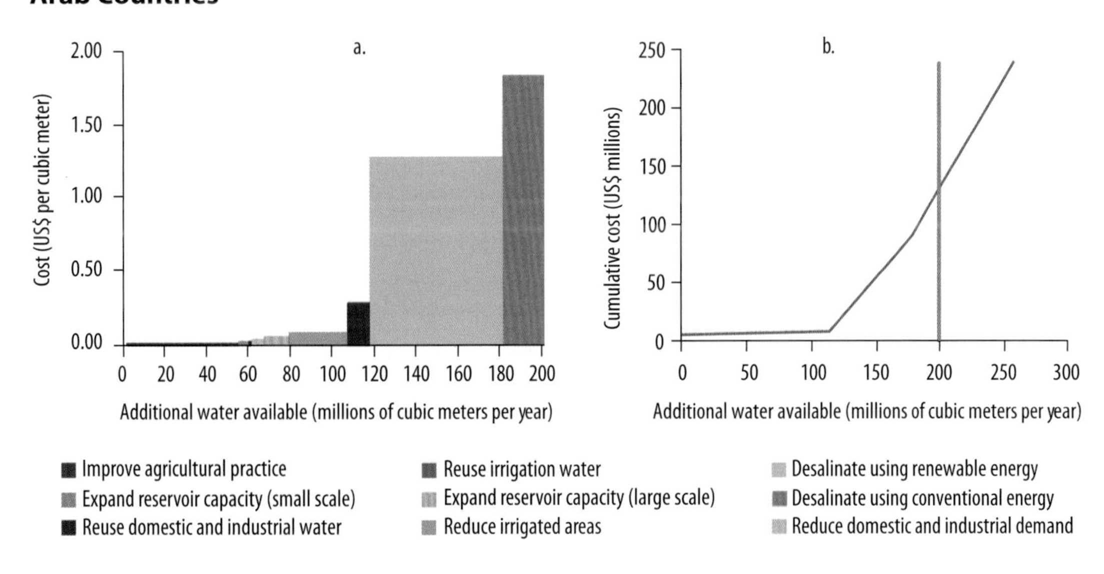

Source: World Bank 2012.

Note: "Desal-CSP" denotes desalination using concentrated solar power

The general assessment presented above should be interpreted with care. Different countries, even in the same region, face different choices and costs in how to close the gap. The World Bank (2012) shows that the majority of Arab countries have already exhausted the water supplies that can be procured at relatively low cost.

The average adaptation costs are US$0.52 per cubic meter in the Middle East and North Africa, but they vary substantially among countries. The per-cubic-meter cost ranges from US$0.02 in Algeria to US$0.98 in the United Arab Emirates (figure 3.10). The per-cubic-meter costs are less than US$0.36 in Algeria, Egypt, the Islamic Republic of Iran, Syria, and Tunisia and more than US$0.64 in Bahrain, the Gaza Strip, Jordan, Kuwait, Oman, Qatar, Saudi Arabia, the United Arab Emirates, and the West Bank.

Long-Term Adaptation Strategies Can Reduce the Water Gap

Water scarcity is a main constraint to socioeconomic development in the Arab region. Along with other stressors, including demographic and land-use changes, climate change will exacerbate the already-precarious high water deficit across the Arab region. In 2040, the region will likely face a reduction in water runoff by 10 percent because of climate change. In combination with the other challenges mentioned, this reduction will result in a 50 percent renewable supply gap. Climate change is also expected to increase the occurrence and intensity of extreme flooding.

The impact of these grim projections can, however, be moderated or avoided if countries take stock of the main challenges and opportunities at the regional, national, and community levels and adopt long-term adaptation strategies to address these challenges. In the previous sections, we have identified key challenges and explored several adaptation options to deal with them. Pulling it all together, we propose the following key messages and policy options that we believe would help the region achieve sustainable water resources management in a changing climate:

1. *Climate change is projected to reduce natural water supplies in most of the region whereas demand is increasing, resulting in a large supply gap.* Most climate models project lower precipitations and higher temperatures for most Arab countries under changing climate conditions—leading to significant reduction in stream runoff. This reduction in runoff will result in an overall desiccation characterized by major reductions in natural water supplies. Over the same period, a parallel increase in water demand of up to 60 percent is projected to result in a 50 percent renewable water supply gap for the region.

FIGURE 3.10

Water Marginal Cost Curves for Algeria and the United Arab Emirates

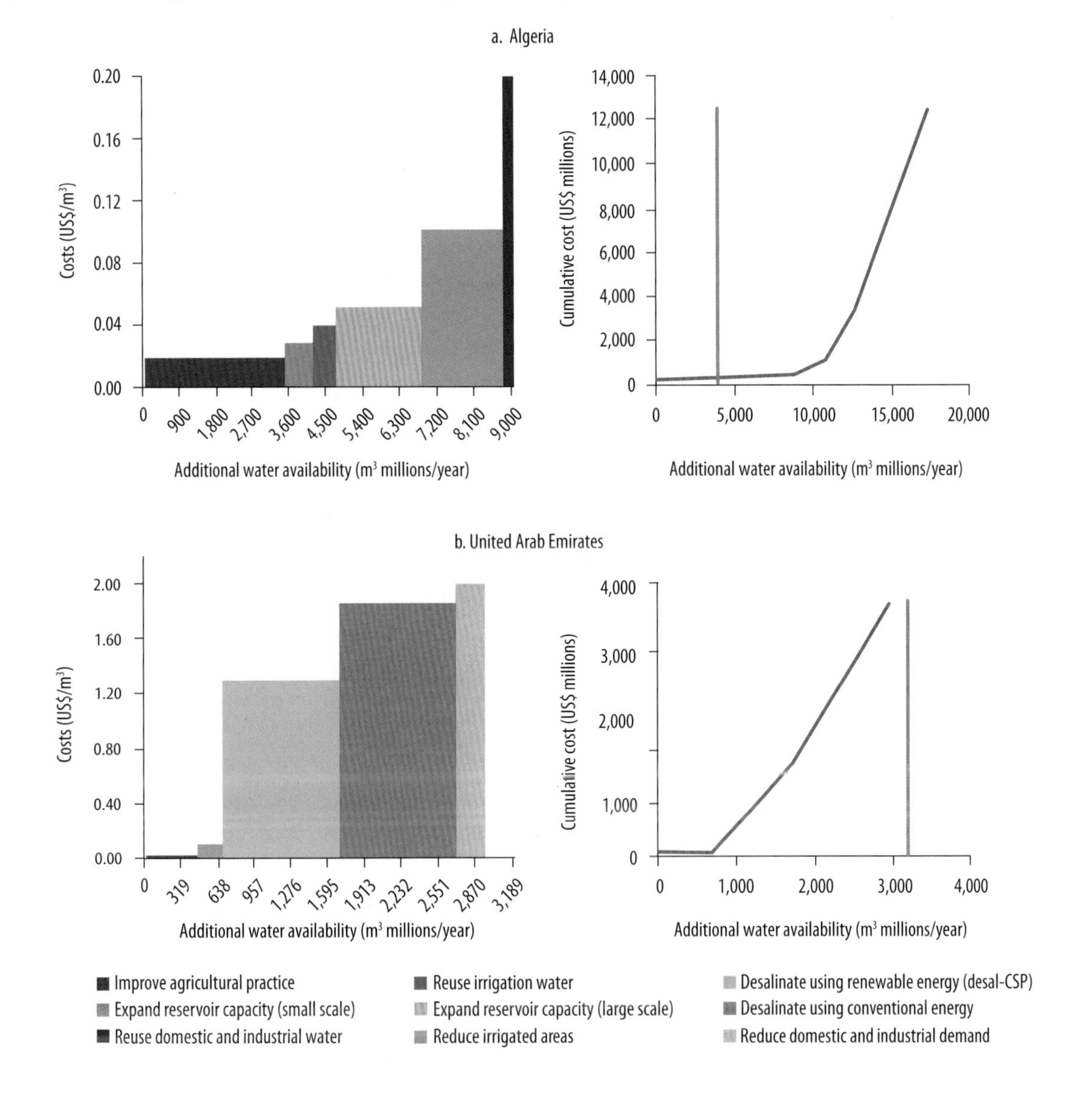

Source: World Bank 2012. [[AQ: correct reference?]]

Note: Not all measures are illustrated because of varying applicability at country level. "Desal-CSP" denotes desalination using concentrated solar power.

2. *The projections call for immediate action, which includes a combination of good water management and climate adaptation.* To meet future demand in a sustainable manner, decision makers have to make choices today to implement integrated water resources management. In addition, adaptation options exist that can be adopted nationally for long-term

planning for the projected additional climate impact on scarcity and increased variability.

3. *Water management requires maintaining the balance among economic efficiency, social equity, and environmental sustainability to enhance overall climate change resilience.* In setting out plans for the future, decision makers should not lose sight of striking a balance among the three pillars of Integrated Water Resources Management: economic efficiency, social equity, and environmental protection.

4. *Integrate water resources management across water and nonwater sectors (agriculture, tourism, and urban development) to create a total resource view with water as a cross-sectoral input to development.* Water use is largely determined outside the realm of the departments responsible for water resources management. Water is better addressed as an integral and high-priority component in national development strategies. Clear directives must be developed for optimal allocation of water across all sectors.

5. *Climate change will require upgrading disaster risk management for floods and drought.* Climate change is projected to increase the frequency and intensity of floods and droughts. In a region with high population growth and economic dependency on agriculture, the necessary institutions and resources must be established to manage these impending risks.

6. *Nonconventional supply options, storage, and conveyance capacity can enhance resilience in the face of droughts and floods.* Notwithstanding the central role of demand-side management, Arab countries may consider developing nonconventional supply options and adequate storage and conveyance capacity to secure minimum level of supplies during extended droughts, which are expected to increase in frequency and intensity under climatic changes. Adequate storage and conveyance capacity is particularly crucial for the least developed countries—such as Mauritania and Somalia—which are extremely vulnerable to rainfall variability.

7. *More aggressive water demand management is required to achieve sustainability and reduce the need for expensive water-supply infrastructure.* Marginal cost analysis for the region shows that water-demand management offers a more cost-effective and sustainable option than capital-intensive water-supply development. Market-based instruments are potentially valuable in increasing water-use efficiency and abating pressure on existing water resources.

8. *Improved water-use efficiency in agriculture will make more water available for use in other needed sectors.* Considering that agriculture consumes more than 80 percent of water resources in the region, even a modest water savings in the sector could significantly increase the amount of water available for other sectors. Water-use efficiency in agriculture can be achieved through more efficient irrigation technologies, changes in cropping patterns and crop mixes, and research and development.

9. *Enhanced regional economic integration will encourage water investment in less developed Arab countries and facilitate trade in water-intensive products from more water-endowed countries.* Arab countries could seek a form of economic integration that facilitates free movement of people, goods, and capital. This measure would both facilitate more efficient use of water resources across the region and mitigate the impact of socioeconomic shocks triggered by severe droughts.

10. *Climate change increases the need for regional and interregional cooperation on managing shared water resources.* Increased stress on shared water resources brought about by climate change necessitates cooperation to facilitate sustainable management of these resources.

11. *Investments should be made in developing an information base and in climate change research and development.* Arab countries need to develop their capacity in research and development with respect to water and climate change issues and invest in data monitoring and information management—a prerequisite for conducting research and policy analysis.

12. *To protect water resources from pollution, governments should enact and enforce water laws and regulations.* Given the increasing value of water resources as they become scarcer in a changing climate, laws and regulations are needed to protect these resources from pollution.

13. *Overall water governance needs improvement at all levels, including the full participation of stakeholders.* Arab countries need to address deficiencies in water governance that hamper efforts to improve water management. The focus should be on enhancing accountability and transparency through establishing clear mandates, authorities, and responsibilities of different institutions.

Notes

1. The World Bank (2012) report does not include the Comoros, Mauritania, Somalia, and Sudan.

2. The 10 percent and 90 percent range in water shortage is between 90 and 280 cubic kilometers per year in the dry and wet scenarios considered in the study.
3. This result, however, may be abated by the reduction in evapotranspiration from the effect of higher carbon dioxide levels. The carbon dioxide effect is still being investigated, and no conclusive results have yet been verified.
4. The 2012 World Bank report does not include the Comoros, Mauritania, Somalia, and Sudan.

References

Agarwal, Anil, Marian S. delos Angeles, Ramesh Bhatia, Ivan Chéret, Sonia Davila-Poblete, Malin Falkenmark, Fernando Gonzalez Villarreal, Torkil Jonch-Clausen, Mohammed Aït Kadi, Janusz Kindler, Judith Rees, Paul Roberts, Peter Rogers, Miguel Solanes, and Albert Wright. 2000. "Integrated Water Resources Management." Global Water Partnership/Technical Advisory Committee Background Paper 4, Stockholm.

AHT Group. 2009. "Identification and Removal of Bottlenecks for Extended Use of Wastewater for Irrigation or for Other Purposes: Summary Report." Report prepared for EUROMED, European Investment Bank, Essen, Germany.

Allan, J. A. Tony. 2007. "Rural Economic Transitions: Groundwater Use in the Middle East and Its Environmental Consequences." In *The Agricultural Groundwater Revolution: Opportunities and Threats to Development*, ed. Mark Giordano and Karen G. Villholth, 63–78. Wallingford, U.K.: CAB International.

Al-Turbak, Abdulaziz S. 1991. "Effectiveness of Recharge from a Surface Reservoir to an Underlying Unconfined Aquifer." In *Hydrology of Natural and Manmade Lakes*, ed. Gerhard Schiller, Risto Lemmelä, and Manfred Spreafico, 191–96. Wallingford, U.K.: IAHS Press.

Assaf, Hamed. 2009. "A Hydro-economic Model for Managing Groundwater Resources in Semi-arid Regions." In *Water Resources Management V: WIT Transactions on Ecology and the Environment*, vol. 125, ed. Carlos A. Brebbia and Viktor Popov, 85-96. Southampton, U.K.: WIT Press, Southampton.

———. 2010. "Water Resources and Climate Change." In *Water: Sustainable Management of a Scarce Resource*, ed. Mohamed El-Ashry, Najib Saab, and Bashar Zeitoon, 25–38. Beirut: Arab Forum for Environment and Development.

———. 2011. "A Framework for Modeling Mass Disasters." *Natural Hazards Review* 12 (2): 47–62. http://dx.doi.org/10.1061/(ASCE)NH.1527-6996.0000033.

Assaf, Hamed, and Mark Saadeh. 2008. "Assessing Water Quality Management Options in the Upper Litani Basin, Lebanon, Using an Integrated GIS-Based Decision Support System." *Environmental Modelling and Software* 23: 1327–37.

Bates, Bryson C., Zbigniew W. Kundzewicz, Shaohong Wu, and Jean P. Palutikof, eds. 2008. "Climate Change and Water." IPCC Technical Paper 6, Intergovernmental Panel on Climate Change, IPCC Secretariat, Geneva.

Brauch, Hans Günter. 2003. "Urbanization and Natural Disasters in the Mediterranean: Population Growth and Climate Change in the 21st Century." In *Building Safer Cities: The Future of Disaster Risk*, ed. Alcira Kreimer, Margaret Arnold, and Anne Carlin, 149–64. Washington, DC: World Bank.

Brekke, Levi D., Julie E. Kiang, J. Rolf Olsen, Roger S. Pulwarty, David A. Raff, D. Phil Turnipseed, Robert S. Webb, and Kathleen D. White. 2009. "Climate Change and Water Resources Management: A Federal Perspective." Circular 1331, U.S. Geological Survey, Reston, VA.

Bucknall, Julia. 2007. *Making the Most of Scarcity: Accountability for Better Water Management Results in the Middle East and North Africa*. Washington, DC: World Bank.

Bushnak, Adil A. 2010. "Desalination." In *Water: Sustainable Management of a Scarce Resource*, ed. Mohamed El-Ashry, Najib Saab, and Bashar Zeitoon, 125–136. Beirut: Arab Forum for Environment and Development.

CAMRE (Council of Arab Ministers Responsible for the Environment). 2009. "Arab Regional Strategy for Sustainable Consumption and Production." 21st Session of CAMRE, Port Ghaleb, Egypt.

Chatila, Jean G., 2010. "Municipal and Industrial Water Demand." In *Water: Sustainable Management of a Scarce Resource*, ed. Mohamed El-Ashry, Najib Saab, and Bashar Zeitoon, 71–90. Beirut: Arab Forum for Environment and Development.

Chebaane, Mohamed, Hazim El-Naser, Jim Fitch, Amal Hijazi, and Amer Jabbarin. 2004. "Participatory Groundwater Management in Jordan: Development and Analysis of Options." *Hydrogeology Journal* 12: (1) 14–32.

Conway, Declan. 2005. "From Headwater Tributaries to International River: Observing and Adapting to Climate Variability and Change in the Nile Basin." *Global Environmental Change* 15: 99–114.

Cullen, Heidi M., and Peter B. deMenocal. 2000. "North Atlantic Influence on Tigris-Euphrates Streamflow." *International Journal of Climatology* 20 (8): 853–63.

Dawoud, Mohamed A. 2009. "Strategic Water Reserve: New Approach for Old Concept in GCC Countries." Paper presented at the 5th World Water Forum, "Bridging Divides for Water," Istanbul, March 16–22.

DGH Rabat. 2002. "Analysis of the Case Study on the Drinking Water Supply in Rabat-Casablanca Coastal Area." Forum Progress toward Water Demand Management in the Mediterranean Region, Plan Bleu for the Mediterranean Centre of Regional Activities, Fiuggi, Italy, October 3–5.

Dyer, Paul D. 2008. "Demography in the Middle East: Implications and Risks." In *Transnational Trends: Middle Eastern and Asian Views*, ed. Amit Pandya and Ellen Laipson, 62–90. Washington, DC: Stimson Center.

Erian, Wadid, Bassem Katlan, and Ouldbdey Babah. 2010. "Drought Vulnerability in the Arab Region: Special Case Study—Syria." United Nations International Strategy for Disaster Reduction, Geneva.

ESCWA (Economic and Social Commission for Western Asia). 2005. "Regional Cooperation between Countries in the Management of Shared Water Resources: Case Studies of Some Countries in the ESCWA Region." E/ESCWA/SDPD/2005/15, United Nations, New York.

———. 2009. "Shared Waters—Shared Opportunities: Transboundary Waters in the ESCWA Region." United Nations, New York.

Evans, Jason P. 2009. "21st Century Climate Change in the Middle East." *Climatic Change* 92 (3): 417–32.

GWP (Global Water Partnership). 2003a. "Egypt: Improving Public Sector Performance: Institutional Strengthening of the Alexandria General Water Authority." Case 162, GWP, Stockholm.

———. 2003b. "Jordan: From Water Service Provision to Planning and Management in the Jordan Valley Authority." Case 161, GWP, Stockholm.

———. 2003c. "ToolBox for IWRM." GWP, Stockholm.

Majzoub, Tarek. 2010. "Water Laws and Customary Water Arrangements." In *Water: Sustainable Management of a Scarce Resource*, ed. Mohamed El-Ashry, Najib Saab, and Bashar Zeitoon, 137-52. Beirut: Arab Forum for Environment and Development.

McCauley, David S., Robert Anderson, Richard Bowen, Ibrahim Elassiouty, Elsayed Mahdy, Ibrahim Soliman, and Hisham Shehab. 2002. "Economic Instruments for Improved Water Resources Management in Egypt." U.S. Agency for International Development–Egypt.

Mutlu, Servet. 2011. "Political Economy of Water Regulation and the Environment in Turkey." In *The Political Economy of Regulation in Turkey*, ed. Tamer Çetin and Fuat Oğluz, 215–45. New York: Springer.

Parry, Martin L., Osvaldo F. Canziani, Jean Palutikof, Paul van der Linden, and Clair E. Hanson, eds. 2007. *Climate Change 2007: Impacts, Adaptation and Vulnerability—Contribution of Working Group II to the Fourth Assessment Report of the Intergovernmental Panel on Climate Change*. Cambridge, U.K.: Cambridge University Press.

Renault, Daniel. 2002. "Value of Virtual Water in Food: Principles and Virtues." Paper Presented at the UNESCO-IHE Workshop on Virtual Water Trade, Delft, Netherlands, December 12–13.

Rogers, Peter, and Alan W. Hall. 2003. "Effective Water Governance." TEC Background Paper 7, Global Water Partnership Technical Committee, Stockholm.

Saadeh, Mark. 2008. "Seawater Intrusion in Greater Beirut, Lebanon." In *Climatic Changes and Water Resources in the Middle East and North Africa*, ed. Fathi Zereini and Heinz Hötzl, 361–71. Berlin and Heidelberg: Springer.

Sadek, M. F., Mamdouh M. Shahin, and C. J. Stigter. 1997. "Evaporation from the Reservoir of the High Aswan Dam, Egypt: A New Comparison of Relevant Methods with Limited Data." *Theoretical and Applied Climatology* 56 (1–2): 57–66.

Sivapragasam, C., G. Vasudevan, J. Maran, C. Bose, S. Kaza, and N. Ganesh. 2009. "Modeling Evaporation-Seepage Losses for Reservoir Water Balance in Semi-Arid Regions." *Water Resources Management* 23 (5): 853–67.

Sowers, Jeannie, and Erika Weinthal. 2010. "Climate Change Adaptation in the Middle East and North Africa: Challenges and Opportunities." Working Paper 2, Dubai Initiative, Cambridge, MA.

Stephan, Raya Marina. 2010. "Trans-boundary Water Resources." In *Water: Sustainable Management of a Scarce Resource*, ed. Mohamed El-Ashry, Najib Saab, and Bashar Zeitoon, 153–70. Beirut: Arab Forum for Environment and Development.

Subah, Ali, and Armin Margane. 2010. "Water Resources Protection in Jordan." Paper presented at the World Water Week Workshop 5, "Management of Groundwater Abstraction and Pollution," Stockholm, September 5–11.

Syvitski, James P. M. 2008. "Deltas at Risk." *Sustainability Science* 3: 23–32.

Ueda, Satoru, and Mohammed Benouahi. 2009. "Accountable Water and Sanitation Governance: Japan's Experience." In *Water in the Arab World: Management Perspectives and Innovations*, ed. N. Vijay Jagannathan, Ahmed Shawky Mohamed, Alexander Kremer, 131–56. Washington, DC: World Bank.

USAID (U.S. Agency for International Development). 1997. "Lake Manzala Engineered Wetland." Project document prepared by the Tennessee Valley Authority, Knoxville.

Ward, Frank A., and Manuel Pulido-Velazquez. 2008. "Water Conservation in Irrigation Can Increase Water Use." *Proceedings of the National Academy of Sciences* 105 (47): 18215–220.

World Bank. 2010. "Oum Er Rbia Irrigated Agriculture Modernization." Project appraisal document, World Bank, Washington, DC.

———. 2012. "Renewable Energy Desalination: An Emerging Solution to Close MENA's Water Demand Gap." World Bank, Washington, DC.

Zeitoun, Mark, and Jeroen Warner. 2006. "Hydro-hegemony: A Framework for Analysis of Transboundary Water Conflicts." *Water Policy* 8 (5): 435–60.

Biodiversity and Ecosystem Services Have a Role in Climate Adaptation

Biodiversity in the Arab Region Is of Global Importance for Agriculture and Adaptation

Without biodiversity and ecosystem services, life and human societies would not exist. Climate variability and change constitute an additional stress that ecosystems are subjected to in this arid and water-scarce region. It is estimated that 20 to 30 percent of species will face a higher risk of extinction if there is 1°C to 2°C warming by the end of the century (Boko et al. 2007). Direct impacts include changes in seasonality, changes in growing seasons, and prolonged droughts. Indirect impacts include increased frequency and intensity of fires. Some of these impacts will result in changes in the structure and composition of vegetation, will lead to longer-term changes in nutrient cycling and runoff, will further exacerbate nitrogen limitations, and will affect plant and forage growth (Bullock and Le Houérou 1996; Gitay and Noble 1998). Increasing sea levels and storm surges are leading to intrusion into the rivers and aquifers and are affecting water quality and food production in the coastal and low areas (Boko et al. 2007; Elasha 2010; Lal, Harasawa, and Murdiyarso 2001; Tolba and Saab 2009).

Although the common image of much of the Arab region is of deserts and rangelands, the ancestors of 80 to 100 crop, fruit tree, and livestock species used today were domesticated there. Such species include varieties of wheat such as durum and spelt, lentils, alfalfa, argan, pomegranates, and sheep, all of which are critical to the livelihoods and well-being of people in the region and are also an important part of global food production systems. These wild species are a genetic resource that has the potential to increase water-use efficiency, as well as drought and disease resistance (Bioversity International 2006; FAO 2010; GTZ 2006). Their value as a source of adaptive traits is even greater under climate change. Some of these species are found in transition zones (such as between deserts and rangelands). They have high genetic variability and resilience to climatic extremes (UNEP 2007). However, much of the traditional knowledge about growing these food species and managing these ecosystems is being lost, and a greater effort in the region and globally is needed to conserve and maintain this important knowledge (World Bank 2009).

Many countries have ecosystems that are of critical value for tourism, for fisheries, and for cultural heritage. Incorporating risks of climate change in the management of these systems is essential. A Touareg proverb in the Arab region highlights this need: "The difference between paradise and desert is not water, but man"—or human management of the environment.

Biodiversity and Ecosystem Services Are Important for Economies and Livelihoods

The importance of biodiversity and ecosystem services (see table S2.1) in livelihoods and economies is often not recognized, nor is it included in national development planning and sectoral strategies. Better valuation of ecosystem services and market and nonmarket mechanisms can help policy makers recognize the importance of biodiversity and ecosystem services and aid them in the discussion of trade-offs.

Global data show that poor people depend on ecosystems for about 20 to 40 percent of their income (UNDP 2008). In the Arab countries, the people in the least developed countries (LDCs) that are likely to be the most dependent include pastoralists and fisherpeople. Grasslands, rangelands, and deserts are dominant in the region; forest cover is generally low (see table S2.2). Many deserts and rangelands are important for grazing livestock of nomadic communities and for nature-based tourism. Land degradation, overgrazing, fuelwood gathering, increasing invasive species, deforestation, hunting, infrastructure development, and the overuse of freshwater are major threats to these systems (Lal, Harasawa, and Murdiyarso 2001; UNEP 1997, 2007).

The economic costs of environmental pollution have been estimated to range from 2.6 percent of the gross domestic product (GDP) in Tunisia to about 4.5 percent in the Arab Republic of Egypt (Croitoru and Sarraf 2010). For example, environmental damage and water degradation from irrigated agriculture results in increased salinity and waterlogging at the cost of about 0.6 percent of GDP in Tunisia and 2.1 percent of GDP in Egypt (Croitoru and Sarraf 2010; World Bank 2007).

TABLE S2.1

Classification of Ecosystem Services as in the Millennium Ecosystem Assessment Framework and Examples from the Chapters in This Volume

Ecosystem service	Examples
Provisioning: People obtain products from ecosystems, such as food, fiber, fuel, forage and rangeland grazing materials, and agrobiodiverse products.	Freshwater for human, livestock, and irrigation needs (chapter 3); sand, clay, and wood for building (chapter 5); nature-based tourism (chapter 6); and medicinal products (chapter 8).
Regulating and supporting: People obtain benefits from the regulation of ecosystem processes, which is necessary for the production of all other ecosystem services.	Clean air (chapter 7); climate regulation (chapter 2); flood control (chapters 3, 5, 6, and 7); erosion control (chapters 3 and 4); regulation of human diseases (chapter 8); and water purification (chapters 3 and 7).
Cultural: People obtain nonmaterial benefits from ecosystems through spiritual enrichment, cognitive development, reflection, recreation, and aesthetic experiences.	Many of the services mentioned in chapter 6.

Source: MA 2003.

TABLE S2.2

Biodiversity and Ecosystems in Arab Countries

Country	Share of land area under conservation (%)	Share of total land classified as drylands (%)	Share of land classified as forests (%)	Number of conservation areas (number in marine)	Wetlands of international importance. Ramsar list (ha)	Number of biosphere reserves (size in thousands of ha)
Maghreb						
Algeria	5.1	21	0.6	18 (4)	1,866	6 (7,312)
Libya	0.1	23	0.1			
Morocco	0.7	92	11.5	12 (4)	14	2 (9,754)
Tunisia	0.3	94	6.5	7 (2)	13	4 (74)
Mashreq						
Egypt, Arab Rep.	9.7	8	0.1	35 (12)	106	2 (2,456)
Iraq		100	1.9	8 (3)		
Jordan	3.4	72	1.1	11 (7)	7	1 (31)
Lebanon	0.5	59	13.4	3 (1)	1	
Syrian Arab Rep.	0.7	98	2.7		10	
West Bank and Gaza			1.5			
Gulf						
Bahrain			1.3			
Kuwait	1.5	92	0.3	5 (2)		
Oman	14.0	14	0.0	6 (2)		
Qatar	0.0		0.0			
Saudi Arabia	38.3	24	0.5	78 (3)		
United Arab Emirates	0.3		3.8	2		
Least developed countries						
Comoros			1.6			
Djibouti			0.3			
Mauritania	1.7	46	0.2	9 (3)	1,231	
Somalia	0.8		10.8	10 (1)		
Sudan	5.2	67	29.4	27 (1)		
Yemen Rep.		30	1.0			1

Sources: UNDP et al. 2003; World Bank's World Development Indicators database; World Bank, various years. Forest Area data accessed from http://data.worldbank.org/indicator/AG.LND.FRST.ZS.

Inland wetlands—many of which are ephemeral—and coastal and marine areas are important economically, culturally, and for their biodiversity. Fish from coastal, marine, and inland waters contribute to the economy and provide up to 20 percent of the protein for people in some parts of the region (see table S2.2). The Dead Sea and Red Sea have unique, endemic, and globally threatened reef and coastal species. Dredging for urban and infrastructure development has extensively altered about half of the coastlines in countries that have marine coasts and has had a negative impact on biodiversity and livelihoods. Other major threats to coastal areas include oil spills, chemical contamination, ballast water spills, and military conflicts. In some countries, organic-rich discharge from livestock and agriculture production and the water discharged from desalination plants are additional threats (Birot, Garcia, and Palahi 2011; UNEP 2007; World Bank 2007). Warmer sea temperatures from climate change add an additional stressor to the reef, and periods of coral bleaching are becoming more common.

Mammals and bird species		Higher plant species		Annual marine catch, 1998–2000 (thousands of metric tons)	Fish protein as a share of animal protein (%)
Number	Number threatened	Number	Number threatened		
275	19	3,164	2	98	6
152	9	1,825	1	33	7
311	25	3,675	2	782	17
243	16	2,196		90	12
221	20	2,076	2	156	19
221	22			12.5	8
188	18	2,100		0.1	5
173	12	3,000		3.6	7
208	12	3,000		2.6	2
56	8	234		5.8	5
165	19	1,204	6	110.0	
202	23	2,028	3	49.0	6
	59	11		112.5	12
	13		5		
	14		2		
233	12	1,100	0	33	11
	26		21	20.7	2
547	29	3,137	17	5.7	2
159	17	1,650	52	122	22

Few studies have been conducted on the economic valuation of ecosystem services in the region, and even fewer on the impacts of climate change on these services. A rare example is a study of forests in Morocco (Birouk et al. 1996). According to the study, traditional forestry accounts for about 2 percent of agricultural GDP, but if nontimber and other services for rural people are included, the figure increases to 10 percent. In Tunisia, 35 percent of forest value is related to forest soil and water conservation in the north (GTZ and MARH 2007). The ecosystems along the Nile and the Nile Delta cover about 5.5 percent of the total area of Egypt, but they provide benefits for 95 percent of the people of that area, with 25 percent of the population living within the Delta, according to a 2011 government review document by the Coastal Research Institute. The *Arab Human Development Report 2009* indicated that high pressure on ecosystems in many Arab countries exceeds their capacity for renewal, thus limiting their resilience to climate change and other stresses (UNDP 2009). Analyses based on the *ecological footprint*, a measure of ecosystem service consumption, show that the

TABLE S2.3

Ecological Capacity and the Deficit or Reserve in Arab Countries, 2007

Country	Ecological footprint consumption (global ha/person)	Total biocapacity (global ha/person)	Ecological deficit/reserve (global ha/person)
Maghreb			
Algeria	1.60	0.60	−1.00
Libya	3.10	0.40	−2.70
Morocco	1.22	0.61	−0.61
Tunisia	1.90	0.98	−0.92
Mashreq			
Egypt, Arab Rep.	1.66	0.62	−1.04
Iraq	1.35	0.30	−1.05
Jordan	2.05	0.24	−1.81
Lebanon	2.90	0.40	−2.50
Syrian Arab Rep.	1.52	0.70	−0.82
West Bank and Gaza	0.74	0.16	−0.58
Gulf			
Bahrain	10.04	0.94	−9.10
Kuwait	6.32	0.40	−5.92
Oman	4.99	2.14	−2.85
Qatar	10.51	2.51	−8.00
Saudi Arabia	5.13	0.84	−4.29
United Arab Emirates	10.68	0.85	−9.83
Least developed countries			
Comoros	1.42	0.29	−1.13
Djibouti			
Mauritania	2.61	5.50	2.89
Somalia	1.42	1.40	−0.02
Sudan	1.73	2.42	0.69
Yemen Rep.	0.94	0.62	−0.32

Source: WWF 2010.

Note: Ecological footprint is a measure of resource consumption, *biocapacity* is a measure of ecosystem production, and *ecological deficit/reserve* is the difference between consumption and capacity of the ecosystems. Ha = hectare.

richer countries, mostly in the Gulf, generally have a large ecological deficit. The LDCs generally have a lower ecological deficit, and two have a surplus (see table S2.3 and figure S2.1).

Managing Risks from Climate Change and Other Pressures Is Imperative

Integrated land and water management that incorporates more efficient and reduced water use, environmental flows to support riverine and wetland ecosystems, and conservation is often an effective approach that can help address multiple stresses, including climate variability and change (CAWMA 2007). Examples from Morocco, the United Arab Emirates, and the Republic of Yemen illustrate how integrated land and water management can help improve ecosystem resiliency and provide other benefits (see box S2.1). Climate variability or climatic extreme events are likely to disproportionately affect populations that rely on ecosys-

FIGURE S2.1

The Ecological Footprint of Arab Countries

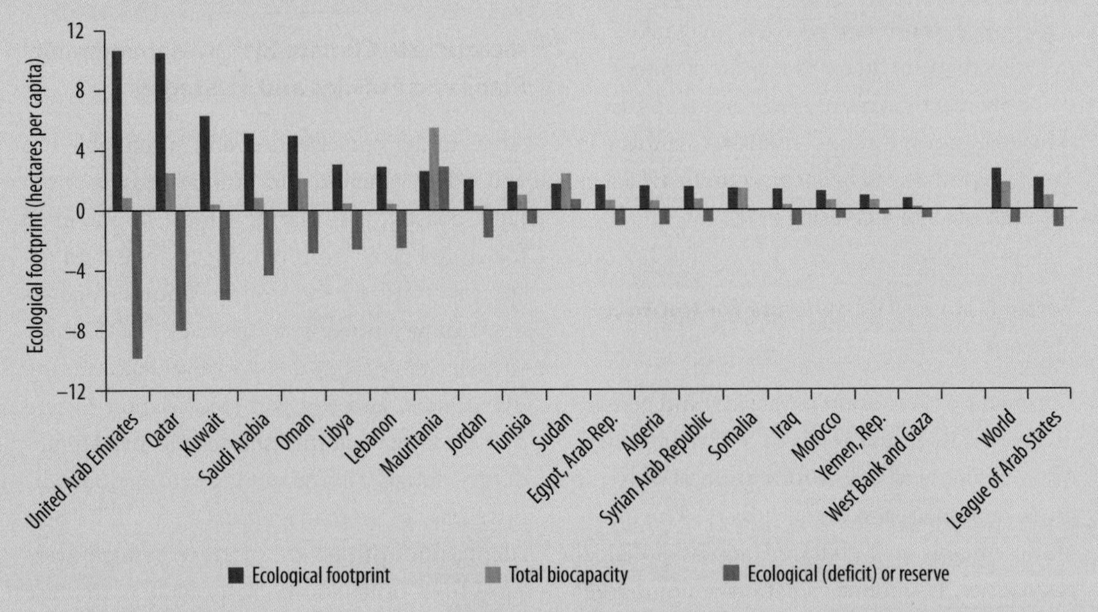

Source: Authors' compilation, based on Global Footprint Initiative 2010.

BOX S2.1

Integrated Land and Water Management in Arab Countries

The Lakhdar watershed in Morocco and the *wadis* in the northern part of the Republic of Yemen incorporate natural forests and woodlands as part of microcatchment vegetation management with active involvement of local communities. Actively preventing and controlling the spread of invasive and alien species to reduce land and water degradation has helped improve food security, increase water supplies, and provide biofuels.

An example of an integrated approach is one used by Masdar Institute in the United Arab Emirates. The system consists of a shrimp and fish production facility, a mangrove forest, wetland systems, and a field of the salt-tolerant *Salicornia* species, which can also be used as a biofuel. Seawater or saline water is channeled into pans for growing commercially profitable shrimp or fish species. The water from the shrimp and fish farms, enriched with organic matters, is channeled into large farms to grow salt-tolerant *Salicornia* species and mangroves. The leaves and small branches can be fed to livestock such as goats and camels, whereas the roots help to stabilize the soil and reduce erosion. Meanwhile, water evaporation from vast stretches of seawater farms will increase humidity. The long-term objective of this farming system is to transform the coastal region through sustainable and natural means: increased humidity and enriched soil will eventually be able to support freshwater farming.

Sources: MIT 2011; World Bank 2009.

tem services or live in ecologically and economically marginal areas in rangelands and coasts. With increasing pressures and a changing climate, ecosystem recovery may take longer and may further reduce the capacity of communities to manage these increasing risks. The adaptation options discussed in the sections that follow can help communities manage the multiple and increasing risks.

Better Data and Knowledge for Improved Management

Information on ecosystem services and genetic diversity is weak and needs to be strengthened. Most of the available information is only for birds, mammals, and higher plants. The information needs to be shared across political boundaries. Economic valuation of nonmarket benefits and off-site effects on soil erosion and water resources would help highlight the values and costs to ecosystems and ecosystem services and would provide a mechanism for incorporating biodiversity and ecosystem services into the national budget (through "green accounting").

Enhance Conservation and Sustainable Use

There is a need to conserve genetic resources and knowledge and practices used to grow ancestral and older varieties of crops and fruit trees. Of the region's known species, 10 percent are listed as threatened (UNEP 2007), and unique ecosystems that are vulnerable to climate change need to be conserved and used sustainably. Unique ecosystems include desert oases, cedar forests (such as those in Lebanon and the Syrian Arab Republic), mangroves (such as those in Qatar), inland reed marshes (such as those in Egypt and Iraq), and high mountain ranges (such as those in Jordan, Oman, Syria, and the Republic of Yemen) (UNEP 2007).

Incorporate Climate Risk in Management Plans and Policies and Strategies

Areas under conservation and sustainable use and other valuable and unique ecosystems—land, coastal, and marine—must be screened for added risks from climate change, and responses must be incorporated into management plans. Biosphere reserves and Ramsar sites (see table S2.2) that are of value to culture, livelihoods, and tourism (see chapter 6) and that are important for fisheries also need to undergo climate risk screening and include climate change in management plans. The best adaptation options are often to reduce pressures from other sources, especially in the case of degraded lands and polluted or overexploited freshwater, coastal, and marine areas. Sustainable management under climate change should be included in strategic environmental assessments (SEAs). Tools and approaches, such as valuation of ecosystem services and payments for environmental services in SEAs, could help integrate biodiversity, ecosystem services, and adaptation in decision making (TEEB 2010).

Use Ecosystem-Based Adaptation to Increase Ecosystem Resiliency

Ecosystem-based adaptation (EBA) is the use of ecosystems to protect and sustain human lives and livelihoods. EBA can be a cost-effective measure and a substitute for more expensive infrastructure investments. EBA can also benefit poor, marginalized, and indigenous communities by maintaining ecosystem services and access to resources during prolonged droughts (GTZ and MARH 2007; World Bank 2009). Conserving floodplain forests, coastal and inland wetlands, mangroves, coral reefs, barrier beaches, and sand dunes as protection from storm surges and flooding are examples of EBA. Some of these services will provide other benefits, such as helping maintain nursery, feeding, and breeding grounds for fisheries, wildlife, and other species on which human populations depend.

Blend Finance to Meet the Added Costs

Clearly, additional funding will be necessary to manage climate risks and develop the resilience or adaptive capacity of ecosystems. Market-based instruments, such as payments for ecosystem services, voluntary carbon markets, and environmental services markets, are potential options for some ecosystems (Brauman et al. 2007).

Management of Biodiversity and Ecosystems Is Essential

Without biodiversity and ecosystem services, life and human societies would not exist. Managing these systems, including risks from climate change, is essential.

- *Arab countries have biodiversity and ecosystems of global importance that are already at risk from climate variability and change and other pressures.* The Arab region played host to the wild ancestors of many of today's crops, fruit tree species, and livestock that are critical to world food production systems. These species are important as present and future adaptation options, especially as stores of wild genes. Also, critical ecosystems in many countries are of value to major economic sectors such as tourism and fisheries, as well as for cultural heritage.
- *Biodiversity and ecosystem services and their contribution to livelihood, and economies need to be recognized and included in decision making.* National development plans and sectoral strategies need to consider biodiversity and ecosystem services in present and future adaptation options. Better valuation of ecosystem services and market and nonmarket mechanisms can help promote recognition of biodiversity and ecosystem services and inform the discussion on trade-offs.
- *National and international efforts and integrated land and water management are needed for the conservation of globally important biodiversity and threatened or unique species and ecosystems.* Conservation and sustainable use efforts that are part of integrated land and water management will contribute to improving ecosystem resilience, reducing the vulnerability of livelihoods, and enhancing national and local economies in many Arab countries.

References

Bioversity International. 2006. "Crop Wild Relatives." Bioversity International, Rome.

Birot, Yves, Carlos Garcia, and Marc Palahi, eds. 2011. *Water for Forests and People in the Mediterranean Region: A Challenging Balance*. Joensuu, Finland: European Forest Institute.

Birouk, Ahmed, Mohammed Tazi, Hamdoune Mellas, and Mohammed Maghnouj. 1996. "Morocco Country Report." Prepared for the Food and Agriculture Organization of the United Nations International Technical Conference on Plant Genetic Resources, Leipzig, Germany, June 17–23.

Boko, Michel, Isabelle Niang, Anthony Nyong, Coleen Vogel, Andrew Githeko, Mahmoud Medany, Balgis Osman Elasha, Ramadjita Tabo, and Pius Yanda. 2007. "Africa." In *Climate Change 2007: Impacts, Adaptation, and Vulnerability—Contribution of Working Group II to the Fourth Assessment Report of the Intergovernmental Panel on Climate Change*, ed. Martin L. Parry, Osvaldo F.

Canziani, Jean P. Palutikof, Paul J. van der Linden, and Clair E. Hanson, 433–67. Cambridge, U.K.: Cambridge University Press.

Brauman, Kate A., Gretchen C. Daily, T. Ka'eo Duarte, and Harold A. Mooney. 2007. "The Nature and Value of Ecosystem Services: An Overview Highlighting Hydrologic Services." *Annual Review of Environment and Resources* 32: 67–98.

Bullock, Peter, and Henri Le Houérou. 1996. "Land Degradation and Desertification." In *Impacts, Adaptations, and Mitigation of Climate Change: Scientific-Technical Analyses—Contribution of Working Group II to the Second Assessment Report of the Intergovernmental Panel on Climate Change*, ed. Robert T. Watson, Marufu C. Zinyowera, and Richard H. Moss, 170–89. Cambridge, U.K.: Cambridge University Press.

CAWMA (Comprehensive Assessment of Water Management in Agriculture). 2007. *Water for Food, Water for Life: A Comprehensive Assessment of Water Management in Agriculture*. London: Earthscan; Colombo: International Water Management Institute. http://www.iwmi.cgiar.org/assessment.

Croitoru, Lelia, and Maria Sarraf. 2010. *The Cost of Environmental Degradation: Case Studies from the Middle East and North Africa*. Washington, DC: World Bank.

Elasha, Balgis Osman. 2010. "Mapping Climate Change Threats and Human Development Impacts in the Arab Region." Arab Human Development Report Research Paper, United Nations Development Programme, Regional Bureau for Arab States, New York.

FAO (Food and Agriculture Organization of the United Nations). 2010. *The State of the World's Plant Genetic Resources for Food and Agriculture*. Rome: FAO.

Gitay, Habiba, and Ian R. Noble. 1998. "Middle East and Arid Asia." In *Regional Impacts of Climate Change: An Assessment of Vulnerability*, ed. Robert T. Watson, Marufu C. Zinyowera, and Richard H. Moss, 231–52. Cambridge, U.K.: Cambridge University Press.

Global Footprint Initiative. 2010. "Data and Results." Global Footprint Initiative, Oakland, CA. http://www.footprintnetwork.org.

GTZ (German Agency for Technical Cooperation). 2006. "Value Chains for the Conservation of Biological Diversity for Food and Agriculture: Potatoes in the Andes, Ethiopian Coffee, Argan Oil from Morocco, and Grasscutters in West Africa." GTZ, Eschborn, Germany.

GTZ (German Agency for Technical Cooperation) and MARH (Ministry of Agriculture and Water Resources). 2007. "Tunisian Adaptation Strategy to Climate Change for Agriculture and Ecosystems." GTZ, Eschborn, Germany.

Lal, Murari, Hideo Harasawa, and Daniel Murdiyarso. 2001. "Asia." In *Climate Change 2001: Impacts, Adaptation, and Vulnerability—Contribution of Working Group II to the Fourth Assessment Report of the Intergovernmental Panel on Climate Change*, ed. James J. McCarthy, Osvaldo F. Canziani, Neil A. Leary, David J. Dokken, and Kasey S. White, 533–90. Cambridge, U.K.: Cambridge University Press.

MA (Millennium Ecosystem Assessment). 2003. "Millennium Ecosystem Assessment Framework." Island Press, Washington, DC.

MIT (Massachusetts Institute of Technology). 2011. "Mission 2014: Feeding the World." MIT, Boston. http://12.000.scripts.mit.edu/mission2014/solutions/seawater-farming.

TEEB (The Economics of Ecosystems and Biodiversity). 2010. "The Economics of Ecosystems and Biodiversity: Mainstreaming the Economics of Nature: A Synthesis of the Approach, Conclusions and Recommendations of TEEB." Progress Press, Mriehel, Malta.

Tolba, Mostafa K., and Najib W. Saab, eds. 2009. *Arab Environment Climate Change: Impact of Climate Change on Arab Countries*. Beirut: Arab Forum for Environment and Development.

UNDP (United Nations Development Programme). 2008. *Human Development Report 2007/2008: Fighting Climate Change—Human Solidarity in a Divided World*. New York: UNDP.

———. 2009. *Arab Human Development Report 2009: Challenges to Human Security in the Arab Countries*. New York: UNDP.

UNDP (United Nations Development Programme), UNEP (United Nations Environment Programme), World Bank, and World Resources

Institute. 2003. *World Resources 2002–2004: Decisions for the Earth*. Washington, DC: World Resources Institute.

UNEP (United Nations Environment Programme). 1997. *Global Environmental Outlook*. Oxford, U.K.: UNEP and Oxford University Press.

———. 2007. *Global Environmental Outlook: GEO 4—Environment for Development*. Mriehel, Malta: Progress Press.

World Bank. 2007. "Assessment of the Tunisian Water Degradation Cost." World Bank, Washington, DC.

———. 2009. *Convenient Solutions to an Inconvenient Truth: Ecosystem-Based Approaches to Climate Change*. Washington, DC: World Bank.

———. Various years. *Little Green Data Book*. Washington, DC: World Bank.

WWF (World Wildlife Fund). 2010. *Living Planet Report 2010: Biodiversity, Biocapacity, and Development*. Gland, Switzerland: WWF. http://wwf.panda.org/about_our_earth/all_publications/living_planet_report/2010_lpr/.

CHAPTER 4

Agriculture, Rural Livelihoods, and Food Security Are Stressed in a Changing Climate

Agriculture Is Essential for the Economy, Food Security, and Livelihoods in the Region

This chapter describes how farming systems in the water-scarce Arab region are vulnerable to climate change with mixed, largely negative impacts until 2050 and more pronounced negative impacts after that. This vulnerability has major implications for food security as well as for the lives and livelihoods of the 150 million people in rural areas whose incomes rely predominantly on natural resources.

Agriculture has played a fundamental role in the history of the Middle East and North Africa. Arab countries contain important sites of early settled agriculture and are centers of the origin and diversity of several major cereal and legume crops. The Middle East is also known for the early domestication of sheep and goats and for innovation in agricultural, water storage, and transport methods to facilitate agriculture between the 4th and 11th century CE, when many new crops and technologies were introduced from the Far East. To effectively adapt to increasing climate variability and change while maintaining productive agriculture for both food security and livelihoods, Arab countries will once again need to invest significantly in agricultural innovation and technology.

The adaptation challenge in rural livelihoods will be complex as more than 34 percent of the region's rural population is poor—ranging from 8 percent in Tunisia to over 80 percent in Sudan. With rural unemployment averaging about 13 percent—although rates are much higher for youth, ranging from 26 to 53 percent—finding alternative income sources will be difficult (Christensen 2007, 67, 73; Dixon and Gulliver 2001, 84;

Photograph by Dorte Verner

IFAD 2010b). In the poorest nations of the region (Mauritania, Somalia, Sudan, and the Republic of Yemen), rural poverty is already chronic and widespread. Elsewhere in the Arab region, poverty mainly affects three high-risk categories: households headed by women (see chapter 7), the landless, and farm laborers. Given the limited assets of these vulnerable groups, adaptation strategies will need to include different forms of safety nets to avoid any further deterioration in conditions that have already led to over one-fifth of the region's children under five years old being stunted, and rural children are almost twice as likely to be underweight as urban children (Christensen 2007, 66; IFAD 2010b).

Food security concerns continue to dominate discussions by policy makers even under current climate conditions, with questions being raised as to whether an increasing reliance on international markets is sufficient to ensure stable access to affordable food supplies for their people. The issue brings geopolitical as well as economic issues to the forefront. Any global rises in food price or volatility under a changing climate scenario will affect all Arab countries because of their dependence on imported food, but in particular the oil-importing Arab countries because food and oil prices tend to rise in parallel.

Although these major concerns seem challenging, this chapter will outline a wealth of adaptation options that can also contribute to sustainable and equitable agricultural growth and rural poverty reduction. More marginal systems have fewer options, and some may become less productive or, even in the longer term, go out of production. Governments need to support the development of agricultural production under a changing climate and to prepare segments of the rural population for transition to alternative livelihoods. This agenda of adaptation governance can strengthen household-level food security for many and if combined with improvements in markets, safety nets and risk management can help meet broader concerns in this field.

Agriculture Is under Stress from Water Scarcity and Population Growth

Climate and Farming Systems

As described in chapter 2, most of the region is arid to hyperarid, with both rainfed and irrigated agricultural systems contributing to overall productivity. The temperate, higher rainfall areas have a Mediterranean climate, which typically has long dry summers and mild wet winters. These areas account for less than 10 percent of the land area but nearly

TABLE 4.1

Principal Farming Systems of the Arab Region

Farming system	Share of the region's land area (%)	Share of the region's agricultural population (%)	Main livelihoods	Prevalence of poverty
Irrigated	2	17	Fruit, vegetables, cash crops	Moderate
Highland, mixed	7	30	Cereals, legumes, sheep, off-farm work	Extensive
Rainfed, mixed	2	18	Tree crops, cereals, legumes, off-farm work	Moderate
Dryland, mixed	4	14	Cereals, sheep, off-farm work	Extensive
Pastoral	23	9	Camels, sheep, off-farm work	Extensive
Agropastoral: millet or sorghum	—	—	Cereals, pulses, livestock	Extensive
Cereals or root crops, mixed	—	—	Cereals, root crops, cattle	Limited
Arid zones	62	5	Camels, sheep, off-farm work	Limited

Source: Annex 4A.
Note: — = not available.

half of the agricultural population. The dry areas, with rainfall under 300 millimeters a year, account for 90 percent of the land but less than 30 percent of the agricultural population. Large-scale irrigated areas cover less than 2 percent of the land area but account for 17 percent of the agricultural population. Farming systems are diverse, determined largely by geography, climate, and natural resource endowments (table 4.1 and annex 4A) (Dixon and Gulliver 2001, 83–84, 87–91).

As precipitation falls over most of the Maghreb, Mashreq, and many of the least developed countries in winter, rainfed crops are grown in the winter months, maturing for harvest generally in spring and early summer. The main rainfed crops are wheat, barley, legumes, olives, grapes and other fruits, and vegetables. Grain production accounts for two-thirds of the cultivated area (against a world average of 46 percent). Yields for rainfed crops vary widely depending largely on the interaction of the crops' genetic makeup (genotype), soil, and cultural and environmental factors.

Irrigated areas are cultivated year-round, with peak demand for irrigation water during the dry summer months in the Mashreq and Maghreb, and in the cooler winters in the Gulf countries. Under irrigation, yields can be very good, for example, with yields of irrigated wheat in the Arab Republic of Egypt averaging 6.5 metric tons per hectare. A wide range of higher-value crops is grown. Fresh fruit and vegetable production account for about 10 percent of the cropped area throughout the region, but the percentage is much higher in countries practicing intensive irrigated agriculture (Egypt 20 percent, Jordan 28 percent, Lebanon 37 percent) and much less in the largely subsistence agricultural economies of Somalia (1 percent) and Sudan (1 percent) (Christensen 2007, 47, 49).

Livestock is integrated in all farming systems, thus providing important synergies and complementarities between and within systems—from extensive pastoralism to feedlots in peri-urban agriculture. Milk, from both cows and goats, is an increasingly important commodity often based on feedlot production systems.

Historically, agriculture has been an engine of growth in the region. With modernization and urbanization, the share of agriculture in regional gross domestic product (GDP) has dwindled, but the sector remains critical to primary production and is the mainstay of the rural economy. In 2005, agriculture contributed 13 percent to regional GDP, ranging from 2 percent in Jordan to over 30 percent in Sudan. Annual agricultural GDP per capita of the agricultural population averages about US$720, ranging from US$133 in the Republic of Yemen to US$1,100 in Tunisia. Throughout the region, agriculture contributes about 20 percent of exports (Christensen 2007, 72; Dixon and Gulliver 2001, 87). With regard to people, farming is still an important occupation in some countries with 48 million (38 percent) engaged overall in the region; however, this figure masks a wide range: from less than 5 percent in Lebanon and the Gulf countries to more than 50 percent in Somalia, Sudan, and the Republic of Yemen (Christensen 2007, 70; Dixon and Gulliver 2001, 83; FAO 2010, chapter 1).

Recent Trends: Agriculture Already under Stress

Cultivable land is abundant but the main constraint is water. Aridity has meant that irrigation has been the principal path to intensification and is the largest water user (see chapter 3 for more details). Further expansion of irrigated production is likely to come largely from productivity gains, especially gains in water-use efficiency rather than from new supply development (Dixon and Gulliver 2001, 91ff).

In general, pressure on land and water resources is expected to grow as populations continue to expand. Population pressure on resources is already leading to environmental degradation and water shortages, and these trends will worsen if nothing changes. This demographic pressure will also increase dependence on imported food; the region already imports half its cereal calories, and this share will likely increase (Dixon and Gulliver 2001, 91).

Rangelands have come under considerable stress. Growth in demand for fresh meat, linked to income growth, has been met by swelling livestock populations, often supported by feed subsidies on imported grains. As a result, livestock populations are today well beyond the carrying capacity of the rangelands. Older systems of rangeland management have not adapted.

Productivity, Technology, and Innovation

After two decades of little or no growth in agricultural productivity (1960–80), the subsequent decades (1980–2000) witnessed a strong average growth rate of 2 percent per annum (see figures 4.1 and 4.2). Al-

FIGURE 4.1

Average Cereal Yield

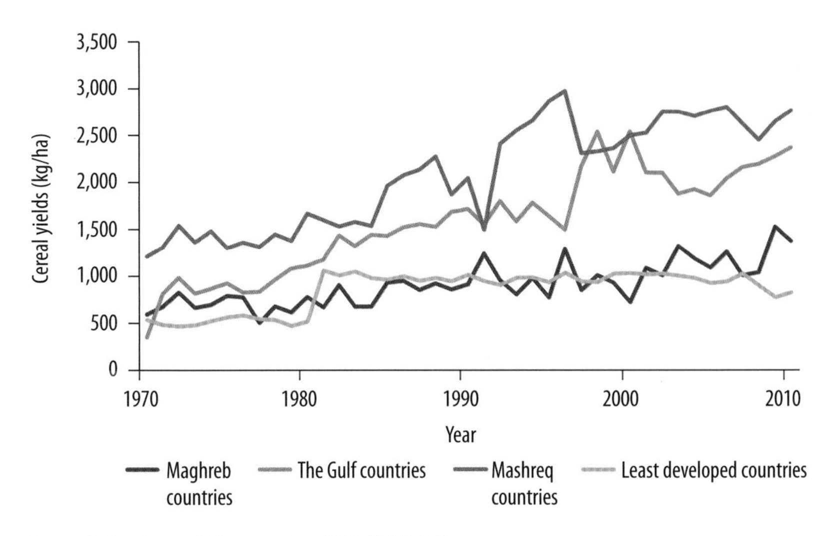

Source: Authors' compilation, based on AQUASTAT 2011.
Note: Ha = hectare.

FIGURE 4.2

Average Vegetable/Melon Yield

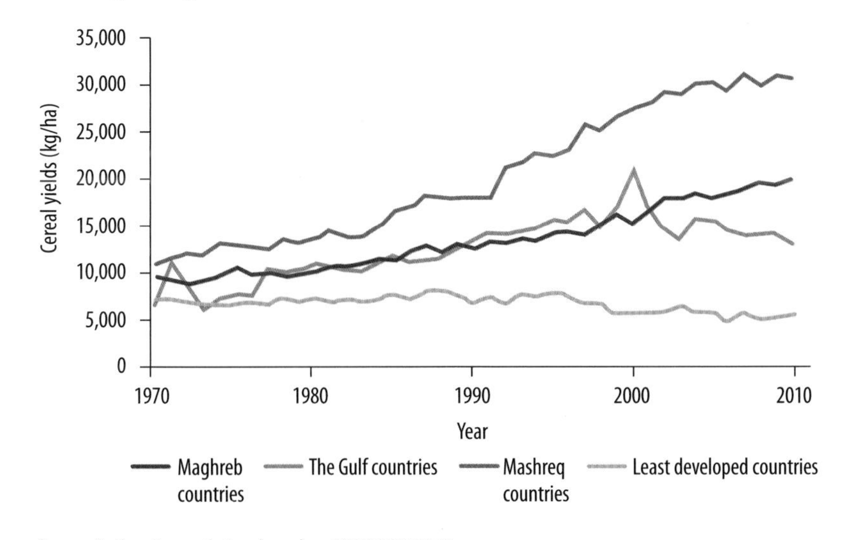

Source: Authors' compilation, based on AQUASTAT 2011.
Note: Ha = hectare.

though the causes are complex and vary by country, variables attributed to this strong growth include more intensive irrigation systems (particularly notable in the Gulf countries) and an increase in production of higher-value crops following the start of trade liberalization in the 1980s. In recent years, cereals production has accelerated with improvements in the terms of trade for cereals and because of rapidly expanding livestock production. This growth is not, however, shared across the entire region (figures 4.1 and 4.2). In some countries, productivity improvements have virtually ceased because access to improved technologies and support services has dwindled, which is particularly noticeable in the least developed countries. Research has made a substantial contribution to yield improvements, but an increased focus is needed on efficient water use and vulnerable systems.

Market orientation

Agriculture in the Arab countries has become predominantly market oriented and commercialized as it responds to fast-growing demand from urban markets for higher-value products. In the Mediterranean countries, market linkages with demand from Europe and formal trade arrangements with the European Union have provided profitable market outlets for fresh fruits and vegetables. Many households have diversified into related off-farm business lines, such as catering, tourism, and so forth.

Policies, institutions, and public goods

Countries across the region have made considerable investment in irrigation, rural infrastructure, and farmer services, such as research and extension. Agriculture has responded with the rapid growth rates noted above; however, some components of public policy have introduced structural distortions in the sector that have reduced its resilience and sustainability, notably the following:

- Earlier water policies allocated water to agriculture. Now water is becoming more valuable for other purposes, but mechanisms for reallocating water between sectors are lacking. At the same time, the lack of demand management through market mechanisms or rationing has led to low water-use efficiency in many agricultural uses.
- Lack of regulation of groundwater extraction has led to depletion of the resource.
- Food security policies have necessitated agricultural production in some areas that has resulted in negative impacts on land and water resources.

- Incentive structures favoring commercial and irrigated production have disfavored research and investment in rainfed farming.

Recently, countries in the region have taken a more integrated appreciation of agriculture's role in the economies, ecologies, and societies of the region, including (a) the value of conserving ecosystems; (b) the environmental services provided by rural areas, such as water filtration and soil conservation; and (c) sociocultural services, such as cultural heritage or traditional agriculture.

Climate Change Affects Farming Systems and Rural Livelihoods

The Arab region will likely be highly vulnerable to climate change (as chapters 2 and 3 have illustrated), which will accentuate the already-severe water scarcity and increase the existing high levels of aridity. The main climate change impacts on agriculture are summarized in annex 4B, with the most important exposures given in table 4.2.

Expected Impacts on Agricultural Water and Farming

All farming systems will be exposed to increased aridity and to declines in water availability (table 4.3). All systems are sensitive to these changes, especially rainfed systems without access to reliable irrigation sources.

TABLE 4.2

Exposure: Expected Climate Changes

Maghreb	Mashreq	Gulf countries	Least developed countries
• Overall a hotter, drier region • Temperature increase of up to 5°C • Decrease in precipitation, fewer rainy days • More droughts, especially in summer • Overall increase in aridity, with 20% less rainfall • Seawater intrusion	• Overall a hotter, drier region • Higher temperatures in both summer and winter • Generally drier, especially in the rainy (winter) season • Rainfall drop below growth threshold for some areas • Seawater intrusion and salinization, particularly in Egypt	• Relatively uniform warming • Possible increase in summer precipitation, but highly uncertain and localized • More severe rainfalls • Seawater intrusion	• Changes in river flows • Variable changes in wetness and aridity, with areas nearer the tropics becoming wetter • More severe rainfalls

Source: Authors' compilation.

TABLE 4.3

Climate Change Impacts on Farming Systems of the Arab Region

Farming system	Exposure: expected climate-related changes	Sensitivity: likely impacts on farming systems
Irrigated	• Increased temperatures • Reduced supply of surface irrigation water • Dwindling groundwater recharge • Loss of production in low-lying coastal areas	• More water stress • Increased demand for irrigation and water transfer • Reduced yields when temperatures are too high • More difficulty in agricultural planning • Salinization from reduced leaching • Reduction in cropping intensity
Highland mixed	• Increased aridity • Greater risk of drought • Possible lengthening of the growing period • Reduced supply of irrigation water	• Reduction in yields • Reduction in cropping intensity • Increased demand for irrigation
Rainfed mixed	• Increased aridity • Greater risk of drought • Reduced supply of irrigation water • Loss of production in low-lying coastal areas	• Reduction in yields • Reduction in cropping intensity • Increased demand for irrigation • More difficulty in agricultural planning
Dryland mixed	• Increased aridity • Greater risk of drought • Reduced supply of irrigation water	• System very vulnerable to declining rainfall; some lands may revert to rangeland • Increased demand for irrigation
Pastoral	• Increased aridity • Greater risk of drought • Reduced water for livestock and fodder	• Very vulnerable system, where desertification may reduce carrying capacity significantly • Increase in nonfarm activities, exit from farming, migration

Source: Annex 4B.

More marginal systems will likely be pushed further to or beyond their margins; some of these lands may revert to pasture or may simply abandon production.

Water resources and irrigation

Water availability is the key determinant of agricultural potential throughout the region, and climate change will affect that availability. In all regions, lower rainfall and higher temperatures will likely lead to a decline in soil-moisture availability to the plant roots and to decreased infiltration and runoff, so that groundwater recharge and river flows will likely diminish. This reduction will affect both rainfed and irrigated production with net crop irrigation requirements estimated to increase by 5 to 20 percent by 2080 (Fischer et al. 2007). Demand for increased irrigation may push up the ratio of irrigation withdrawals to available renewable water resources.

Crop and livestock yields and production

These changes in agroclimatic conditions will affect production, although the pace and direction of change are far from certain, and they will inevitably vary considerably across locations. For example, in the longer run, yields of key rainfed cereals may drop, with projected maize yields falling by 15 percent and wheat yields by 5 percent over much of the region by 2050, particularly in the Mashreq and Maghreb countries (Iglesias and Rosenzweig 2010). Another study (Breisinger et al. 2010) suggests that cereal yields may still increase on average at least until 2050 (see figure 4.3), but by less than what would have been the case in the absence of climate change.

Overall, the most likely picture that emerges for cereals is threefold: (a) a combination of improved incentives because of rising commodity prices together with application of productivity-enhancing measures will keep yields rising to midcentury, after which they will begin to drop off;

FIGURE 4.3

Average Cereal Yields in the Middle East and North Africa: Historic Climate and Alternative Scenarios, 2025 and 2050

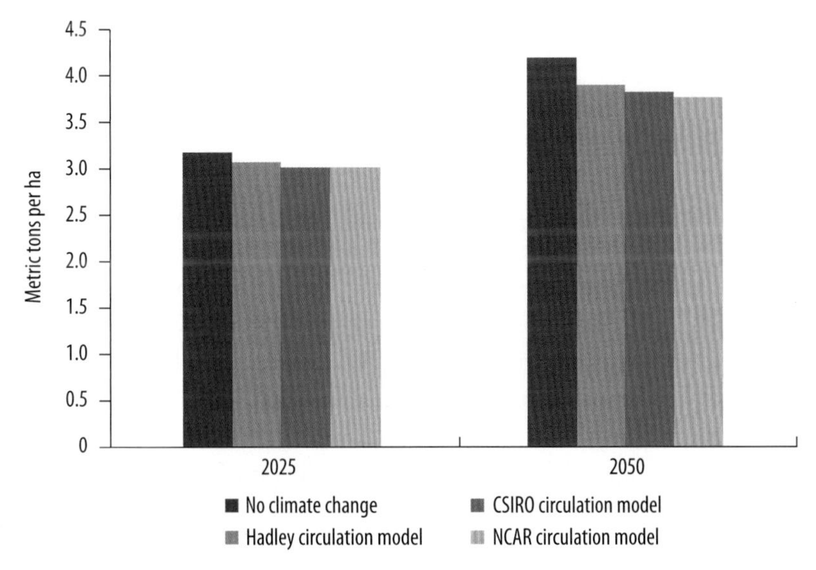

Source: Breisinger et al. 2010.

Note: Hadley, CSIRO, and NCAR are different circulation models. CSIRO = Commonwealth Scientific and Industrial Research Organization; NCAR = National Center for Atmospheric Research; ha = hectare.

TABLE 4.4

Projected Rate of Annual Change for Rainfed Wheat under One Climate Change Scenario

	Morocco	Egypt, Arab Rep.	North Africa	Syrian Arab Republic	Iraq
Yield: annual rate of increase or decrease (%)					
Baseline, 2010	5.4	1.4	5.4	1.3	4.6
2020	1.2	2.4	2.0	0.7	0.7
2050	0.2	1.4	0.0	−0.3	−0.3
Area: annual rate of increase or decrease (%)					
Baseline, 2010	0.0	0.5	0	−0.1	−0.1
2020	−0.1	0.1	−0.5	−0.2	−0.2
2050	−1.0	−0.7	−1.0	−1.1	−1.1
Forecast production (thousand metric tons)					
Baseline, 2010	4,634	4,847	n.a.	4,585	1,851
2020	5,322	6,511	n.a.	5,106	2,040
2050	5,622	10,848	n.a.	5,162	2,133

Source: International Food Policy Research Institute crop models (using CSI A1B), taken from the following International Food Policy Research Institute website on December 19, 2011: http://www.ifpri.org/publication/food-security-farming-and-climate-change-2050.

Note: n.a. = not applicable.

(b) natural resource and climate change pressures will contribute to a slow fall in the production area from 2010; and (c) overall output will rise in most major producing countries, but after 2050 production of wheat and maize across the region will start to decline (Breisinger et al. 2010). The International Food Policy Research Institute's series of global crop models supports this assessment (see table 4.4 for the example of rainfed wheat in a number of Middle East and North African countries) (Breisinger et al. 2010; World Bank 2007).

An element of uncertainty is introduced by the expected increased variability and increased frequency of extreme events, especially of drought but also of destructive storms, floods, and heat waves. These factors will reduce yields in the year of incidence and may also create risk aversion and disincentives to investment, thereby making planning at household, local, and national levels more challenging. Yields of several economically important fruit species (olives, apples, pistachios and other nuts, and pomegranates) may also suffer reduced yields or crop failure if winter temperatures are too high.

Climate change will also increase risks for livestock production, particularly for the intensive feedlots that are becoming more common. The likely rise of animal pests and diseases under climate change could affect all animal production, but especially intensive systems. The lack of prior conditioning to extreme weather can result in major losses in confined livestock feedlots. In addition, animal nutrition models have shown that higher temperatures can limit dairy milk yield in relation to feed. The following section explores the adaptation options available for rural households, both in agriculture and in the broader food security and livelihood contexts.

A Range of Adaptation Options Are Available

Farmers Have Many Adaptation Options

This section examines a range of likely farmer responses to specific climate change and variability impacts. At least some technology and institutional options that can help maintain productivity are known and either are accessible to farmers or could be made accessible. However, the strength of these adaptive measures declines in inverse proportion to the sensitivity of the system to climate change. The most marginal and affected systems—dryland and pastoral systems—are those for which the fewest solutions are available. Without policy and program intervention, these systems will most likely have the greatest changes to agricultural production and rural impoverishment.

Likely Farmer Adaptations in Agriculture[1]

Farmers are likely to respond to rising temperatures by a mix of changes in varieties of crops and, where available, by recourse to supplementary irrigation. Climate change is expected to result in warmer temperatures and increased aridity across most Arab countries. Farmers may be able to manage risks with more drought-tolerant or shorter-cycle varieties and (where available) supplementary irrigation, or by switching their cropping pattern. Soil moisture conservation techniques, such as changed tillage and mulching practices, intercropping, and shade planting, will also be important (see figure 4.4).

In predominantly rainfed systems, farmers may use conservation tillage and rainwater harvesting techniques to compensate for reduced soil moisture. Rainwater harvesting techniques have been practiced in the rainfed systems of the region for centuries. Technologies range from simple in-field structures diverting water to a planting pit, through structures in the catchment that divert runoff to storage or run-on fields, to permanent terraces, or to dams (FAO 2011a). Rainwater harvesting can boost yields by two to three times more than conventional rainfed agriculture, especially when combined with improved varieties and minimum tillage methods that conserve water.

FIGURE 4.4

Mixed Crops

Source: ICBA.
Note: Shown are annual pearl millet and sorghum varieties that are salt- and drought-tolerant and provide both forage and food for animals and humans, respectively, with many of them multicut.

Increasing unpredictability of rainfall suggests that farmers may adapt by using supplementary irrigation, growing drought-tolerant or shorter-cycle crops, or lengthening the growing season. Unpredictable rainfall may translate into delayed planting, with a negative impact on yields. If planting is delayed by more than a few weeks, crops may not mature or they may fail altogether or produce reduced yields. In addition, unpredictable rainfall may lead to spells of drought during the growing season, resulting in yield losses.

Farmers can use more supplementary irrigation and drought-tolerant or shorter-cycle crops or change their cropping calendar. The impact of delayed rains could be partly offset by using supplementary irrigation early in the season, by growing shorter-cycle crops or varieties, or by taking advantage of warmer average temperatures to extend the growing season into autumn, especially where supplementary irrigation is available. Farmers may also switch to fast-growing crops, such as maize, that can be planted later. Where crops encounter stress in dry periods after planting, farmers may seek to plant drought-tolerant varieties.

Where rainfall becomes more concentrated, farmers may practice more surface irrigation and water harvesting. Concentration of rainfalls reduces water available in the soil and therefore reduces evapotranspiration and plant growth, but it also increases runoff and surface-water availability. Such concentration may thus reduce the productivity of rainfed agriculture but may increase the availability of water for irrigation and water harvesting. Farmers can seek to develop more supplementary surface irrigation and water-harvesting infrastructure to capture the increased runoff.

Climate change impacts may reduce soil fertility and increase soil erosion, but farming practices can mitigate these impacts and maintain soil health. Soils throughout the region are generally low in natural fertility, and climate change will further deplete fertility through erosion and a decline in organic matter. A wide range of soil conservation measures is available. Soil fertility can be restored through integrated soil fertility management, including manuring and crop rotations. Inclusion of nitrogen-fixing legumes in the rotation improves the nutrient balance in the soil (see figure 4.5). Farmers may seek to further diversify their mixed farming systems with crop rotation, intercropping, and agroforestry. This diversity will reduce risks and will also allow restoration of soil nutrients. Chemical fertilizers can also play a role. To conserve moisture and prevent erosion through runoff, farmers may also combine structural measures like terraces with vegetative or agronomic measures (FAO 2011b, 5.3.1).

Declining groundwater availability may cause farmers to return to traditional agricultural and water-harvesting techniques. Climate change may bring reduced groundwater recharge. However, reserves are already being run down, and adaptation options are essential whatever the climate change outcome. Improvements to the productivity and profitability of traditional rainfed and terrace cultivation and of water-harvesting schemes may offer some possibilities.

Declining water availability and unpredictable rainfall may sharpen the need for efficient groundwater and surface irrigation, especially supplementary irrigation. Farmers' likely response to increasing aridity is to improve the productivity of water use through more efficient groundwater and surface irrigation, especially supplementary irrigation. Groundwater will continue to play a key buffer role in maintaining optimal soil moisture, and this role will grow with increasing climatic variability. Groundwater productivity can be improved by conjunctive use and by adopting precision techniques such as drip irrigation, combined with agronomic measures such as *fertigation*—the application of water-soluble nutrients through irrigation—and protected greenhouse agriculture. Switches in the cropping pattern can also increase the profitability of groundwater use. The scope for improving efficiency in surface irrigation is enormous. A conventional gravity-fed small-scale irrigation system is

FIGURE 4.5

Drought-Tolerant Plants

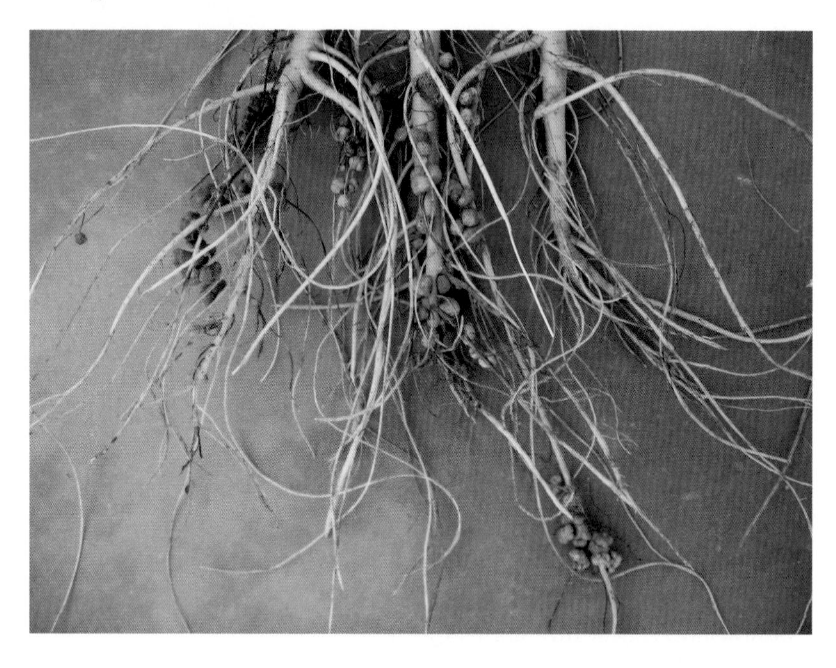

Source: ICBA.
Note: Drought-tolerant plants, such as cowpea, guar, and lablab, enrich the soil by nitrogen fixation in root nodules through symbiotic association with rhizobia.

typically 35–45 percent efficient. Reducing those losses through lined canals or piped conveyance and through drip, bubbler, and sprinkler irrigation can increase an irrigation system's efficiency up to 70–80 percent. Farmers can do much to improve in-field water efficiency through better irrigation scheduling and improvements in agronomic efficiency, for example, changes in crops, cropping calendar, and husbandry practices (FAO 2011b, 5.2.2).

Combined changes in water availability and temperature may encourage farmers to switch to more adaptable cropping patterns; to "conjunctive management" of rainfall, surface water, and groundwater; and to efficient protected agriculture and pressurized irrigation. If aridity increases, farmers can switch to more adaptable crops (figure 4.6). A first level of response could be to switch between crops with differing responses to climate change within an agroecologically homogeneous "crop group," for example, switching between fava beans and lentils within the legume crop group. The most prevalent crop switch will likely be from wheat to barley. This switch will coincide with a switch in farming systems from cereals production for human consumption to production of barley and

FIGURE 4.6

Various Date Palm Species

Source: ICBA.
Note: There are date palm species that are salt- and drought-tolerant and produce plentiful and delicious dates.

straw as part of an integrated, semi-intensive production system of sheep and goat meat to satisfy rising domestic demand. The role of barley as an adaptation strategy is enhanced by the fact that even during very dry years when grain yields are minimal, straw production or "green grazing" for flocks remains a viable production alternative.

Alternatively, farmers can switch to a different crop group. For example, where temperate fruit yields are affected by failure to meet vernalization requirements, farmers may begin to plant subtropical fruits like citrus in new zones and elevations that are less exposed to frost occurring from climate change. Water stress may best be handled by conjunctive management of rainfall and surface water and groundwater irrigation, and by the precision irrigation techniques described above.

Growing salinization will prompt changes in cropping patterns and soil and water management. Salinization will likely increase in coastal areas because of the effect of rising sea levels and seawater intrusion into aquifers. Farmers generally may also use more saline water as depleting aquifers become saltier and water scarcity drives development of more saline water sources. Farmers will seek out more salt-tolerant crops, use freshwater to blend with saline sources, and use off-season freshwater sources to leach salt residues in the soil profile (box 4.1).

BOX 4.1

Brackish Water Sources for Biosaline Agriculture

Research shows the value of brackish water sources for biosaline agriculture (Taha and Ismail 2010; Taha, Ismail, and Dakheel 2005). In research commissioned by the International Fund for Agricultural Development, the International Center for Biosaline Agriculture analyzed the potential use of saline and brackish water resources in Arab countries for animal feed production. For example, sorghum, shown here, has been screened and evaluated for drought and salt tolerance and is being used as forage and food from central Asia to North Africa.

The study showed that sufficient saline and brackish water resources exist to irrigate up to 330,000 hectares. These findings are particularly relevant to the Arab region, given the increasing salinization of groundwater.

Source: ICBA.

The likely farmer responses discussed above concern the application of existing technology in order to adapt farming systems to climate change effects. Empirical evidence shows that these adaptations are already under way (see box 4.2).

BOX 4.2

Yemeni Farmers Grasp the Challenge of Adapting to Climate Change

A study found that most Yemeni farmers (77 percent) knew about climate change and thought that it could affect their farms (64 percent). Most farmers thought that climate change was manifested through increases in average temperature, variability and irregularity of rainfall, and higher frequency of extreme events, such as droughts. These views concur with those of climate scientists. Many farmers (37 percent) said that they had changed their agricultural practices in the past to cope with adverse climate conditions, and more than half (54 percent) thought they should start now to adjust their farm activity to cope with possible future adverse climate conditions. The options they proposed consisted of a mixture of traditional and modern farming practices, such as changing cultivation practices, rehabilitating terraces and spate irrigation systems, switching to higher-value crops, and increasing the scale of farm operations. The study concludes: "Climate change may be a fruitful source of innovation in that it inspires old and new remedies, more respectful of the environment, more preoccupied with recovering traditional land management techniques and plant varieties, and in general more attentive to the needs and the capabilities of rural communities."

Source: Scandizzo and Paolantonio 2010.

Institutional and Technological Developments Should Advance Together

Institutional adaptation needs to accompany technological innovation as they are mutually supportive. Table 4.5 offers some examples from the different farming systems of how they might go hand in hand in developing adaptation strategies. Some institutional adaptation can occur spontaneously at the local level through, for example, farmers' organizations for better catchment management, collaborative approaches to groundwater management, and community management of pasture. Some adaptation strategies require partnerships with government, decentralization of irrigation management to water users' associations, water demand management, payments for ecosystem ser-

TABLE 4.5

Adaptive Capacity to Climate Change Impact in the Main Farming Systems of the Arab Region

Farming system	Selected examples of likely farmer adaptations available (access to technology, institutional adaptive capacity, and a further research agenda)
Irrigated	**Technology:** (a) increased (or substituted) supply through treated wastewater, saline, and drainage water reuse (and possibly further diversions); (b) improved water-use efficiency at both system level and field level; (c) amended soil fertility by integrated nutrient management; (d) improved water management to reduce salinity; and (e) agronomic and postharvest improvements for more income per drop **Institutional adaptation measures:** (a) decentralization of irrigation and drainage management to water users' associations and demand management measures (including the use of financial measures) and (b) participatory approaches to groundwater management **Further research agenda:** (a) more sustainable land and water management techniques and (b) integrated nutrient management
Highland mixed	**Technology:** (a) watershed management, (b) conservation tillage, (c) better integration of crops and livestock, and (d) agronomic and postharvest improvements **Institutional adaptation measures:** (a) participatory approaches and equitable sharing of benefits; (b) compensation for externalities and downstream benefits (payment for ecosystem services); and (c) reduction in overgrazing through more equitable regulation and control of common grazing resources—with participation, plus investment in water points, and elimination of subsidies on animal feed
Rainfed mixed	**Technology:** (a) improved management of water, (b) terrace restoration and soil contouring, and (c) agronomic and postharvest improvements **Institutional adaptation measures:** (a) land consolidation, (b) community-based watershed management, and (c) support mechanisms like payments for ecosystem services **Further research agenda:** (a) technologies on crop-livestock integration, and (b) risk management
Dryland mixed	**Technology:** (a) windbreaks, water harvesting, water management and conservation, (b) zero tillage, (c) together with agronomic and postharvest improvements **Institutional adaptation measures:** (a) communal land and water management, (b) participatory research and development, and (c) financial support mechanisms, like payments for ecosystem services **Further research agenda:** (a) varieties with shorter growing period, drought resistance, and improved grain and straw quality; (b) new varieties and techniques, such as intercropping; and (c) systems research on crop-livestock interaction and resource conservation, with a focus on risk reduction and sustainability
Pastoral	**Technology:** (a) intensified livestock productivity and (b) diversification **Institutional adaptation measures:** (a) communal grazing management and (b) financial support mechanisms, like payments for ecosystem services

Sources: Christensen 2007; Dixon and Gulliver 2001, 87–91.

vices to compensate for externalities, or land consolidation. Box 4.3 illustrates the combination of technology, institutional development, and community outreach in Abu Dhabi that results in a more water-sustainable agricultural system.

BOX 4.3

Abu Dhabi Government Has Already Taken Steps Toward More Water-Sustainable Agricultural Production

In Abu Dhabi, the agriculture sector, which uses more than 70 percent of the water resources, is being remodeled. Subsidies supporting traditional agricultural production are being phased out; greenhouse-based agriculture is being encouraged; new well drilling is being limited; and farmers are being redirected to grow crops with less water demand. In October 2010, the Abu Dhabi Food Control Authority withdrew subsidies for growing Rhodes grass, a forage crop with high water demand, in the western agricultural region. Alternate forage crops are being introduced and adapted to meet local demands. Extension support has been set up through the Farmers' Services Centre, which is demonstrating model farms based on salt- and drought-tolerant species and efficient irrigation methods to ensure sustainable agriculture. Farmers' markets have been established at various sites in the urban centers of Abu Dhabi and Dubai, and value-added products such as weekly vegetable boxes are made available for sale.

Source: Authors' compilation.

Adopting Strategies to Support Food Security

For many countries, food security policies have promoted both the development of agricultural production and import strategies. The drivers are as much political as they are economic, with policies that encourage domestic food production by tilting the incentive structure in favor of this. This causes economic losses to the nation and possibly increases food insecurity of producing households by reducing their potential incomes. Although no Arab country maintains food self-sufficiency today, decision makers have the perception that a basic level of internal food security is required to limit the shocks of global food price volatility. Such policies have various trade-offs, particularly with respect to the use of scarce water resources. Considering the region's environmental conditions, the impact of food shocks cannot be completely isolated. Box 4.4 illustrates the findings of a recent study of such trade-offs in Morocco.

Following the global food price shocks of 2008, policy analysis for the region has emphasized the need for the following:

- Improving data and strengthening the capacity for evidence-based decision making
- Protecting vulnerable households nationwide by strengthening safety nets (via cash transfer programs, labor-intensive employment programs, and health and nutrition interventions, with a particular focus on women and children), education, and family planning
- Protecting rural households and contributing to the national food supply and price stability by formulating a rural livelihoods strategy focused principally on enhancing agricultural and off-farm incomes and production through investment in research and development, rural infrastructure, and market development; and
- Reducing exposure to market supply and price risks by
    1. *Improving supply-chain efficiency* by improving trade in agricultural commodities through global, regional, and bilateral agreements and promoting efficient domestic food distribution and retailing
    2. *Introducing cost-effective risk management instruments calibrated to the risk assessed*, for example, by (a) establishing food reserves or buffer

BOX 4.4

Trade-Offs between Self-Sufficiency and Food Security in Morocco

A recent study shows that Morocco could, in theory, achieve 85 percent self-sufficiency in cereals at current yield levels and that full self-sufficiency could be achieved if yields were to rise by 40 percent. However, this self-sufficiency would come at a high cost—about US$10 billion between 2008 and 2022—through revenue forfeited by not producing higher-value crops. If Morocco produced instead high-value crops, the US$10 billion could be used to purchase a much greater quantity of imported cereals. In addition, production of higher-value crops would create far more agricultural employment for landless laborers than cereals production would.

Sources: Lampietti et al. 2011; Magnan et al. 2011.

stocks to help stabilize prices and smooth consumption fluctuations, (b) forward contracting, (c) employing financial hedging products, and (d) building in risk management provisions to bilateral and multilateral agreements

3. *Promoting and supporting regional and global responses to protect against price volatility*, including (a) pressure to curtail and reform biofuel policies and subsidies, (b) establishing regional or global grain reserves for rapid market intervention, and (c) setting up an international working group to monitor trade and trigger action (Breisinger et al. 2010; Lampietti et al. 2011; McDonnell and Ismail 2011).

Food security strategies have to be adapted to the nature of the risks. Table 4.6 provides an outline for risk assessment under climate change. Annex 4F provides a checklist of possible options.

TABLE 4.6

Food Security Strategy Options under Climate Change

Country characteristics	Examples	Main food security risks	Possible impact of climate change	Principal strategic responses
Poorer countries with vulnerable populations dependent on farming	• Mauritania • Somalia • Sudan • Yemen, Rep.	• Rural malnutrition and famine	• Intensified risk	• Safety nets, education, family planning • Rural livelihoods strategy focused on agricultural productivity and risk management • Bilateral and multilateral agreements for food aid
Middle-income countries that want moderate food prices for their citizens and to maintain a viable rural sector	• Maghreb and Mashreq countries	• Price spikes • Difficult access and affordability for poorer rural areas and households	• Long-term increase in global food prices • Reduced production and incomes for rural people	• Safety nets, education, family planning • Rural livelihoods strategy focused on agricultural productivity and risk management, off-farm, and so on • Improving supply-chain efficiency
Better-off countries requiring assurance of food supplies	• Oil-exporting countries of the Gulf region	• Geopolitical risk	• Limited	• Improving supply-chain efficiency • Risk management instruments

Source: Authors' compilation.

Adopting Strategies to Support Rural Livelihoods

Agricultural income is by far the largest source of overall rural household income. Rural household-level livelihood strategies will need to adapt to climate impact on agriculture and agricultural income. Table 4.7 lists some options at the household level, whereas annex 4E gives examples of national-level possibilities, which are discussed later in this section.

Farm incomes

Where water, working capital, and markets are available, farmers will likely continue the trend away from production of staples toward higher-value crops like fruits, vegetables, and flowers, which give the highest return to the scarcest factor—water. Fortunately, markets for these products expand with urbanization, demographic growth, and rising incomes. However, rising preference for meat and dairy products will pose a challenge since they are water intensive.

TABLE 4.7

Strategies and Actions for Improved Rural Livelihoods at the Household Level

Measure	Action	Potential result
Increase returns from agriculture	• Develop access to markets and improve marketing • Grow high-value crops • Develop downstream value added, for example, through food processing and packaging	• Maximization of income generated for food produced • Increased income from agriculture • Further income for rural livelihoods
Increase additional income sources	• Develop household industries • Develop forestry resources	• Additional income from manufacturing of craft items or other small-scale industry • Increased income and restoration or preservation of land
Increase skills and education levels	• Support attendance at schools and other learning opportunities • Increase use of information and communication technology	• Increased possibilities of secure income from other employment opportunities • Increased access to current climate/price information and increased skills levels

Sources: IFAD 2008, 2010b; McDonnell and Ismail 2011.

As markets develop, commercial farmers should be able to mange risks and to maintain—or even increase—their incomes, but smaller farmers and food deficit households may be affected negatively. The challenges for smaller farmers will be to become commercial farmers and to link into market circuits. The many millions of rural households with a food deficit may be affected negatively if food prices rise (see below), and consequently they will need coping strategies.

Off-farm rural activities

Clearly, not all farmers will be able to maintain their income from farming, and many nonfarm rural households will also be vulnerable, so diversification of livelihoods will be an important adaptation option. Developing new sources of income that are less climate dependent can improve household resilience and reduce risk. Such sources might include adding value to agricultural production by processing and marketing, as well as new household industries, such as crafts or tourism, or light manufacturing. These options may encounter barriers: (a) low rural demand for products and services, (b) difficulty of access to markets and information, (c) poor infrastructure base, and (d) inadequate access to finance. Public policies need to lift these barriers and create an enabling environment, and education and skills development could help support off-farm employment (Christensen 2007, 50).

Exit from farming and rural to urban migration

In areas where rural livelihood options are inadequate, people will likely migrate, particularly during periods of climate stress like droughts and floods. This probability will be true for farm laborers, especially those who are economic migrants. Households will need to adapt roles and responsibilities, strengthen local social networks in both the receiving and exiting communities, and invest in skills and education to prepare for future environmental conditions. Public policies and programs will be needed to ease the transition.

Effective Adaptation Requires Policy Reforms and Support Programs

An important consideration is that farmers are appropriately supported and that farmer reactions contribute to—and do not undermine—larger national objectives of sustainable, equitable, and efficient development. Climate change serves to highlight a number of opportunities for better

FIGURE 4.7

A Conceptual Framework for Considering Integrated Adaptation Strategies for Addressing Agriculture, Water, Food Security, Rural Livelihoods, Gender, and Environment Issues

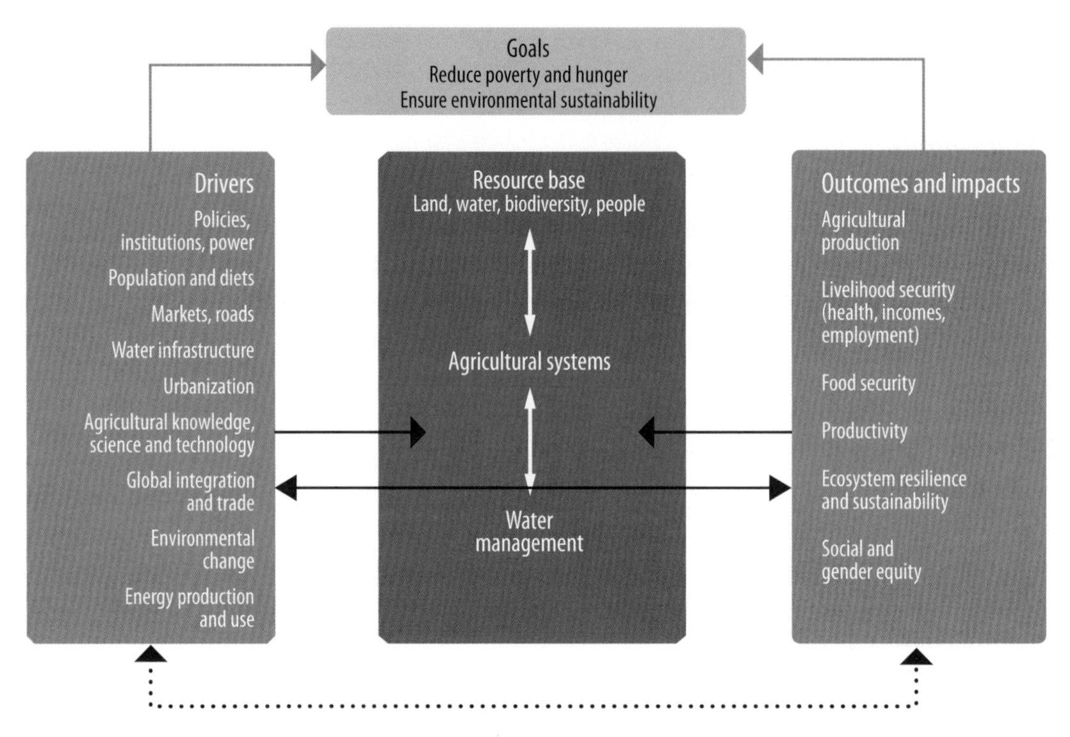

Source: Molden 2007.

integration between farmers, systems, and national levels of concern and action (see figure 4.7). Strategies can combine many different interventions ranging from on-ground local initiatives to advances in national infrastructure and developments in governance and institutional arrangements.

Annex 4E provides a summary checklist of actions that governments may consider. Strategies should be developed with both technical and institutional measures (Clements et al. 2011; FAO 2010). Key areas of government support could include the following:

- Ensuring that the policy and incentive framework facilitates adaptation
- Providing information on likely climate change effects and impacts
- Ensuring that adaptive technology is available

- Ensuring that farmers have access to the technical and financial resources necessary to change
- Promoting appropriate institutional change, such as participatory local planning, farmer cooperatives, and farmer field schools
- Integrating changes from the local level into higher-level planning, for example, for water resources planning and allocation at the basin scale, for development of irrigation, and so forth
- Ensuring that externalities—particularly environmental impacts and upstream and downstream impacts—are managed equitably and sustainably
- Conducting research to ensure that optimal productivity and risk management measures are available
- Facilitating development of appropriate forms of crop and weather insurance
- Improving meteorological capacity and services geared toward farmers' needs

Aligning agriculture and water policies

The agricultural sector affects and is affected by policies in other areas. Future policy and government program development will need to ensure that adaptation strategies are aligned to maximize benefits not only in food production but also in ecosystems services and rural incomes. A key area will be between water and agriculture policies (see figure 4.7). The frequent misalignment among government objectives in these two areas needs to be addressed.

In Arab countries, water is widely overallocated, particularly to agriculture, and competition and conflict between agricultural uses and higher-value municipal and industrial uses, already common, will intensify under climate change (see chapter 3 for greater analysis and potential adaptation strategies). Mechanisms may need to be devised for transferring water from agriculture. A need exists to review allocative efficiency at the basin scale and to evolve the regulatory and market-based tools that will allow reallocation of water in a way that is adaptable under climate change (World Bank 2006b).

By adjusting the incentive structure to recognize the scarcity of water and the need to use it to obtain the highest returns, demand management is an imperative (World Bank 2006b, 13). In effect, increasing supplies will also be essential, either through increased water productivity or use of nonconventional water to satisfy the increasing demands for irrigation water. Particularly in water-scarce Arab countries, investment in reuse of treated wastewater and drainage water can offset water scarcity. Institutional arrangements, including a legal and regulatory framework, are needed for allocation and safe use, and protecting both humans and the environment. Programs have to be assessed at the level of overall basin efficiency and socioeconomic benefit. Egypt, Tunisia, and many of the Gulf countries are already managing successful programs (FAO 2011b; McDonnell and Ismail 2011; World Bank 2006a, 177).

Finance

The lack of access to finance by governments has limited the implementation of adaptation strategies, although implementation has been improving with the establishment of a number of international funds to support programs. A greater awareness is needed among Arab countries of how these funds may be accessed and used.

Finance is important for adaptation strategies by helping households widen their economic opportunities, increase their asset base, and diminish vulnerability to shocks. Such help will require the development of rural financial services, which can enhance social protection, particularly microcredit and savings and loans approaches (McDonnell and Ismail 2011; Saab 2009). Considerable success has been made with investments in communal infrastructure supported by social funds, such as those in Egypt and the Republic of Yemen. INDH (National Initiative for Human Development) in Morocco is a successful homegrown version of the same approach that is supporting agricultural projects, as well as community investments. Extension of these activities to communal investment in natural resource management, watersheds, irrigation development, and so forth, could support adaptation (Christensen 2007, 57).

The trade in environmental services through payment for ecosystem services mechanisms has attracted interest and financing both within countries and from international investors. Payment for ecosystem services could be a key instrument for supporting climate change adaptation in the highland mixed and rainfed mixed systems (compensation for externalities and downstream benefits of watershed management). It could also support conservation investments, like terracing, biodiversity conservation, or the maintenance of traditional agricultural heritage in a number of systems, including dryland mixed and pastoral systems (FAO 2011b).

One important innovation is the development of carbon markets with Dubai acting as a hub for a number of enterprises in this area. Another possibility is to work within the Clean Development Mechanism of the Kyoto protocol. That mechanism has been successful in financing some types of agricultural projects, such as methane capture or use of agricultural by-products as an energy source. Morocco and the United Arab Emirates already have agricultural projects in the agriculture sector under this system (Larson, Dinar, and Frisbie 2011).

Research, extension, and information: learning from and with farmers

Farmers are already innovative and proactive in adapting to climate constraints, and an understanding of their behavior will help match services to needs and assist in broader adoption and dissemination of best practices. Traditional agricultural systems, usually characterized by a high degree of complexity and plant biodiversity, can provide valuable knowledge for adaptation. Much can be learned from the very specific use of environmental knowledge and natural resources in these systems (FAO 2010).

Their efforts should be supported by innovation and experimentation in the many international and national centers of excellence in agriculture and water research. Organizations such as the Arab Center for the Study of Arid Zones and Drylands, the International Center for Agricultural Research in the Dry Areas, and the International Center for Biosaline Agriculture each bring specialist knowledge that is based on field and laboratory trials in the Arab region.

Governments and farmers need supporting data to develop adaptation strategies, programs, and policies. Satellite-generated data allow the measurement of changes in land cover, the forecast and monitoring of crop yields, crop stress, production, stream flows, groundwater levels, soil moisture and water storage, and pollution plumes in water or in the soil. These data can be combined inputs into hydrological and crop modeling to anticipate likely impacts of climate change. The MAWRED (Modeling and Monitoring Agriculture and Water Resources Development) program at the International Center for Biosaline Agriculture and the related WISP (Water Information Systems Platform) projects, at key centers in Egypt, Jordan, Lebanon, Morocco, and Tunisia combine state-of-the-art water modeling with remotely sensed data to help answer key regional and national policy questions. These programs are examples of local centers working with international organizations such as the U.S. National Aeronautics and Space Administration (through funding from the U.S. Agency for International Development and the World Bank's Global Environment Facility program) to bring innovative information generation to the region.

Adapting institutions: decentralization and community collaboration

Many adaptations to climate change will involve communal approaches to natural resource management. Institutional changes across the region have, in recent years, brought increased emphasis on "subsidiarity"—decentralizing decision making to the lowest possible level, encouraging participation of stakeholders, and fostering local-level community or interest groups as primary agents of development and as counterparts to public services. Many adaptations will require this kind of local collaboration, including the following (Dixon and Gulliver 2001, 121):

- Revival and adaptation of older systems of rotational grazing and land management that involve all stakeholders in planning, managing, and monitoring
- Watershed-based (rather than individual) soil and water management systems
- Development of collective groundwater management
- Water-demand management through communal regulation
- Conservation of ecosystems, environmental services, and sociocultural heritage

The challenge will be for public services to organize themselves cross-sectorally, so that support is forthcoming not only for agriculture but also for water management, marketing, downstream processing, off-farm activities, the environment and so forth, as well as human development aspects that are key to sustaining livelihoods. Bottom-up community organizations gaining support from top-down "convergent" public services and community development funds make up a key set of capabilities that can achieve a substantial impact on rural livelihoods. New business lines, such as "green agriculture," ecotourism, and landscape and cultural heritage management, require these public-community partnership approaches. It will be important to factor in participation of both women and youth and to promote the engagement of civil society organizations at all levels from local to national (Christensen 2007, 61).

Ways Forward to Reduce Climate Stress

Climate change in the Arab region is likely to act as a "threat multiplier" rather than a "game changer." It will increase many of the complex challenges that already exist, such as water scarcity, growing populations, environmental degradation, and unemployment. Many governments have taken steps toward developing climate change adaptation strategies for the key areas of agricultural production, rural livelihoods, and food security. Often, strategies for climate change adaptation reinforce or complement existing programs and projects.

Adaptation plans will differ among countries, reflecting the region's varying environmental, socioeconomic, and political conditions. A range of policy measures (regulatory, financial, and information instruments) will be needed within strategic frameworks to address both the institutional and economic barriers to adaptation, as well as the changing natural environment. Transformation of the agriculture sector under adaptation strategies can yield important economic results, as well as enhance food security. When allied with economic diversification programs, the development returns might be substantial. There are thus important opportunities to be grasped as well as challenges to be faced in developing adaptation strategies.

Given the current political complexities in the region, responding to the impacts of climate change may seem daunting. But if future social and economic risks are to be managed, the agricultural production systems of Arab countries must incorporate climate resilience. Communities in this region have survived changing climates in the past, but current population and other pressures are so great that national governments need to be actively involved in adaptation plans so that people can sustain their existence in rural areas and so that food is available to all. If adaptation to climate change can be properly managed, the agriculture sector and the rural economy not only could cope with many of the risks of climate change but also could strengthen the structure of the sector through more sustainable land and water management practices and higher productivity agriculture (McDonnell and Ismail 2011). See annex 4C for measures and potential results in farm-level adaptation to climate change, and annex 4D for strategies and actions for agriculture and water-related adaptation at the national level.

Summary of Key Messages

Agricultural production

- Farming systems in general will likely be negatively affected by increasing aridity, greater unpredictability, and growing water scarcity.
- Some farming systems may benefit from warmer temperatures that extend the growing season or increase productivity of winter crops.

- Until midcentury, favorable terms of trade and available productivity improvements and adaptation strategies could allow continued increases in output in most systems if the institutional and incentive frameworks are favorable. Thereafter, negative climate change effects may override these potentials and production may decline.
- Marginal farming systems in dryland and pastoral systems are particularly vulnerable, and some lands may shift to less intensive production or go out of production altogether.
- A wide range of both technological and institutional adaptation measures is available for most farming systems. Farmers are already applying many of these measures.
- The challenge will be to ensure that knowledge-based approaches combine governments' ability to guide adaptation through the incentive structure; support research; and provide technical and institutional support for the knowledge, skills, and adaptive capacity that farmers need.
- Responses to climate change should generally coincide with best-practice agendas on sustainable land and water management, as well as with institutional decentralization and empowerment of local stakeholders.
- Because aridity will increase and water is the binding constraint to agriculture in the region, governments will need to evaluate trade-offs between supporting climate change responses in agriculture and preparing parts of the rural economy for transition away from agriculture.
- Key areas for government action include (a) developing climate modeling and adaptation strategies, emphasizing robust "no regret" options, especially those that combine adaptation and mitigation; (b) preparing adaptation strategies; (c) adapting water management and agricultural services; (d) decentralizing and strengthening local participatory governance, land tenure, and environmental stewardship; (e) advancing trade liberalization and market development; (f) strengthening cooperation on climate change; and (g) working out options for financing adaptation, especially through global funds.

Food security

- Arab countries with vulnerable populations dependent on farming can improve food security by a rural livelihoods strategy principally focused on enhancing agricultural and off-farm incomes and production. This approach will help rural household food security and will also contribute to the national food supply and price stability.

- In evaluating the trade-offs between promoting domestic food production and promoting production that gives the highest return to water, governments will need to take account of the realities of food security at both the household and national level. At the household level, farmers will need to maximize income to ensure household-level food security and escape from poverty.
- All Arab countries can work to improve supply-chain efficiency and to cooperate with global efforts to stabilize food markets and prices. Poorer countries may also seek bilateral and multilateral agreements for food aid.
- All Arab countries can strengthen safety nets (cash transfers, labor-intensive employment programs, and health and nutrition interventions) and education.
- Better-off countries requiring assurance of food supplies can reduce exposure to market supply and price risks not only by improving supply-chain efficiency but also by introducing cost-effective risk management instruments calibrated to the risk assessed.

Rural livelihoods

- Existing rural community resilience will be impaired by climate change impacts on natural resources and on financial and social capital.
- Adaptation strategies at the household level need to develop income opportunities less dependent on natural resources, alongside programs for agricultural intensification.
- Governments have a major role to play in developing and implementing rural livelihood adaptation strategies, including (a) establishing a policy and incentive framework favorable to diversified rural enterprises and livelihoods; (b) supporting integrated rural development programs, especially for infrastructure, health, and education; (c) decentralizing public services, empowering local government, and empowering community-based and nongovernmental organizations; (d) facilitating the inevitable rural-urban migration; (e) ensuring that social safety nets are in place in rural areas; and (f) developing disaster reduction and risk management systems.

ANNEX 4A

Characteristics of the Arab Region's Farming Systems

Farming system	Countries where found	Characteristics
Irrigated: (1) *large scale*	Arab Republic of Egypt, Iraq, Morocco, Syrian Arab Republic	• 8.1 million ha of irrigated cropland • Agricultural population: 16 million • High population density and small farm size • Largely run-off from the river, but also *karaz* or *qanat* systems drawing on underground water • Large center pivot schemes dependent on pumped groundwater • Generally, moderate water-use efficiency • Cash crops, including cotton, sugar beet, vegetables, fodder • Cropping intensities: 120–160%
Irrigated: (2) *small scale*	Throughout the region, including the Arab Republic of Egypt, the Maghreb, Oman, the Syrian Arab Republic, and the Republic of Yemen. Often complementary to other systems, such as highland mixed (see below)	• Dependent on streams and springs, on oases, and on spate flows and flood recession • Supplementary irrigation from groundwater is practiced
Irrigated: (3) *groundwater*	Throughout the region	• Individual groundwater irrigation developed rapidly in recent years, often associated with protected agriculture (such as greenhouses) and with pressurized irrigation (drip, sprinkler)
Highland mixed	Morocco (Atlas), Republic of Yemen	• Cultivated area: 22 million ha (5 million ha of which are irrigated) • Agricultural population: 27 million • Annual rainfall: 200-800 mm • Dominated by rainfed cereal and legume crops with tree crops (olives, fruits, khat, coffee) on terraces • In Morocco, livestock transhumance also important
Rainfed mixed	Algeria (coastal plain), Iraq (Kurdistan), Lebanon, Morocco (Rif), Syrian Arab Republic (coastal plain)	• Cultivated area: 14 million ha • Agricultural population: approximately 16 million • 8 million cattle • Rainfall: 300–1,000 mm • Approximately 600,000 ha now irrigated by tube well, increasing yields and allowing summer cropping • Main winter crops: wheat, barley, chickpeas, lentils, and fodder crops • Tree crops (olives, fruit, nuts), melons, and grapes grown in areas with > 600 mm of rainfall • Some protected irrigation of vegetables and flowers for export
Dryland mixed	Algeria (inland plains), Libya, Morocco (Atlantic coastal plain), northern Iraq, West Bank and Gaza, Syrian Arab Republic, Tunisia	• Dry, subhumid zones • Annual rainfall: 150–300 mm • Cultivated area: 17 million ha • Agricultural population: 13 million • Approximately 3 million ha receive some level of irrigation • Main crops: barley and wheat, with fallow years • 6 million cattle; innumerable sheep and goats

Farming system	Countries where found	Characteristics
Pastoral	Desert margins of Algeria, Arab Republic of Egypt, Iraq, Jordan, Libya, Morocco, Oman, Syrian Arab Republic, Republic of Yemen	• Extensive, on 250 million ha of semiarid steppe • Low population densities • Rainfall: < 150 mm • Approximately 2.9 million ha of irrigated cropland scattered through the system • Agricultural population: 8 million • 60 million sheep and goats • 3 million cattle • Seasonal migration; strong links to peri-urban feedlots
Agropastoral (millet, sorghum)	Somalia, South Sudan, Sudan	• Sorghum, millet, pulses, cattle, sheep, goats
Cereal/root crop mixed	South Sudan, Sudan	• Maize, sorghum, millet, root crops, cattle
Arid zones	Deserts of Algeria, Arab Republic of Egypt, Gulf states, Iraq, Jordan, Libya, Morocco, Syrian Arab Republic, Tunisia	• Extensive area • Agricultural population: approximately 4 million, largely in oases and irrigation schemes (notably in the Maghreb and Libya) • 2.7 million cattle, plus camels, sheep, and goats

Sources: Christensen 2007; Dixon and Gulliver 2001, 87-91.

Note: Percentages of the region's land area and agricultural population are not calculated for the agropastoral and cereals/root crop mixed farming systems. Ha = hectare.

ANNEX 4B

Predictive Impacts of Climate Change on Agriculture and Livestock

Climate change effects	Impacts	Areas affected in Arab countries
Agroclimatic conditions		
• Increased temperature during growing season	• Reduction in yield resulting from shorter duration of various physiological development stages • Possible benefit in some areas from longer growing season	• Reductions possible across all areas • Possible benefits in higher-elevation agriculture where water is available
• Increase in days above threshold temperatures and key plant development times	• Nonlinear decline in yields as above-threshold temperatures are met (thresholds: wheat 26°C, maize 30°C, rice 34°C) • For some crops (such as olives in Morocco), higher annual low temperatures will be problematic since some plants need cold in winter	• Across all areas
• Increase in length of dry season	• Reduction in (a) yields, (b) number of crops per year, (c) length of time that rangelands can be grazed, and (d) number of animals supported	• Large parts of Iraq and the Syrian Arab Republic by the end of the 21st century
• Unpredictable rainfall patterns	• Increase in fires leading to loss of grazing land and forests	
Hydrological conditions		
• Decrease in groundwater recharge	• Less water available for irrigation	• Most countries • Decrease of more than 70% by 2050 along the southern Mediterranean
• Decrease in surface flows	• Less water available for irrigation	• River-based systems (such as the Arab Republic of Egypt, Iraq, and Jordan)
• Increase in heavy rainfall	• Floods and greater soil erosion removing natural resource base, where land-use practices, such as bunds or terraces, are not in place	• Sloped areas across region
Soil conditions		
• Decrease in soil moisture	• Reduced plant growth and yield, with plant foliage often irreparably damaged • Increased need for supplemental or full irrigation • Restricted grazing	• Rainfed agriculture areas • Irrigated and mixed rainfed areas • Pasture and rangelands
• Decrease in soil carbon	• Reduced nutrient levels and soil water holding capacity, leading to decline in yields, further reducing soil carbon content	• Carbon content in the Arab region tends already to be low
Atmospheric conditions		
• Increase in carbon dioxide atmospheric levels	• Increased photosynthesis (such as, for C3 plants like wheat) could increase yields • Reduced plant uptake of nitrogen may cause lower nutritional value of crops	• Across all areas, increased yield potential constrained by water stress and soil nutrient levels

Climate change effects	Impacts	Areas affected in Arab countries
Sea-level rise		
• Inundation of agricultural land	• Loss of productive land in low-lying areas	• Bahrain, Comoros, Kuwait, Nile Delta, Qatar, and United Arab Emirates
• Salinization of soils	• Decrease of productive land or reduction in yields	• Coastal zones, particularly in the Gulf countries
• Salinization of aquifers	• Decrease in water available for irrigation or reduction in yields	• Coastal zones, particularly in the Gulf countries
Plant and animal health conditions		
• Pests and diseases	• Different pest and disease vectors will affect growth and yields in both plants and animals	• All areas; Egypt already seeing bluetongue disease and Rift Valley fever
• Humans	• Increased disease vectors and heat stress will affect labor activities	• Across all areas

Sources: Bengtsson, Hodges, and Roeckner 2006; Black, Brayshaw, and Rambeau 2010; Boko et al. 2007; Chenoweth et al. 2011; Christensen 2007; Dasgupta et al. 2007; Dixon and Gulliver 2001; Döll and Flörke 2005; Easterling et al. 2007; Elasha 2010; Evans 2009, 2010; Giorgi 2006; Jin, Kitoh, and Alpert 2010; Kitoh, Yatagai, and Alpert 2008; Lobell et al. 2011; McDonnell and Ismail 2011, based on Ghassemi, Jakeman, and Nix 2005; Milly, Dunne, and Vecchia 2005; UNDP 2009; World Bank 2006b, 2007, 2011; WFP 2010.

ANNEX 4C

Measures and Potential Results in Farm-Level Adaptation to Climate Change

Measure	Action	Potential results
Choice of crops	• Adopt drought- and heat-resistant varieties	• Reduction of risks of yield loss and lower irrigation requirements
	• Adopt salt-tolerant varieties	• Increased production on salinized land or with salinized water
	• Adopt pest-resistant varieties	• Reduction of crop loss when climate conditions are favorable for increased weeds and pests
	• Adapt maturation time to new conditions	• Improved yields by maturation in shortened or lengthened growing season
	• Use altered mix of crops	• Risk-hedging leading to reduction of overall production variability
	• Adopt agroforestry	• Increase in overall production using trees with crops/shrubs that are drought or salt tolerant for both enhancing shade and water retention for other crops as well as providing trees to harvest
Choice of livestock	• Raise more drought-hardy animals (camels, goats)	• Fewer animal and production losses
	• Increase use of pest-resistant breeds	• Fewer animal and production losses
	• Improve livestock breeding	• Development of animal species that are more adapted to new environmental conditions
Tillage and time of operations	• Change planting dates	• Adaptation to altered precipitation and temperature patterns
	• Increase protected agriculture	• Reduction in water lost through evapotranspiration and wind
	• Increase land leveling	• Better water dispersion and filtration, and reduction in erosion
	• Introduce deep plowing	• Break up hard pan to increase infiltration
	• Change tillage and mulching practices	• Increases retention of moisture and organic matter and reduces greenhouse gas emissions
	• Change mulching	• Increases moisture retention and minimizes weeds
	• Switch from spring to winter crops	• Fewer losses from lengthened summer dry periods
	• Change fallow period	• Increases retention of moisture and organic matter
	• Increase carbon additive input	• Compensate for loss of carbon through increased decomposition rates of soil carbon
Crop husbandry	• Alter row and plant spacing	• Extension of roots to groundwater
	• Expand intercropping	• Increase shading to reduce moisture losses, increase overall land productivity and crop diversity
Livestock husbandry	• Increase supplementary feeding, feed blocks	• Less impact from lower-nutrient fodder
	• Increase disease management	• Fewer losses from new disease vectors and pests
Irrigation and water harvesting	• Use supplemental irrigation in dry farming areas	• Fewer losses in drier conditions
	• Improve irrigation technology, management	• Lower water use; less moisture stress and higher yields for crops
	• Irrigate with marginal water	• Reduction in the overexploitation of freshwater; more water available for crops
	• Monitor water use, reduce overwatering	• Efficient use of water for optimizing crop yield
	• Introduce rainwater harvesting	• Increase in available water
Soil fertility	• Adapt fertilizer to specific conditions	• Maximization of yields and minimization of input cost
	• Adapt timing of fertilizer application	• Increase in mineral nutrients available to crops by applying when needed
Fisheries	• Improve fisheries management	• Conserve stocks and reduce environmental stresses

Sources: Cai and Sharma 2010; Cai et al. 2011; Haddad et al. 2011; McDonnell and Ismail 2011, based on AOAD 2007; Molden et al. 2010; Sulser et al. 2011, table 4.4.

ANNEX 4D

Strategies and Actions for Agriculture and Water-Related Adaptation at the National Level

Measure	Action	Potential result
Financial instruments	• Improve targeted subsidies	• Better support for use of new crop and livestock species, conservation fishing practices, agroforestry, agricultural infrastructure, and land conservation measures
	• Increase investment	• Better equipment and infrastructure to support increased agricultural and water productivity
	• Obtain funding, such as through payments for ecosystem services or from international climate financing programs	• Support for sustainable management practices of both land and marine environments
	• Increase microfinance	• Support for new equipment, land, seeds, and other requirements needed for adaptation
	• Improve social safety nets	• Increased support for the most vulnerable to protect their asset bases during extreme events
	• Introduce climate-related insurance	• Reduction (over a limited period) of financial risks from climate-induced production failures and short-term losses from moves to new crops, livestock, or irrigation practices
Information services	• Increase funding and support of research	• Increased knowledge about crop and agricultural management possibilities; development of on-farm research projects to gain local support; better access to seeds and new breeds; increased coordination between Arab, regional, and international climate change and agricultural research centers
	• Improve farmer extension services	• Increased farm-level knowledge and access to new practices and possibilities for developing climate-resistant agriculture
	• Improve drought monitoring	• Early warning to producers and supporting-government institutions of potentially difficult conditions
	• Improve disease and pest monitoring	• Early warning to producers and supporting government institutions of potential yield losses
Institutional and legal development	• Establish cooperatives for marketing and supplies	• Spreading of risk; maximization of market opportunities
	• Further develop land tenure	• Improved access to subsidies and credit and better investment opportunities
	• Regulate environments, such as land, water bodies, and marine areas	• Reduced degradation of natural resource base from climate change and human actions

Sources: Alwang and Norton 2011; Knutson and Bazza 2008; McDonnell and Ismail 2011, based on AOAD 2007; Sulser et al. 2011; UN and LAS 2010; World Bank 2009.

ANNEX 4E

Strategies and Actions for Improving Rural Livelihoods and Adaptation at the National Level

Measure	Action	Potential result
Financial instruments	• Target subsidies	• Support for economic diversification; increased investment incentive to private sector to establish businesses in rural areas
	• Access various payment schemes, such as for ecosystems services and from international climate funds	• Increased income through supporting activities, such as ecosystems services or climate mitigation/adaptation measures to improve living standards and environment
	• Increase investment	• Better education, health, and transport infrastructure and programs to support economic development
	• Improve safety nets	• Increased support for the most vulnerable during periods of disaster and enhanced risk
Information services	• Support education and skills development initiatives	• Increased possibilities for employment away from agriculture
	• Support improvements in health provisions	• Protection of rural populations from increased health risks to ensure ability to earn incomes
	• Develop information and communication technologies	• Increased access for rural communities to information that can improve skills/education and their preparedness for current climate conditions
	• Provide support for local entrepreneurship	• Increased local businesses for income and employment opportunities
	• Provide pricing information	• Maximization of income for rural producers
Institutional and legal development	• Strengthen role of rural institutions and governance	• Greater voice for concerns and needs of rural communities
	• Develop and support women's groups	• Increased support for women in developing and implementing adaptation strategies
	• Support local nongovernmental organization development	• More practical hands-on support to rural communities to prepare adaptation and increase climate resilience
	• Increase support of migrating households	• Help for migrants to establish themselves in new areas and to limit the impacts of loss of social capital

Sources: Alwang and Norton 2011; IFAD 2010a, 2010b; McDonnell and Ismail 2011, based on AOAD 2007; UN and LAS 2010.

ANNEX 4F

Strategies and Actions for Food Security Adaptation at the National Level

Measure	Action	Potential result
Financial instruments	• Target food price controls/subsidies	• Reduced impact of food price rises to vulnerable groups without leakage to higher-income groups
	• Access climate finance payments	• Development of agricultural production and food chains
	• Use hedging instruments	• Lower impact of price spikes by spreading risk
	• Increase investment	• Development of food-chain transport and storage systems; better economic development in poor areas and increased incomes
	• Improve safety nets	• Increased income to vulnerable groups to maintain their food access
	• Implement climate-related insurance	• Reduction of risk of local climate extreme impacts on agricultural production and income failure
Information services	• Monitor food availability	• More information about global and local food supplies so that actions can be taken before crises emerge
	• Increase research	• Increased knowledge about the complex interactions of the dimensions of food security to develop better adaptation strategies
	• Increase information about food consumption and nutrition	• Healthier and more sustainable food consumption patterns and reduction of waste
Physical stock developments	• Build strategic food stocks at national and local levels	• Reduction of the impacts of variations in supply and price of food in global markets
Institutional and legal development	• Improve food security policies	• Better strategic policies for food security as opposed to food production
	• Improve bilateral and multilateral food supply chains	• Establishment of food supply agreements with producing areas to establish reliable incomes and investment for food producers, and secure supplies for consumers

Sources: Alwang and Norton 2011; Godfray et al. 2010; IFAD 2010a, 2010b; Lampietti et al. 2011; McDonnell and Ismail 2011, based on AOAD 2007; Nelson et al. 2010; Sulser et al. 2011; UN and LAS 2010; World Bank 2009.

Note

1. This discussion draws on a paper written by some of the authors for FAO's (2011b) *State of Land and Water*.

References

Alwang, Jeffrey, and George W. Norton. 2011. "What Types of Safety Nets Would Be Most Efficient and Effective for Protecting Small Farmers and the Poor against Volatile Food Prices?" *Food Security* 3 (Suppl. 1): S139–48.

AOAD (Arab Organization for Agricultural Development). 2007. "Strategy for Sustainable Arab Agricultural Development." League of Arab States AOAD, Cairo.

AQUASTAT. 2011. "FAO's Global Information System on Water and Agriculture." http://www.fao.org/nr/water/aquastat/main/index.stm.

Bengtsson, Lennart, Kevin I. Hodges, and Erich Roeckner. 2006. "Storm Tracks and Climate Change." *Journal of Climate* 19: 3518–43.

Black, Emily, David J. Brayshaw, and Claire M. C. Rambeau. 2010. "Past, Present, and Future Precipitation in the Middle East: Insights from Models and Observations." *Philosophical Transactions of the Royal Society A* 368 (1931): 5173–84.

Boko, Michelle, Isabelle Niang, Anthony Nyong, Coleen Vogel, Andrew Githeko, Mahmoud Medany, Balgis Osman Elasha, Ramadjita Tabo, and Pius Yanda. 2007. "Africa." In *Climate Change 2007: Impacts, Adaptation, and Vulnerability—Contribution of Working Group II to the Fourth Assessment Report of the Intergovernmental Panel on Climate Change*, ed. Martin L. Parry, Osvaldo F. Canziani, Jean P. Palutikof, Paul J. van der Linden, and Clair E. Hanson. Cambridge, U.K.: Cambridge University Press.

Breisinger, Clemens, Teunis van Rheenen, Claudia Ringler, Alejandro Nin Pratt, Nicolas Minot, Catherine Aragon, Bingxin Yu, Olivier Ecker, and Tingju Zhu. 2010. "Food Security and Economic Development in the Middle East and North Africa: Current State and Future Perspectives." IFPRI Discussion Paper 00985, International Food Policy Research Institute, Washington, DC.

Cai, Xueliang, David Molden, Mohammed Mainuddin, Bharat R. Sharma, Mobin-ud-Din Ahmad, and Poolad Karimi. 2011. "Producing More Food with Less Water in a Changing World: Assessment of Water Productivity in 10 Major River Basins." *Water International* 36 (1): 42–62.

Cai, Xueliang, and Bharat R. Sharma. 2010. "Integrating Remote Sensing, Census, and Weather Data for an Assessment of Rice Yield, Water Consumption, and Water Productivity in the Indo-Gangetic River Basin." *Agricultural Water Management* 97 (2): 309–16.

Chenoweth, Jonathan, Panos Hadjinicolaou, Adriana Bruggeman, Jos Lelieveld, Zev Levin, Manfred A. Lange, Elena Xoplaki, and Michalis Hadjikakou. 2011. "Impact of Climate Change on the Water Resources of the Eastern Mediterranean and Middle East Region: Modelled 21st Century Changes and Implications." *Water Resources Research* 47: W06506.

Christensen, Ida. 2007. *The Status of Rural Poverty in the Near East and North Africa*. Rome: Food and Agriculture Organization of the United Nations.

Clements, Rebecca, Jeremy Haggar, Alicia Quezada, and Juan Torres. 2011. *Technologies for Climate Change Adaptation: Agriculture Sector*. Roskilde, Denmark: UNEP Risoe Centre.

Dasgupta, Susmita, Benoit Laplante, Craig Meisner, David Wheeler, and Jianping Yan. 2007. "The Impact of Sea Level Rise on Developing Countries: A Comparative Analysis." Policy Research Working Paper 4136, World Bank, Washington, DC.

Dixon, John, and Aidan Gulliver. 2001. *Farming Systems and Poverty: Improving Farmers' Livelihoods in a Changing World*. Rome and Washington, DC: Food and Agriculture Organization of the United Nations and World Bank.

Döll, Petra, and Martina Flörke. 2005. "Global-Scale Estimation of Diffuse Groundwater Recharge." Frankfurt Hydrology Paper 3, Institute of Physical Geography, Frankfurt.

Easterling, William E., Pramod Aggarwal, Punsalmaa Batima, Keith Brander, Lin Erda, Mark Howden, Andrei Kirilenko, John Morton, Jean-François Soussana, Josef Schmidhuber, and Francesco Tubiello. 2007. "Food, Fibre, and Forest Products." In *Climate Change 2007: Impacts, Adaptation, and Vulnerability–Contribution of Working Group II to the Fourth Assessment Report of the Intergovernmental Panel on Climate Change*, ed. Martin L. Parry, Osvaldo F. Canziani, Jean P. Palutikof, Paul J. van der Linden, and Clair E. Hanson, 273–313. Cambridge, U.K.: Cambridge University Press.

Elasha, Balgis Osman. 2010. "Mapping Climate Change Threats and Human Development Impacts in the Arab Region." Arab Human Development Report Research Paper, United Nations Development Programme, Regional Bureau of Arab States, New York.

Evans, Jason P. 2009. "21st Century Climate Change in the Middle East." *Climatic Change* 92: 417–32. doi:10.1007/s10584-008-9438-5.

———. 2010. "Global Warming Impact on the Dominant Precipitation Process in the Middle East." *Theoretical and Applied Climatology* 99 (3–4): 389–402. doi:10.1007/s00704-009-0151-8.

FAO (Food and Agriculture Organization of the United Nations). 2010. "Climate Change, Water, and Food Security." FAO, Rome.

———. 2011a. *Conservation Agriculture Training Manual for Extension Agents and Farmers*. Damascus: Arab Center for the Study of Arid Zones and Dry Lands and FAO.

———. 2011b. *State of Land and Water*. Rome: FAO.

Fischer, Günther, Francesco N. Tubiello, Harrij van Velthuizen, and David A. Wiberg. 2007. "Climate Change Impacts on Irrigation Water Requirements: Effects of Mitigation 1990–2008." *Technological Forecasting and Social Change* 74 (7): 1083–107.

Ghassemi, Fereidoun, Anthony J. Jakeman, and Henry A. Nix. 2005. *Salinisation of Land and Water Resources: Human Causes, Extent, Management, and Case Studies*. Sydney: University of New South Wales Press.

Giorgi, Filippo. 2006. "Climate Change Hot-Spots." *Geophysical Research Letters* 33: L08707.

Godfray, Charles J., John R. Beddington, Ian R. Crute, Lawrence Haddad, David Lawrence, James F. Muir, Jules Pretty, Sherman Robinson, Sandy M. Thomas, and Camilla Toulmin. 2010. "Food Security: The Challenge of Feeding 9 Billion People." *Science* 327 (5967): 812–18.

Haddad, Nasri, Mahmud Duwayri, Theib Oweis, Zewdie Bishaw, Barbara Rischkowsky, Aden Aw Hassan, and Stefania Grando. 2011. "The Potential of Small-Scale Rainfed Agriculture to Strengthen Food Security in Arab Countries. *Food Security* 3 (Suppl. 1): S163–73.

ICBA (International Center for Biosaline Agriculture).

IFAD (International Fund for Agricultural Development). 2008. "The Role of High-Value Crops in Rural Poverty Reduction in the Near East and North Africa." IFAD, Rome.

———. 2010a. "Climate Change Strategy." IFAD, Rome.

———. 2010b. "Rural Poverty Report 2011." IFAD, Rome.

Iglesias, Ana, and Cynthia Rosenzweig. 2010. "Effects of Climate Change on Global Food Production under SRES Emissions and Socio-economic Scenarios." Socioeconomic Data and Applications Center, Columbia University, Palisades, NY. http://sedac.ciesin.columbia.edu/mva/cropclimate/.

Jin, Fengjun, Akio Kitoh, and Pinhas Alpert. 2010. "Water Cycle Changes over the Mediterranean: A Comparison Study of Super-High-Resolution Global Model with CMIP3." *Philosophical Transactions of the Royal Society A* 368: 5137–49.

Kitoh, Akio, Akiyo Yatagai, and Pinhas Alpert. 2008. "First Super-High-Resolution Model Projection That the Ancient 'Fertile Crescent' Will Disappear in This Century." *Hydrological Research Letters* 2, 1–4.

Knutson, Cody L., and Mohamed Bazza. 2008. "A Review of Drought Occurrence and Monitoring and Planning Activities in the Near East Region." Regional Office for the Near East, Food and Agriculture Organization of the United Nations, Cairo.

Lampietti, Julian A., Sean Michaels, Nicholas Magnan, Alex F. McCalla, Maurice Saade, and Nadim Khouri. 2011. "A Strategic Framework for Improving Food Security in Arab Countries." *Food Security* 3 (Suppl. 1): S7–22.

Larson, Donald F., Ariel Dinar, and J. Aapris Frisbie. 2011. "Agriculture and the Clean Development Mechanism." Policy Research Working Paper 5621, World Bank, Washington, DC.

Lobell, David B., Marianne Bänziger, Cosmos Magorokosho, and Bindiganavile Vivek. 2011. "Nonlinear Heat Effects on African Maize as Evidenced by Historical Yield Trials." *Nature Climate Change* 1: 42–45.

Magnan, Nicholas, Travis J. Lybbert, Alex F. McCalla, and Julian A. Lampietti. 2011. "Modeling the Limitations and Implicit Costs of Cereal Self-Sufficiency with Limited Data: The Case of Morocco." *Food Security* 3 (Suppl. 1): 49–60.

McDonnell, Rachael, and Shoaib Ismail. 2011. "Climate Change Is a Threat to Food Security and Rural Livelihoods in the Arab Region." International Center for Biosaline Agriculture, United Arab Emirates.

Milly, P. Christopher D., Kathryn A. Dunne, and Aldo V. Vecchia. 2005. "Global Pattern of Trends in Stream Flow and Water Availability in a Changing Climate." *Nature* 438: 347–50.

Molden, David, ed. 2007. "Water for Food, Water for Life: A Comprehensive Assessment of Water Management in Agriculture." IWMI/Earthscan, London.

Molden, David, Theib Oweis, Pasquale Steduto, Pren Bindraban, Munir A. Hamjra, and Jacob Kijne. 2010. "Improving Agricultural Water Productivity:

Between Optimism and Caution." *Agricultural Water Management* 97 (4): 528–35.

Nelson, Gerald C., Mark W. Rosegrant, Amanda Palazzo, Ian Gray, Christina Ingersol, Richard Robertson, Simla Tokgoz, Tingju Zhu, Timothy B. Sulser, Claudia Ringler, Siwa Msangi, and You Liangzhi. 2010. "Food Security, Farming, and Climate Change to 2050: Scenarios, Results, Policy Options." International Food Policy Research Institute, Washington, DC.

Saab, Najib. 2009. "Arab Public Opinion and Climate Change." In *Arab Environment Climate Change: Impact of Climate Change on Arab Countries*, ed. Mostafa K. Tolba and Najib Saab, 1–12. Beirut: Arab Forum for Environment and Development.

Scandizzo, Pasquale, and Adriana Paolantonio. 2010 "Climate Change, Risk, and Adaptation in Yemeni Agriculture." World Bank, Washington, DC.

Sulser, Timothy B., Bella Nestorova, Mark Rosegrant, and Teunis Rheenen. 2011. "The Future Role of Agriculture in the Arab Region's Food Security." *Food Security* 3 (Suppl. 1): S23–48.

Taha, Faisal, and Shoaib Ismail. 2010. "Potential of Marginal Land and Water Resources: Challenges and Opportunities." *Proceedings of the International Conference on Soils and Groundwater Salinization in Arid Countries*. Oman: Sultan Qaboos University, 99–104.

Taha, Faisal, Shoaib Ismail, and Abdullah Dakheel. 2005. "Biosaline Agriculture: An International Perspective within a Regional Context of the Middle East and North Africa (MENA)." Paper presented at the International Conference on Water, Land, and Food Security in Arid and Semi-arid Regions, Bari, Italy, September 6–11.

UN (United Nations) and LAS (League of Arab States). 2010. "The Third Arab Report on the Millennium Development Goals 2010 and the Impact of the Global Economic Crises." UN, New York.

UNDP (United Nations Development Programme). 2009. "Arab Human Development Report 2009: Challenges to the Human Security in the Arab Countries." Regional Bureau of Arab States, UNDP, New York and Cairo.

WFP (World Food Programme). 2010. "Identifying Adaptation Interventions to Climate Change: WFP's Role in Egypt, Syria, oPt, and Tajikistan—Preliminary Study." WFP Regional Bureau for Middle East, Central Asia and Eastern Europe.

World Bank. 2006a. "Directions in Development: Reengaging in Agricultural Water Management. Challenges and Options." World Bank, Washington, DC.

———. 2006b. "Making the Most of Scarcity: Accountability for Better Water Management in the Middle East and North Africa." World Bank, Washington, DC.

———. 2007. "Middle East and North Africa Region (MENA): Regional Business Strategy to Address Climate Change." World Bank, Washington, DC.

———. 2009. "Improving Food Securities in Arab Countries." World Bank, Washington, DC.

———. 2011. "Middle East and North Africa Water Outlook." World Bank, Washington, DC.

CHAPTER 5

Climate Change Affects Urban Livelihoods and Living Conditions

The aim of this chapter is to discuss the relative vulnerability to climate change of urban areas and their inhabitants in the Arab countries, to demonstrate some actual responses to these challenges, and finally, to propose additional policy options to strengthen the capacity for cities to adapt to climate change. The chapter will argue that climate change is currently threatening cities. It will further make the case that rapid and uncontrolled urban expansion exacerbates climate change impacts, especially the construction of informal human settlements and urban drainage systems. The chapter proposes that urban poverty reduction measures that share cobenefits with climate change adaptation should be prioritized; urban greening efforts should be enhanced to match current urbanization rates; the informal settlement of land and building of homes should be regulated; specific measures should be taken to improve the availability of water resources; urban water drainage systems should be improved to meet the projected challenges associated with flooding; and local governments' capacities should be developed in climate change adaptation and given a greater role in adaptive planning once their capacity for such responsibilities has been strengthened.

The world's first cities were located near the coastlines and waterways of the Arab countries. In exploring the historic areas of cities such as Algiers, Byblos, Damascus, or Jericho, it can be seen that traditional building techniques still demonstrate how cities were built to cope with the hot, dry environment. Buildings were built closely together to provide shaded walkways from the intense sun. Walls were thick and solid to absorb the daytime heat, both for shielding the interior spaces during the day and for emitting that same heat at night. Fountains of water were used

Photograph by Dorte Verner

to cool interior spaces through evaporation while wind towers cooled through convection. The cities themselves were placed near water, but above flood plains. Over time, globalization has led to the construction of cities that are less responsive to local climate and environmental conditions. Meanwhile, the climate itself is changing, and these changes must be considered in the planning and construction of cities. Urban areas concentrate a large proportion of people and assets, and many cities are in coastal and low-lying areas, with high exposure to climate change risks. In the Arab countries, it is anticipated that over 75 percent of the total population will live in urban areas by 2050, the majority of which are located in coastal areas (Mirkin 2010).

Climate Change Threatens Cities

Climate change is a major global development challenge, which threatens both human development and efforts toward achieving the United Nations (UN) Millennium Development Goals. In urban areas, the negative effects of drought, flooding, temperature increases, and sea-level rise are exacerbated by existing challenges, which include poverty, rapid urbanization, a lack of safe and affordable housing, inadequate urban drainage systems, and governance structures that lack the local capacity needed to effectively adapt to climate change (Satterthwaite et al. 2007; UNDP 2008).

Climate change affects urban areas differently from rural areas. People's livelihoods within cities tend to be more diverse and less reliant on the natural environment than in rural communities and are, therefore, more resilient to climate exposure. Still, urban populations are not insulated from climate risks. Cities do not exist independently from the countryside and largely depend on broad trade networks to supply food, water, and other resources. For example, drought may decrease agricultural yields on farms, which will, in turn, lead to an increase in food prices. This disproportionately affects urban residents, who have less capacity to grow food for subsistence. Furthermore, climate change can impact the availability and transport of these resources to cities. As will be shown, cities located in coastal areas are more exposed to sea-level rise and storm surges. Many cities are also built in a way that encourages the development of low-income or informal housing in areas vulnerable to floods and landslides. Finally, cities simply have more assets and more people than do rural areas, so in sheer volume, more humans and resources are threatened by climate change in urban settings.

Climate change impacts are not felt uniformly across cities. It has been shown in other regions that the poor are disproportionately vulnerable because they lack the resources to adapt, they live in areas exposed to weather events, and they are more likely to continue suffering indirect losses after a disaster, through displacement and the disruption of livelihoods and social networks (Hardoy and Pandiella 2009; Mearns and Norton 2010). Other studies show that poor women are particularly vulnerable. As livelihoods are disrupted and men seek new productive alternatives, women are left to assume the responsibilities abandoned by their male counterparts in addition to their traditional domestic responsibilities (Ashwill and Blomqvist 2011).

In urban areas, financial assets are critical to livelihood security because most work is wage based and most resources are accessed through cash exchanges (Sanderson 2000). Human and social assets are also at risk in that climate change has been shown to adversely impact health and contribute to migration (Raleigh and Jordan 2010; Verner 2010). Table 5.1 demonstrates many of the ways that climate change has been shown to affect the assets, services, and health of urban residents.

Cities Are Changing and Becoming More Exposed

Arab cities are changing in ways that worsen the impacts from climate change. Urbanization, which can be understood as the physical growth of urban areas and their populations, was said to contribute to vulnerability in the Fourth Assessment of the Intergovernmental Panel on Climate Change (Parry et al. 2007). In addition, rapid urbanization is often either unplanned or planned without considering climate change. This results in the creation and expansion of informal settlements, which are highly vulnerable to climate events; the destruction of natural environments, which act as a buffer to climate change impacts; and the design of inadequate drainage and wastewater management systems, which can make the consequences of extreme weather events more severe.

Urbanization

Over the past 40 years, the population of Arab countries has grown nearly threefold to the current population of 340 million people. It is projected that the region will be home to some 395 million people by 2020 and to 598 million by 2050 (El-Batran 2008; Mirkin 2010).

As the population grows, it is rapidly urbanizing. Cairo, for example, gains an estimated 1,000 new residents every week (Ghoneim 2009). This urbanization, which is not only occurring in large metropolises but also in small and medium-size towns (Kharoufi 2011), is occurring for a number of reasons. These reasons include displacement from wars and

TABLE 5.1

Possible Climate Change Impacts in Urban Areas in the Arab Countries

Natural hazards and extreme events	Incremental impacts on urban systems	Impacts on urban residents
Increased temperature		
• Heat waves • Fires	• Increased heat island effect • Increased outdoor pollution • Reduced interior air quality • Increased interior temperatures • Reduced groundwater table • Changing disease vectors • Stress on storm water system • Increased energy demand • Increased road surface damage • Increased demand for water	• Asthma • Heat stress and stroke • Thirst • Illness • Property losses • Housing instability • Disruptions in access to power, transport systems, and supplies
Decreased precipitation		
• Drought • Fires	• Groundwater depletion • Subsidence • Stress on building foundations • Reduction in green space and growing conditions • Reductions in urban agriculture • Changes in fish populations • Increased runoff contamination • Changing disease vectors	• Water shortages • Food shortages • Exposure to new disease vectors • Inability to fish and farm • Higher food prices • Disruptions of hydroelectricity • Housing instability
Increased precipitation		
• Flooding • Mudslides • Epidemics	• Stress on storm water and sewage systems • Stress on building foundations • Stress on building envelope • Slope instability • Road washouts • Changing disease vectors	• Exposure to flood-related toxins and wastes • Illness • Substandard construction • Disruption of basic services • Provision and access to supplies • Housing instability • Property loss and relocation • Community fragmentation
Sea-level rise		
• Storm surges • Flooding	• Coastal erosion • Altered coastal ecosystems • Impacts on wetlands • Salinization of water sources • Stress on water treatment systems • Stress on storm water systems • Disruptions to shipping and ports	• Exposure to flood-related toxins and wastes • Disruption in availability of potable water, food, and other supplies • Property loss and relocation • Community fragmentation

Sources: Adapted from Carmin and Zhang 2009; Dickson et al. 2010; Dodman and Satterthwaite 2008; Wilbanks et al. 2007; World Bank, forthcoming.

conflicts, high fertility rates, substantial rural-to-urban migration (also see box 5.1) and some international immigration (mainly between Arab countries but also from outside the region) (El-Batran 2008). In fact, the annual urbanization rate exceeds the population growth rate in most Arab countries (see table 1.1, chapter 1 of this volume). Figure 5.1 shows urbanization rates for the world, the Arab region, and select Arab countries from the past 50 years.

In Dubai, developed areas nearly tripled over the 20-year period between 1984 and 2003, from 78.5 square kilometers to 266.1 square kilometers (Ghoneim 2009). Map 5.1 shows how urban areas expanded in Beirut from 1984 to 2006. Somalia, Sudan, and the Republic of Yemen, are the most rapidly urbanizing of the three least developed countries, with rates of urban growth at 3.5 percent, 4.5 percent, and 4.9 percent, respectively (see table 1.1, chapter 1 of this volume).

The expansion of cities into flood plains, *wadis*, wetlands, and water catchments affects not only local residents, but also the city as a whole, because these areas are the natural safeguards against flooding. Hillside deforestation and building development increases the amount of water runoff and can lead to landslides, a secondary hazard. In addition, there is a problem with homes and roads being constructed entirely in natural water passages, blocking natural water flows. With increased paving and construction, more rainwater will be lost to sheeting rather than reabsorbed into local aquifers. Across the region, intense seasonal rains, which used to occur once every 20–30 years, are now occurring on a nearly annual basis (Al Khan 2010). The loss of vegetative ground cover, deforestation, the excessive drying of soil, and the laying of asphalt all decrease the land's natural ability to absorb and cleanse this moisture.

Global urbanization is predominantly coastal, and this trend is amplified in Arab countries, where most of the 37,000 kilometers of coastline are developed (Ghoneim 2009). Map 5.2 illustrates the population density in the region, with the highest densities located along the coastlines and near rivers. This contributes to vulnerability as it exposes the densest populations to sea-level rise, lowland flooding, and storm surges.

The Arab countries are particularly vulnerable to coastal flooding given the high concentration of populations in these areas. Coastal flooding refers to any type of flooding from the sea, because of high tides, storm surges, sea-level rise, and high river levels. According to Ghoneim (2009), a conservative 1-meter sea-level rise scenario would inundate 41,500 square kilometers of coastal areas in the Arab countries, directly affecting more than 37 million people. By contrast, a more extreme 5-meter rise in sea level would flood 113,000 square kilometers, with impacts spread unequally across many countries (Ghoneim 2009).

BOX 5.1

Climate Change Contributes to Urbanization: The Example of Rural-to-Urban Bedouin Migration in the Syrian Arab Republic

While urbanization increases vulnerability to climate change, climate change itself can contribute to urbanization. In Syria, for example, the UN estimates that a recent four-year drought drove nearly 800,000 rural villagers to makeshift camps around the cities of Aleppo, Damascus, and Homs (Sinjab 2010). In Damascus, former wheat farmers now live in dusty tents made of plastic and cotton sheets, with no plans to return to their former villages (Assaf 2010).

For a recent World Bank study, the team interviewed Bedouin migrants who were driven from Syria's Badia rangeland to the outskirts of Palmyra and other cities because of drought. These migrants experienced losses to their traditional livelihood of sheep herding and, as a consequence, moved to the city in search of wage labor. Ahmed Mehened, a single 38-year-old man from the community of Satih, said, "We lost our flock due to the continued dry years. We waited until all of the sheep were gone, and when they were, we left." Ahmed and his brothers owned 70 sheep in 2004, but within a few years they were forced to abandon their traditional livelihood. Unlike the more powerful herders, who had sufficient savings and sheep to withstand the drought, small flock owners, such as Ahmed, were forced to leave for the cities. Some have permanently moved to peri-urban areas within Syria, and others have immigrated to different countries in the region, such as Lebanon and Saudi Arabia, in search of work.

Photograph by Dorte Verner

Ahmed, along with his brothers and neighbors, moved to Palmyra, where a tree nursery provided them with three months of work. After that, he moved to Homs to harvest olives for one month, while his brothers moved elsewhere. They have all since returned to the nursery near Palmyra, where Ahmed lives with his sister and earns LS 6,000 (about US$100) per month. He says he is able to cover his expenses but cannot save, and he also says that many others are worse off. The government has provided tents, with a single rug and a pillow, but without any other basic services for the migrants. The UN reports that the dietary intake of these Bedouin migrants has plummeted. Meat and dairy consumption has decreased, and a large number of migrants live exclusively on sugared tea and bread. Education in these communities also tends to be very low.

Ahmed finds it difficult to create contacts and find work in Palmyra and Homs; without a formal education or professional training, Ahmed says he has no hope for a future in the city. He would like to return to the rangeland as a herder.

Source: World Bank 2011a.

FIGURE 5.1

Urban Population Growth in Selected Arab Countries, the Arab Region, and the World

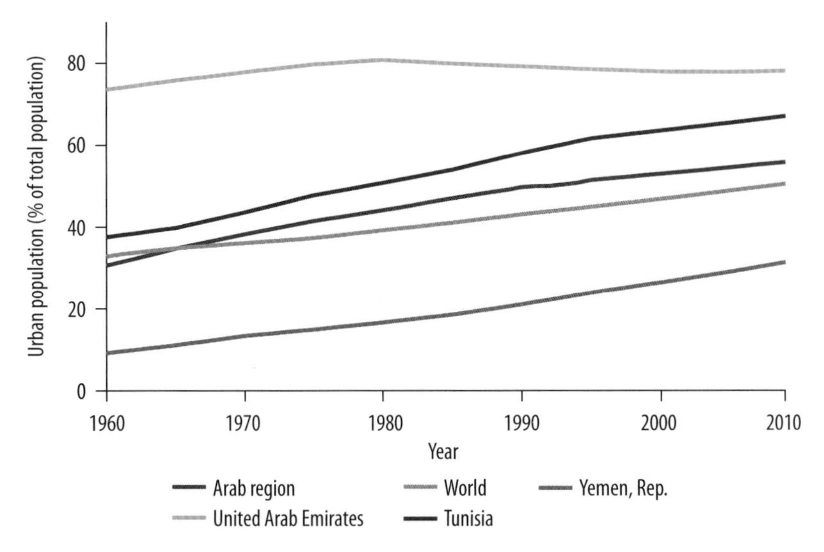

Source: Authors' representation based on data from the World Bank's World Development Indicators database.

MAP 5.1

Urban Expansion in Beirut, 1984–2006

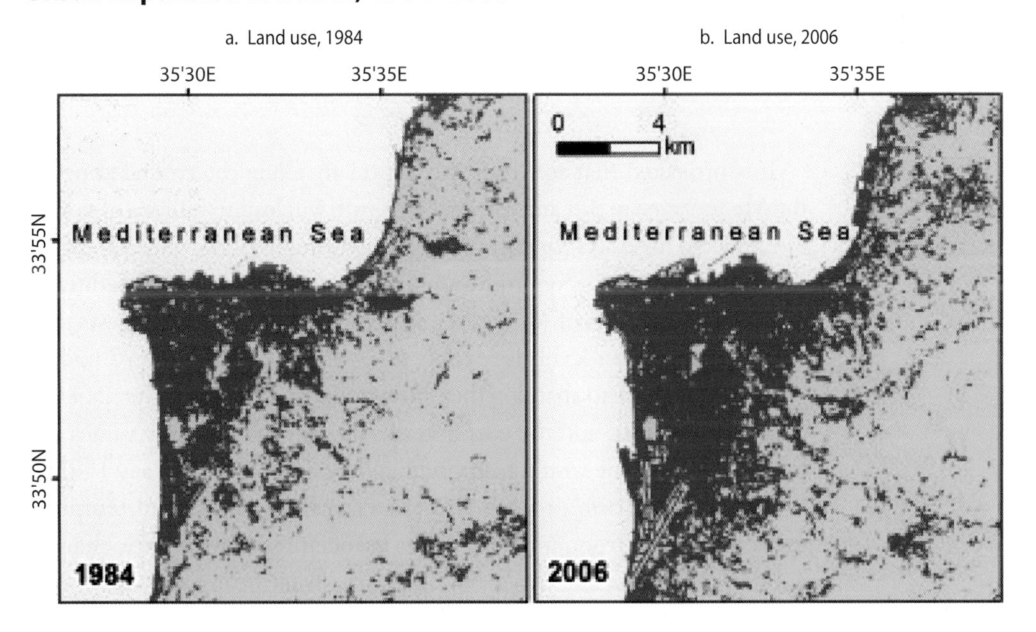

Source: Ghoneim 2009.

Note: Red areas demonstrate the increasing urban footprint of Beirut between 1984 and 2006.

MAP 5.2

Population Density of the Arab Countries, 2005

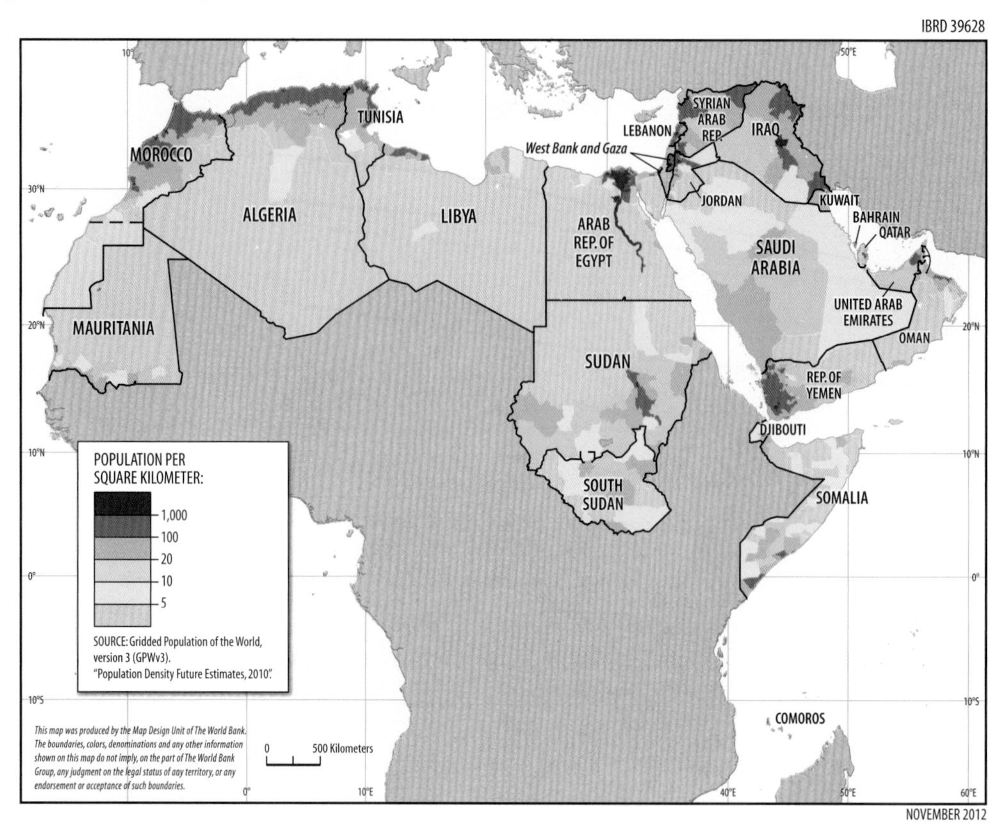

Source: World Bank, from the UNEP GEO (Global Environment Outlook) Data Portal, 2011, as compiled from the Center for International Earth Science Information Network, Food and Agriculture Organization of the United Nations, and *Centro Internacional de Agricultura Tropical*. http://geodata.grid.unep.ch.

It is projected that sea-level rise, especially along the coastal zones of the Mediterranean Sea and the Arabian Gulf, will lead to billions of dollars in estimated losses (Ghoneim 2009). An assessment of the vulnerability of the most important economic and historic centers along the Mediterranean coast (the cities of Alexandria, Rosetta, and Port-Said) suggests that, in the event of a sea-level rise of only 50 centimeters, more than 2 million people would have to abandon their homes (figure 5.2). Moreover, 214,000 jobs would be lost, and the cost in terms of land and property values and lost tourism income would be more than US$35 billion (El Raey 1997).

The urbanization process also exacerbates the increased temperatures and more frequent heat waves associated with climate change. Asphalt, concrete, and other buildings trap solar heat and store it during the day, then radiate this heat at night, preventing a drop in local temperatures. Known as the urban heat island effect, urban areas tend to experience higher temperatures than do nearby rural areas. Poor air

FIGURE 5.2

Estimated Percentage Increase in Storm Surge Zone Areas by Country, in Selected Arab Countries, 2009

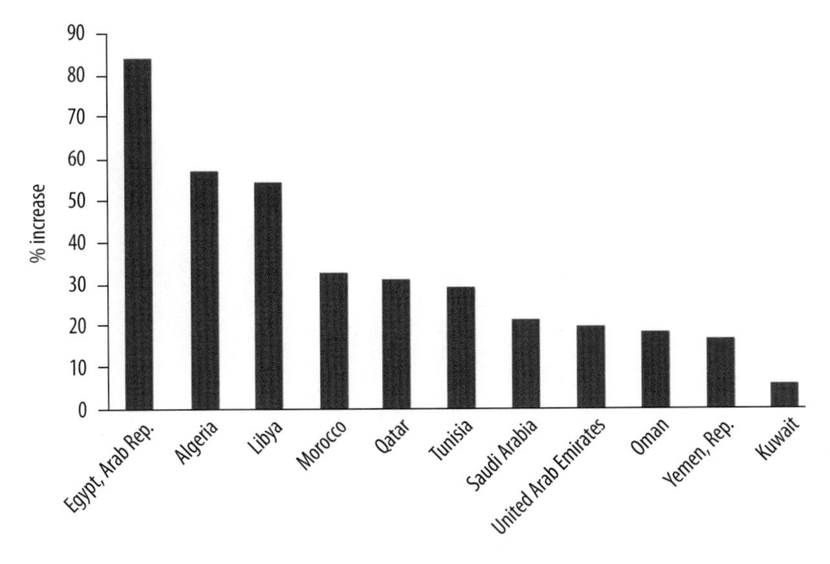

Source: Modified after Dasgupta et al. 2009.

Note: Percentages are drawn from a global study measuring the potential impact of a large (1-in-100-year) storm surge by contemporary standards, compared with the intensification that is expected to occur in this century. In modeling the future climate, the study took into account changes in sea-level rise, geological uplift, and subsidence along the world's coastlines.

quality in cities can also exacerbate this heat retention (see Cairo, for example, map 5.3).

The Arab region is particularly sensitive to the urban heat island effect, and recent modeling suggests that by 2050 nighttime temperatures in many cities could be 3°C to 3.5°C warmer than the surrounding countryside, which itself could be 2°C to 3°C warmer than now (McCarthy, Best, and Betts 2010). This will have a severe impact on those who rely on passive nighttime cooling to keep their houses and business habitable.

Higher temperatures have been associated with many deleterious human impacts. These include greater risk of disease (for example, malaria and dengue); increased water evaporation, which can contribute to water scarcity; greater likelihood of heat stroke; and increased energy consumption for cooling purposes (Verner 2010). (For more information on climate change and health, please see chapter 8 of this volume.)

Informal Housing Leaves the Urban Poor Exposed to Risk

In the Arab countries, unplanned and self-built housing is expanding in many cities because the supply of formal and affordable housing does not meet the demand of growing populations. For example, rapid urban pop-

MAP 5.3

Rise in Cairo Surface Temperature, 1964–2002

a. Cairo, 1964

b. Cairo, 2002

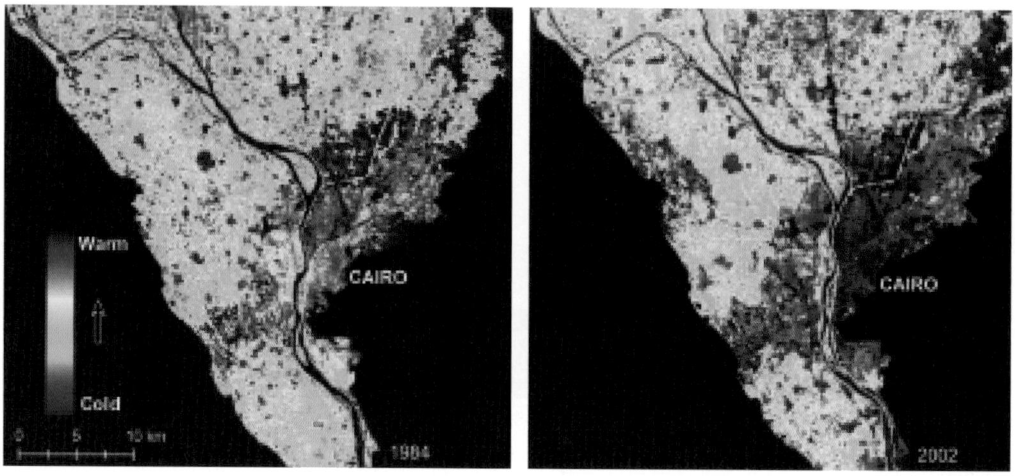

Source: Ghoneim 2009.

Note: Red areas indicate local surface temperatures and urban heat island effect following 20 years of extensive urbanization and development in Cairo.

ulation growth in the Syrian Arab Republic has led to a large housing deficit, estimated to be 687,000 units by the government (Government of Syria 2006). In many Syrian cities, informal housing has become the norm. In Damascus, it is estimated that 50 percent of urban growth is informal (Fernandes 2008).

Housing is critical to the urban poor because 50 percent of poor people's incomes are generated in the home. Still, the urban poor are priced out of more desirable areas and frequently create informal settlements in insecure lands, making them the first to suffer the consequences of a disaster (Huq et al. 2007). Without the resources to settle in desirable and secure locations, and with ill-regulated land use, poor migrants make their homes along waterways, flood-prone lowland areas, and on unstable slopes. This puts not just their homes and families in harm's way, but also their livelihoods. In a tragic example, a major disaster hit Algeria in 2001 when flash floods in slum districts killed at least 600 people and displaced a thousand more (*Economist* 2001).

The inability to meet this housing demand with formal, low-income living options has led to a general tolerance by authorities of the expansion into hazard-prone and environmentally sensitive areas. In 2005, nearly 800,000 homes were built in high-density settlements along the

Damascus fault line, an area prone to earthquakes, or in ecological buffer zones, which perform important environmental functions (IRIN 2005). This underscores the lack of planning and regulation that contributes to these informal settlements.

Inadequate Drainage Will Not Accommodate New Precipitation Patterns

The increased vulnerability of Arab cities to flooding from sea-level rise, more frequent storms, uncontrolled urban growth, and the destruction of natural water drainage systems (for example, wetlands) is exacerbated by poorly designed urban water drainage systems. If there is inadequate drainage, from too few drains or a limited capacity to deal with heavy precipitation, localized flooding occurs. Also, solid waste and sediments can block storm water drains and impede the flow of water from the impacted area. When this happens, particular weather events can quickly become natural disasters with significant human impacts.

To provide an example, inadequate drainage contributed significantly to the 2009 flooding in Jeddah, Saudi Arabia. In this case, more than 90 millimeters of rain, equal to twice the yearly average, fell within four hours. Jeddah's drainage network was not properly designed and was incomplete, with unfinished segments that could not divert the amount of water accumulation (Assaf 2010 and chapter 3 of this volume). The flood tore through the poor neighborhoods of southern Jeddah, damaging more than 7,000 vehicles and 8,000 homes as well as claiming 150 lives. More significantly, these impacts could have been reduced, or even eliminated, had urban construction been avoided in the natural drainage areas, known as *wadis*. As a result, a large number of the victims were migrant workers who lived in poorly constructed, informal shanty houses in the *wadi* area. To make matters worse, the police and civil defense units were poorly prepared to handle large-scale disasters (Assaf 2010).

Despite this, few cities are making investments in storm drainage to accommodate the increased exposure to flood waters, instead choosing cheaper, short-term solutions. Following the 2010 floods in Dubai, the city's Municipality Drainage and Irrigation Network Department relied heavily on mobile pumping units to make up for the lack of adequate storm drainage (Al Khan 2010). Such a solution is not a sustainable answer to a climate that is becoming increasingly unpredictable. These problems are compounded in cities that use the same drainage system for storm waters and waste. Flooding in these cases makes the floodwaters toxic with human and other waste (Assaf 2010).

Cities Struggle to Provide Water for an Increasingly Difficult Scenario

Climate change is exacerbating the existing problem of water scarcity in a region that has 89 percent of its land categorized as arid and semiarid. Droughts are a major risk in the Arab region, affecting more than 40 million people from 1980 to 2010 (Erian, Katlan, and Bahbah 2010). Even under normal conditions, arid regions depend on such few rainfall events that a small deviation can have negative consequences. Refer to box 5.2 for an example of how increased aridity affected a city in Mauritania.

BOX 5.2

Climate Change in Dryland Cities: Nouakchott, Mauritania

Dryland cities are prone to desertification and sandstorms, which affects not only the buildings and infrastructure but also the health and well-being of urban residents, including drinking water shortages, increases in food prices, and decreased air quality from dust contribute to a general reduction in the local quality of life.

Nouakchott, a low-lying coastal city of 550,000, regularly experiences temperatures between 14°C and 37°C and has an average annual rainfall of 600 millimeters. This rainfall has been steadily decreasing since 1968, while the population has been growing. The amount of vegetation in the city has been reduced because of the grazing of livestock, and this has destabilized the 11 coastal dunes that surrounded the city. This has exposed the city to more frequent sand storms, which have major social and economic consequences.

These consequences include damages to infrastructure and increased incidences of respiratory infections, which have subsequent effects on commerce, industry, and the everyday lives of people in the city. The eastern and northeastern portions of the city have also been exposed to increased sand damage. The airport is also surrounded by dunes, which have been encroaching onto the runway and disturbing operations. Areas in L'Aftout es Saheli, a long and narrow coastal lagoon of Nouakchott, remain unaffected, as do other locations where urban growth has paved over the sand. But this increase in asphalt has its own set of negative consequences (described in the earlier section, "Urbanization").

Source: IIED 2009.

The Intergovernmental Panel on Climate Change (IPCC) has projected that climate change will reduce the availability of water for urban water supply systems. Higher temperatures and reduced precipitation are expected to cause supply shortages because of slower aquifer replenishment rates and reduced amounts of surface water (Danilenko, Dickson, and Jacobsen 2010). Arab cities already struggle to supply adequate water resources to their populations, so any further exacerbation to this sensitive sector could have significant impacts. According to Assaf (2010), inadequate urban water services result in lower standards of living. In 2002, the Food and Agriculture Organization reported that piped water supplies in many Arab cities, including Amman, Beirut, Damascus, and Sana'a, were intermittent. This forced many urban residents to use private water vendors, which do not have the same accountability or safeguards in place to provide a quality product (WHO 2005). In many cases, this meant that poor urban residents supplemented their supplies with lower-quality water (Assaf 2010).

Many cities fail to properly maintain their water utility networks. Without routine maintenance, the operational life of an aging water infrastructure may be exceeded, resulting in perpetually increasing costs. Because of poor maintenance, it is estimated that Beirut and Casablanca lose up to 40 percent of their water supplies through leaks, while Damascus is estimated to lose approximately 64 percent (WHO 2005). Leaking water distribution systems can also lead to contamination and ground instability (Wheater and Evan 2009).

In addition to natural water scarcity, other factors can also cause demand to outstrip supply. These factors include unauthorized taps (Sana'a), unmetered usage (Beirut), and water pricing that is below market value (Alavian et al. 2009; Assaf 2010). Because of the unauthorized taps of ground water through private wells, Sana'a may be the first city in the world to run out of water (Hudes 1999).

Wastewater Treatment and Rainwater Catchment Provide Additional Water Sources

With increasing urban populations, the amount of human wastewater is rising. This, combined with the scarcity of water, has already led many Arab countries to embrace wastewater treatment and reuse. In cities, treated wastewater, or "gray water," can be used for sustaining stream flows and wetlands; groundwater recharge; watering for landscaping, gardens, parks, street medians, and golf courses; fire protection; industrial uses; and domestic uses such as laundry, cleaning, toilet flushing, and air conditioning. Dual treatment and distribution systems have even allowed some cities such as Windhoek, Namibia, and Tokyo, Japan, to supply

TABLE 5.2

Wastewater Treatment Capacity of Selected Arab Economies

Economy	Capacity
Bahrain, Kuwait, Oman, Qatar, Saudi Arabia, and United Arab Emirates	• Adhere to strict quality standards • Have the capacity to treat a high percentage of wastewater • Use a high percentage of the treated wastewater in agriculture and landscape irrigation and the additional treated water is disposed of at sea
Arab Republic of Egypt, Iraq, Jordan, Morocco, and Syrian Arab Republic	• Follow moderate regulations that do not meet national or international standards • Because of a lack of capacity to handle large loads of wastewater, effluent is frequently disposed of in surface water bodies for later agricultural use • The governments restrict the dumping of raw wastewater into *wadis*
Lebanon, West Bank and Gaza, and Republic of Yemen	• Have little to no capacity or regulation to treat and reuse wastewater • Large amounts of raw wastewater are released into *wadis* and used in agriculture

Source: Choukr-Allah 2010.

potable water through wastewater treatment (UNEP and GEC 2005). Wastewater reuse also decreases the impact of anthropogenic pollution in the environment from wastewater dumping.

Some cities in the region, such as Amman and Damascus, already have highly effective wastewater treatment facilities, with additional treatment plants under construction in several other cities. Still, the capacity to treat wastewater falls well short of the total amount that could be treated. For example, the Arab Republic of Egypt, as a country, has the capacity to treat nearly 1.6 billion cubic meters per year of municipal wastewater, but the amount of wastewater produced in Alexandria and Cairo alone already exceeds this amount (Choukr-Allah 2010). Arab countries were ranked, in descending order, from greatest capacity to lowest, according to their wastewater treatment capacity and practices (table 5.2) (Choukr-Allah 2010).

Rainwater harvesting, much like wastewater reuse, also increases the amount of water available in urban areas. Historically, rainwater harvesting has been used in the region for 2,000 years. The Nabateans used rainwater both for drinking water and agriculture to support their desert cities (Lange et al. 2011). A recent study in the West Bank shows that nearly 87 percent of the total rainfall in an average rainy season can be harvested. This is equivalent to about 480 millimeters of rainfall (Lange et al. 2011).

Arab Cities Are Responding to Climate Changes

Ancient Arab settlements were built of local materials, such as mud brick and stone, using traditional techniques to keep them comfortable in the desert heat. Reflecting on traditional architecture and planning techniques can make future construction more responsive to local conditions. While designing the modern city of Masdar in the United Arab Emirates, architects researched vernacular architectural techniques from the region and concluded that, "settlements were often built on high ground . . . to take advantage of the stronger winds. Some also used tall, hollow 'wind towers' to funnel air down to street level. And the narrowness of the streets—which were almost always at an angle to the sun's east-west trajectory, to maximize shade—accelerated airflow through the city. . . ." (Ouroussoff 2010).

Building Codes Can Reduce Vulnerability

As a response to the region's housing shortfall in cities and rapid urban growth, building construction has expanded. For example, Beirut, divided and damaged by years of war, expanded significantly to provide housing for internal migrants (refer to map 5.1). More recently, coastal lands have been reclaimed for the development of affluent housing and malls (UN ESCWA 2009a).

Much of this new construction has not followed traditional Arab methods, but instead, uses designs that were popularized in other parts of the world. These designs are often ill-suited to the Arab climate. New urban centers, such as Abu Dhabi, Doha, and Dubai, showcase modern high rises with glass facades—the antithesis of thermal massing. These buildings, used for both residential and commercial purposes, feature inoperable windows; combined with the region's rising temperatures, high rises create a huge energy demand to power air-conditioning systems (Gelil 2009). This leads to greater greenhouse gas emissions, a major contributor to the greenhouse effect and global warming.

As a response, many countries around the world have begun implementing "green" building codes.[1] The term *green building* refers to all processes related to an environmentally responsible structure, throughout the life cycle of a building (U.S. Environmental Protection Agency 2010). Building codes can be used to create more robust structures that will be better suited to the increased temperatures and storms expected with climate change. Additional benefits to green buildings include an increased return on investment through reduced energy, operating, and maintenance costs; increased sales and leasing potential of buildings; better occupant health and productivity; and reduced use of fossil fuels (LGBC 2011).

International assessment and certification systems for green buildings have emerged over the past two decades. These systems include the Building Research Establishment Environmental Assessment Method (BREEAM) in the United Kingdom in 1990, Leadership in Energy and Environmental Design (LEED) in the United States in 1998, and Green Star in Australia in 2003 (IBE 2011). No unified assessment system for green buildings has been developed so far for the Arab region, but individual countries or subregions have formulated their own methods for assessing the environmental impact of building projects.

The Council of Arab Housing and Construction Ministers, formed by the League of Arab States, has been working since the 1990s to prepare unified building codes for the Arab region. Six stages from this plan were finalized and resulted in the production of 14 unified Arab building codes; other stages are planned. Abu Dhabi in the United Arab Emirates has developed a national system based on LEED, called *Estidama* (Arabic for sustainability), that includes aspects from its own culture, environment, and ideas of sustainability. The Egyptian Green Building Council developed the Green Pyramid Rating System in January 2009. The system uses a whole-building approach to sustainability (EGBC 2009, 2011).

A green building rating system in Lebanon was launched in 2011 by the Lebanon Green Building Council with support from the International Finance Corporation. The system is termed the *ARZ Building Rating System*, after the famous Lebanese cedar tree. The ARZ system meets minimum international environmental requirements while taking into account local conditions in Lebanon. A green building audit was conducted during the development of the ARZ that revealed the system would reduce greenhouse gas emissions and would result in a considerable return on investments and an average payback period of two to three years (CWO 2011; LGBC 2011).

The Jordan Green Building Council was established in October 2009 as a nongovernmental organization, and the Jordan National Building Council has recently published the "Guide for Green Buildings in Jordan." Although it is a guide rather than a code, it is expected that it will provide the technical standards to reach sustainability for new buildings and provide the background to produce green building codes for Jordan in the near future.

Urban Greening Can Reduce Heat Impacts

To combat localized climate conditions such as increasing temperatures or severe drought, some Arab cities, such as Amman, Cairo, Casablanca, and Rabat, have begun urban greening projects. These projects typically involve the planting of drought resistant vegetation, which serves to reduce the urban heat island effect, decrease air pollution, and absorb excess rainfall and even CO_2. Oftentimes these projects include irrigation systems, which maintain a level of micromoisture and sustain the new plants. In Amman, these plans include the planting of native species of trees and bushes and using treated wastewater for landscape irrigation (figure 5.3).

Weak Governance Contributes to Lack of Adaptation Action

Despite the modest efforts described previously, many challenges remain in increasing the resilience of Arab cities to climate change. According to Fünfgeld (2010), there are four institutional challenges that prevent cities from effectively adapting to climate change:

- A lack of knowledge on emerging scientific information on climate change and its impact on cities (global models, downscaling, and so on);
- A limited understanding of how broader socioeconomic processes influence urban vulnerability;
- The insufficient integration of information on climate change vulnerability into local planning processes and development agendas; and
- The absence of a suitable governance framework for climate-risk management in cities.

In many Arab countries, long-standing governing structures are changing. The Arab Spring and the independence of the Republic of South Sudan have illustrated that social groups are demanding democratic representation after years of "skewed access to political power and wealth," (UNDP 2009, 56). According to the United Nations Economic and Social Commission for Western Asia, the history of state-centered management is being challenged (UN ESCWA 2009a).

Still, relying for years on powerful national governments has meant that local governments and municipalities have developed a very low capacity to adapt to climate change. This is a problem because local governments are in a much better position to understand the diverse relationships, resources, and vulnerabilities of local communities to climate change (Ashwill, Flora, and Flora 2011). In addition, local authorities are present and accountable during the planning, implementation, and monitoring of projects within their jurisdiction (World Bank, forthcoming). This makes them a much more effective facilitator of climate change

FIGURE 5.3

Climate-Sensitive Streetscape in Amman: An Urban Greening Project, Before and After

a. Existing

b. Planned

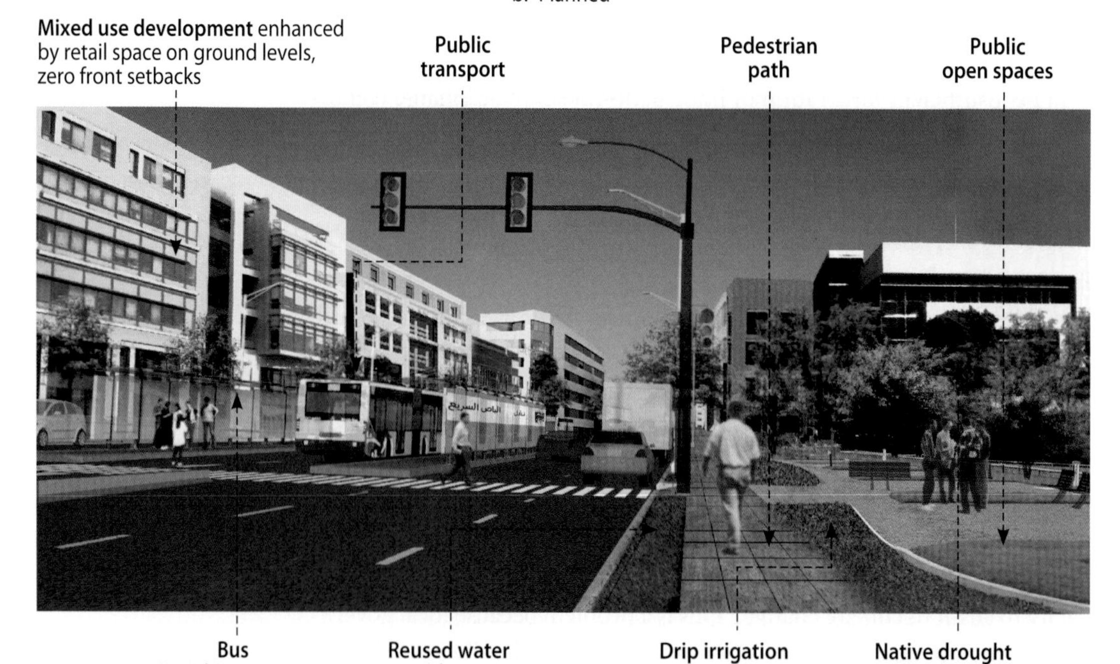

Source: GAM 2011.

adaptation than their national counterparts, who lack the site-specific knowledge to plan efficiently at the local level.

General knowledge related to urban issues and climate change is also lacking. Much of existing climate change research, both globally and within the region, fails to properly account for urban issues. In fact, the IPCC Fourth Assessment Report (AR4) (Wilbanks et al. 2007) recognizes that the urbanization effect is missing from global climate model projections (Christensen et al. 2007). Although they are referenced in AR4, urban planning strategies for adaptation and mitigation are not comprehensively reviewed. Fortunately, it is expected that the next IPCC assessment report will cover more adaptation and mitigation strategies that relate to urban settlements and infrastructure (IPCC 2009).

In some cases, the lack of knowledge and institutional capacity has led to egregious examples of negligence. In 2008, the Al-Duwayqa landslide in Cairo killed 119 slum residents. It was later determined that officials had known the imminent risks to residents but failed to act (Amnesty International 2010).

Policy Makers Have Opportunities to Make Cities More Resilient

From the previous discussion, several conclusions and actionable recommendations emerge. Policy makers, urban planners, and development practitioners should take these into account as they examine the specific ways in which to design responsive and effective urban adaptation strategies. These strategies must go beyond responding to the impacts and disasters associated with climate change and be proactive in addressing the root causes of vulnerability to severe weather events and long-term changes to the climate. Adaptation must look to reduce the relative exposure and sensitivity of urban communities to climate change while aiming to strengthen the adaptive capacity of the populations living in them.

Urban poverty reduction measures should continue. The case can be made that poverty is the single most important indicator in determining human vulnerability to climate change. Poor people have the least resources available to adapt to the climate and often live, or have assets, in the most disaster-prone areas. Sound development strategies are typically sound climate change adaptation strategies. Several studies have shown that climate change impacts are diverse, long term, and difficult to predict, so responses should be designed to be beneficial under all possible climate scenarios (Ashwill, Flora, and Flora 2011; Heltberg, Siegel, and

Jorgensen 2009; World Bank 2010b). This means that many urban development initiatives that aim to reduce poverty should be continued and strengthened. It has been shown that many cities in the Arab region are growing rapidly and already struggle to provide the basic delivery of quality services to their growing populations. Improving basic services, especially for the urban poor, will improve the capacity of these residents to cope with environmental challenges and hazards related to climate change. Disaster risk reduction strategies are also important responses to climate change. Disaster preparedness involves raising public awareness, training police and civil defense emergency units, and developing responsive disaster forecasting systems. Refer to box 5.3 for more specifics on increasing disaster preparedness.

Urban greening projects should be enhanced at a scale that matches urbanization rates. This chapter has demonstrated that urban growth contributes to human vulnerability in cities. The settlement of hazard-prone land and the encroachment onto, and destruction of, natural ecosystems that regulate air temperatures, stabilize hills, absorb excess rainfall, and cleanse water and air supplies increases the vulnerability of all urban residents, particularly the poor. Therefore, the greening of urban areas through urban reforestation projects, vegetative restoration (especially around water sources), and other activities will go a long way in minimizing the worst impacts from climate change. This can be a difficult task in the Arab countries, which are known for their dry and barren landscapes. However, many modern techniques that maximize water retention (for example, rainwater catchment) or efficient water use (for example, wastewater management and reuse) can be applied to minimize the risks associated with rapid urbanization. It is also important that regulations and the enforcement of zoning and building laws are established so as to discourage urban growth into natural areas. When possible, site-responsive architectural and urban design techniques should be used. As seen in traditional architecture, these practices have been proven to work over centuries of implementation in lessening the impact of the climate on daily city life. Many of these urban greening efforts will also function as a carbon sink, thus mitigating the human contribution to climate change.

Urban authorities should regulate the informal housing sector and enforce these regulations. It has been shown that some of the most vulnerable people in the Arab cities are those who have built homes in informal settlements. These areas tend to be the most exposed to climate change and environmental hazards. Because the urban poor often use their homes for income generation, this also entails a risk to not only their homes and possessions but also their livelihoods. The regulation of these settlements encompasses several different policy carrots and sticks. Regulators should

BOX 5.3

Case Study: Climate Change Adaptation and Disaster Risk Reduction in the Coastal Cities of North Africa

North Africa's increasingly populous coastal cities face tangible risks today, and these will multiply as the impacts of global climate change further manifest themselves. To better understand the risks these cities face, and to help them prepare the necessary adaptive responses, the World Bank led a two-year study focused on three cities critical to the region's economic, social, and political life: Alexandria, Casablanca, and Tunis. The study also examined the Bouregreg Valley between Rabat and Salé in Morocco, an area that is undergoing large-scale urban development.

Looking at a mix of factors, including expected sea-level rise, patterns of coastal erosion, frequency of torrential rainfall and flooding, and housing patterns and urbanization, the study found that most risks increased over the 2010–30 period. For this period, each of the three cities faced cumulative losses of well over US$1 billion from all risks (including seismic risk), with the portion of risks associated with climate change increasing. Moreover, looking beyond the 2030 time horizon, sea-level rise would increase further, weather extremes would become more frequent, and climate change would account for a still greater portion of the risks. Climate change scenarios involve uncertainty, and because of this, the authors of the study called for both flexibility and "no-regrets" actions (actions that provide benefits under all future climate scenarios) that will be cost-effective in these scenarios.

Alexandria. The city of Alexandria is home to a population of 4.1 million people, with an expected 65 percent surge to 6.8 million inhabitants by the year 2030. The city's expansion is anticipated to proceed westward along the Al-Bouhayra Governorate border, and to the south. Shoreline areas such as the Abu Quir depression and areas near the Maryut Lake, along with other low-lying spaces, will likely experience an influx of people. A recently expanded seafront highway has steepened the seabed and intensified patterns of coastal erosion and vulnerability to storm surges.

Flooding risks are expected to increase by 2030, particularly if informal settlements multiply in areas that are already below sea level and are vulnerable to inundation. Alexandria's densely urbanized coastline presents an obvious vulnerability, particularly as sea-level rise, coastal erosion, and the retreat of beaches expose more of the city to submersion. To minimize the risks, policy makers can ensure that the 2030 Greater Alexandria Master Plan, currently under preparation, takes into account climate change impacts in directing city growth and setting rules for density and land use. The study also calls for improved early warning systems, possibly including a network of "smart buoys" along the coastline to generate timely data and provide warning of coastal storms. Alex-

(continued)

BOX 5.3 *Continued*

andria could further be made more resilient with investments in key infrastructure, including improved coastal marine defenses and more robust drainage systems.

Casablanca. Casablanca, home to 3.3 million people, with another 300,000 in the surrounding areas, is expected to grow to 5.1 million by 2030. The rapid pace of urbanization along the waterfront and low-lying areas raises major concerns. Slums and crowded areas with poorly built structures add to the risks.

The city's vulnerability to flooding, coastal erosion, and marine inundation will become more pronounced in the years leading up to 2030, with sea levels expected to rise by as much as 20 centimeters while the beaches retreat by as much as 15 meters. The study identifies specific areas of vulnerability, such as a 10-kilometer coastal segment between eastern Casablanca and Mohammedia that is already vulnerable to erosion.

Urban planning priorities would include slum rehabilitation and protection plans for areas that have already been subject to episodic flooding. Moroccan institutions can become more responsive and effective by removing overlapping functions and simplifying operations. Improved information systems would help ensure that people receive adequate warnings of potentially dangerous weather. By strengthening coastal defenses in areas already subject to erosion and sea surges, the city could protect a number of low-lying, densely populated areas. Box Figure B5.3.1 shows the adaptation cost curve for Casablanca, identifying the cost-benefit ratio for one-time investments, institutional measures, and infrastructure investments.

Tunis. Compared with the other cities in this study, Tunis anticipates only a modest population growth of 33 percent over the next two decades, increasing from 2.2 million in 2010 to nearly 3 million by 2030. Despite this, the city faces special risks of its own. In downtown Tunis, land subsidence leaves some buildings tilting dangerously, and seismic risks are amplified by poor soil quality. The coastline is seriously threatened by erosion, requiring reinforced beach defenses.

Topographical data show that specific areas in the lower downtown area are also vulnerable to marine submersion under certain storm scenarios. By 2030, the frequency of torrential rainfall could increase by 25 percent under widely accepted scenarios for climate change. Tunis could face an increased recurrence of extreme weather events with shorter cycles of frequency.

Climate-resilient urban planning will be crucial for Tunis in order to manage the increasing risks. In low-lying areas of the city prone to flooding, upgraded drainage systems will be necessary. Illegal housing development at the periphery will need to be contained. In addition, careful zoning, with allowance for green spaces, along with rigorous enforcement of standards, will be critical.

Source: World Bank 2010a.

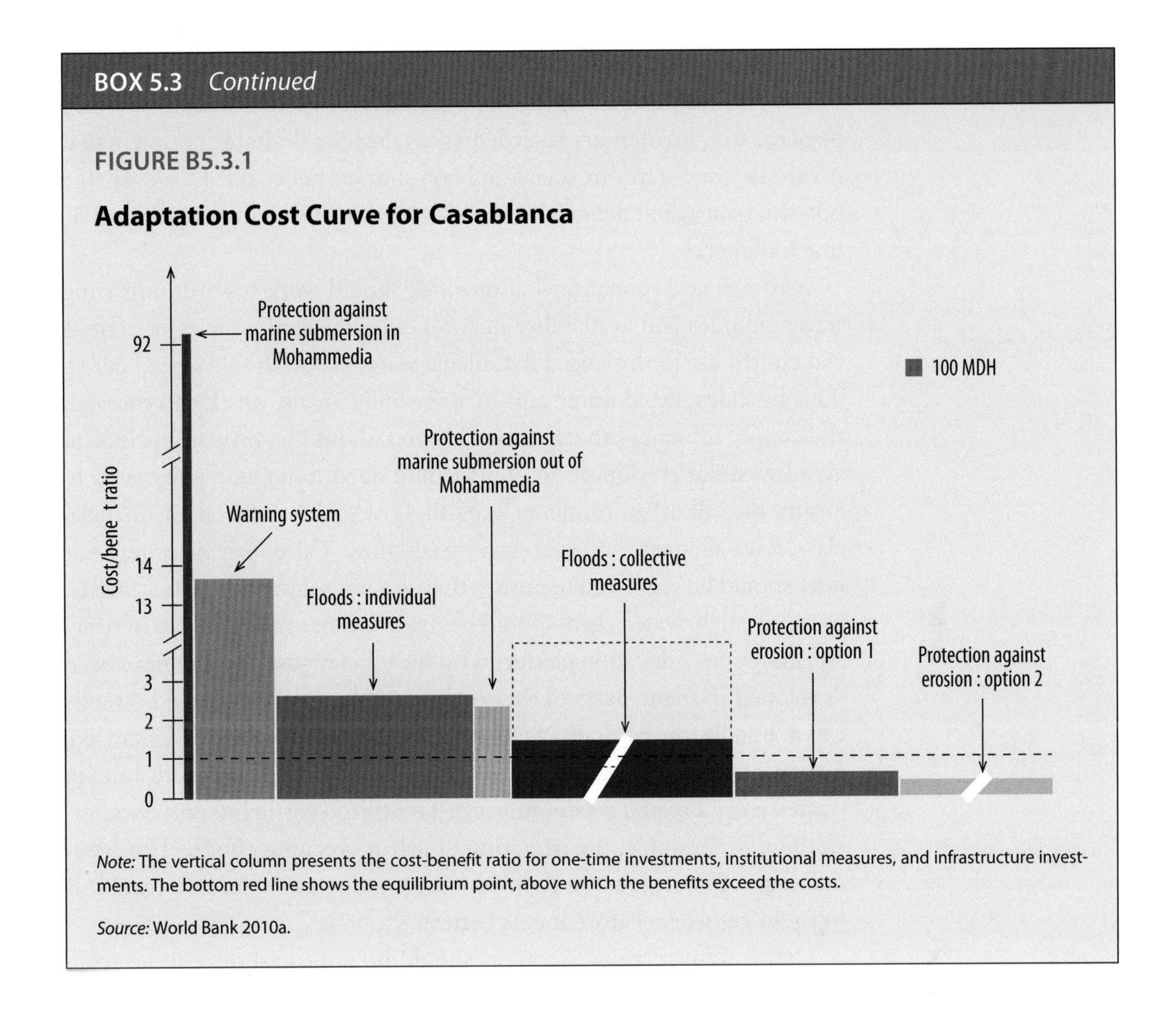

Note: The vertical column presents the cost-benefit ratio for one-time investments, institutional measures, and infrastructure investments. The bottom red line shows the equilibrium point, above which the benefits exceed the costs.

Source: World Bank 2010a.

seek to limit the expansion of settlements into high-risk areas. This would include the delineation of "high-risk" areas (prone to landslides, sea-level rise, or other flooding) and establish clear zones where development is either curtailed or totally banned. Of course, simply writing regulations will not solve the problem, so enforcement mechanisms must also be created with the resources to sustain them. Such a regulatory system must ensure that inhabitants in these zones have access to alternate sites for housing. These alternate sites can be well planned to be climate resilient and "green." Access to this housing should include plans for formally recognizing the land tenure of residents who have moved or settled in designated areas. Once inhabitants have land tenure, they will also then be eligible for basic services and other social safety nets. Such a system would serve the dual purpose of reducing the exposure of urban residents to climate events and strengthening their adaptive capacity. Still, many residents may not want to leave their informal homes, even if threatened by the climate. In these cases, the assisted relocation of households may be necessary. Involuntary resettlement is the least desirable option in that it can do great harm to social and community networks. However, the potential social and financial costs associated with climate disasters can be much higher.

National and subnational authorities should work toward improving water supplies and availability in Arab cities. Several policies or actions can contribute to this end. First, illegal water access should be regulated. This includes illegal water taps by a few individuals, which can threaten the supply of many others. Second, water should be properly priced to avoid wasteful consumption. This would need to include safeguards to ensure that all urban residents, regardless of social, political, or financial class, have adequate access to quality supplies. Third, private water vendors should be regulated to ensure they are not selling cheap, but unsafe, water. Fourth, water users should be made aware of the dangers of overconsumption. Education platforms on the efficient use of water have been developed in many parts of the world, and these efforts could be replicated. Finally, the capacity of cities to treat wastewater must be enhanced. Urban population growth means the increased production of human wastewater. Treated wastewater can be utilized for many purposes, including, for example, the irrigation of urban greening efforts. The accumulation of waste is not a problem that is going away for cities, and it is likely to get worse before it gets better.

Urban water drainage systems should be improved or relieved from the pressure of managing floods. As was mentioned, poorly designed drainage systems can lead to major calamities related to flooding and turn mild floods into major crises. To improve these systems, drainage infrastructure should be designed to handle worst-case flood scenarios. This can be based on proper statistical analyses of hydrological records with full consideration of projected changes to the climate. It is also important to take into account the adverse effect of sea-level rise on the effectiveness of current drainage systems in coastal areas. Table 5.3 outlines a variety of flood-risk management measures that can decrease the pressure of storm water on drainage systems.

Efforts should be made to improve the capacity of local governments to lead climate change adaptation efforts. Because climate change impacts are manifested at the local level, there are major limits as to how well central governments can engage in planning for climate resilience. Broad changes in climate can have very locally specific consequences, and it is important that governments at this level are empowered to respond ef-

TABLE 5.3

Flood-Risk Management Measures in Urban Areas

Size of the urban area	Noncoastal	Coastal
Small	Conveyance: channels, storm drainage, and floodplain restoration Storage: pond/basin and rainwater harvesting Sustainable urban drainage Infiltration Building design, resilience/resistance Wetlands and environmental buffers	Resettlement/retreat Defenses Conveyance: storm drainage and floodplain restoration Storage: pond/basin Sustainable urban drainage Building design, resilience/resistance Wetlands and environmental buffers
Medium	Defenses Conveyance: channels and storm drainage Storage: pond/basin and rainwater harvesting Sustainable urban drainage Infiltration Building design, resilience/resistance	Defenses Conveyance: storm drainage Storage: pond/basin and rainwater harvesting Sustainable urban drainage Building design, resilience/resistance Wetlands and environmental buffers
Large	Defenses Conveyance: channels, storm drainage, and diversion Storage: pond/basin/public square and dam Sustainable urban drainage Building design, resilience/resistance Solid waste management	Defenses Conveyance: storm drainage Storage: pond/basin/public square Sustainable urban drainage Building design, resilience/resistance Barrier and barrage systems Solid waste management

Source: Jha et al. 2011.

fectively. To do this, local governments should be provided the resources, mandate, and trainings needed to respond to climate change. It is also important that they are held accountable for improving climate resilience. This means that clear and attainable targets should be set (in consultation with all relevant stakeholders). It is true that the deficiency of knowledge is a key constraint to local adaptation. Therefore, city and municipal governments should have key roles in ensuring that there is a clear and widely disseminated information base about climate change, its threats, and possible responses. Sharing local knowledge among Arab municipalities on appropriate climate change adaptation measures is encouraged, as many cities have successful, but unrecognized, climate change initiatives under way (box 5.4).

BOX 5.4

The Need for a Climate-Resilient Arab Energy Sector

The Arab energy sector impacts regional development and global markets. The energy sector plays a vital role in the socioeconomic development of the Arab countries, many of which are endowed with vast hydrocarbon resources. The Arab countries hold nearly 58 percent of global oil reserves and nearly 29 percent of global gas reserves (OAPEC 2010). In addition, there is a huge potential for the development of renewable energy resources such as solar and wind. Oil and gas revenues, estimated at US$571 billion in 2008, have been the major source of income in most of the Arab countries, especially in the Gulf Cooperation Council (GCC) region. According to the Arab Monetary Fund, the oil and gas sector makes up about 38 percent of the total Arab gross domestic product (AMF 2010). The Arab region hosts about 7.8 million barrels per day of oil refining capacity to meet domestic demand of petroleum products and exports around 2.9 million barrels a day to international markets. Oil refining is also a large water consumer and is thus affected by water shortages. Climate-induced disruptions in Arab refineries would severely impact the global energy market.

Climate change will impact both energy supply and demand. Extreme weather events can result in devastating economic and social impacts on energy infrastructure. Oil- and gas-producing facilities in some coastal low-lying areas vulnerable to sea-level rise and offshore facilities might also be vulnerable to extreme weather events such as storms, which would lead to breaks in production (API 2008). These disruptions can have far-reaching implications given that energy systems tend to be centralized and serve large populated areas.

Furthermore, storms can also negatively affect the transmission and distribution of power and oil, gas, and other fuels. Distribution systems are vulnerable to extreme weather events such as falling trees, for example, caused by storms. This is also true in the case of oil and gas pipelines, which run for thousands of kilometers in the Arab region, and are exposed to storms, storm-related landslides, erosion processes, and floods.

Renewable energy sources are also vulnerable to the impacts of climate change. Solar, wind, hydropower, and thermal technology are designed for specific climatic conditions. Solar energy generation can be affected by extreme weather events and increased air temperature that can alter the efficiency of photovoltaic (PV) cells and reduce PV electrical generation (World Bank 2011b). Climate-induced water scarcity, drought, and sandstorms would also decrease the efficiency of PV solar systems. Natural seasonal variability of wind speed has a significant impact on the energy produced from wind turbines. Alterations in the wind speed frequency distribution can affect the optimal match between power availability from natural resources and the output of wind turbines. The amount of hydroelectricity that can be generated depends on the variation in water inflows to a plant's reservoirs. A large number of thermal power plants in the Arab region, especially in the GCC region, are located near coastal areas. Lower levels of precipitation and higher temperatures will negatively

BOX 5.4 *Continued*

influence the cooling processes of power plants and even a modest variation in ambient temperature may represent a significant drop in energy supplies.

In relation to supply, the Arab region is expected to be warmer and drier, which will increase the use for domestic air-conditioners and desalination. In turn, this will have unforeseen effects on energy consumption. As seen elsewhere in this chapter, modern building techniques have led many cities in the Arab countries to depend on mechanized cooling systems rather than on natural ventilation. Moreover, much like renewable energy sources, desalination technology depends on very specific climatic conditions and sea temperatures to operate efficiently.

Energy supply affects water availability in the Arab countries. A significant share of energy is used across the Arab countries in groundwater abstraction, desalination, treatment, transfer, and distribution. Projected climate change–induced declines in freshwater supplies and increases in demand in the region would increase energy requirements for all these activities. Projected increases in average air and water temperatures and limited availability of adequate cooling water supplies are expected to affect the efficiency, operation, and development of new power plants.

Fossil-based combined heat and power thermal plants are commonly used for water desalination in the Arab region, which hosts nearly 50 percent of the world's desalination capacity (AFED 2010). Saudi Arabia is already producing 18 percent of the world's desalinated water, and this is projected to double to meet growing demand (KACST 2010). The strong interdependency between energy, water, and climate change makes it imperative that climate adaptation policy formulation be coordinated across these sectors.

Adaptation options need to be implemented. Energy infrastructure and technology are long-term capital investments. Therefore, it is important that energy planners, policy makers, and consumers be well prepared for these climate change impacts so that necessary adaptation measures can be taken. Energy systems must be adapted to withstand anticipated climate change and its impacts. This can be achieved by increasing the resilience of the energy system. Options include diversifying energy supplies to ensure the security of supply, proper siting of energy facilities away from vulnerable geographic areas, promoting regional energy integration to share energy resources during emergency situations, disaster preparedness planning, and risk management.

Climate impact assessments and adaptation planning should be mainstreamed into the environmental impacts and strategic environmental assessments of the energy sector. Infrastructure projects including energy should take climate-proofing into account. In this context, improving energy efficiency and scaling up renewable energy technologies would also be considered to further expand the portfolio of energy options. In addition to improving future energy availability, this could also help to reduce the need among nearly 60 million Arabs who currently lack access to modern energy services (AFED 2011).

Source: Authors' compilation.

Note

1. Building codes can be defined as a collection of rules and regulations adopted by authorities that have appropriate jurisdiction to control the design and construction of buildings, their alteration, repair, quality of materials, use and occupancy, and related factors (Harris 2006).

References

AFED (Arab Forum for Environment and Development). 2010. *Water: Sustainable Management of a Scarce Resource*. Beirut: AFED.

———. 2011. *Green Economy in a Changing Arab World*. Beirut: AFED.

Alavian, Vahid, Halla Maher Qaddumi, Eric Dickson, Sylvia Michele Diez, Alexander V. Danilenko, Rafik Fatehali Hirji, Gabrielle Puz, Carolina Pizarro, Michael Jacobsen, and Brian Blankespoor. 2009. *Water and Climate Change: Understanding the Risks and Making Climate-Smart Investment Decisions*. Washington, DC: World Bank. http://go.worldbank.org/FQYU823WW0.

Al Khan, Mohammed N. 2010. "Dubai Battles Floods Round-the-Clock." *Gulf News*, March 3. http://gulfnews.com/news/gulf/uae/weather/dubai-battles-floods-round-the-clock-1.591192.

AMF (Arab Monetary Fund). 2010. *Annual Report 2010*. Abu Dhabi, United Arab Emirates: AMF.

Amnesty International. 2010. "City Officials Convicted over Deadly Cairo Rockslide." Amnesty International, London, May 28. http://www.amnesty.org/en/news-and-updates/head-2010-05-28.

API (American Petroleum Institute). 2008. "Considering the Offshore Industry's Hurricane Response." API, Washington, DC.

Ashwill, Maximillian, and Morten Blomqvist. 2011. *Gender Dynamics and Climate Change in Rural Bolivia*. Washington, DC: World Bank.

Ashwill, Maximillian, Cornelia Flora, and Jan Flora. 2011. *Building Community Resilience to Climate Change: Testing the Adaptation Coalition Framework in Latin America*. Washington, DC: World Bank.

Assaf, Hamed. 2010. "Water Resources and Climate Change." In *Arab Environment: Water: Sustainable Management of a Scarce Resource*, ed. Mohamed El-Ashry, Najib Saab, and Bashar Zeitoon, 25–38. Beirut: Arab Forum for Environment and Development.

Carmin, JoAnn, and Yan F. Zhang. 2009. "Achieving Urban Climate Adaptation in Europe and Central Asia." Policy Research Working Paper 5088, World Bank, Washington, DC.

Choukr-Allah, Redouane. 2010. "Wastewater Treatment and Reuse." In *Arab Environment: Water: Sustainable Management of a Scarce Resource*, ed. Mohamed El-Ashry, Najib Saab, and Bashar Zeitoon, 107–24. Beirut: Arab Forum for Environment and Development.

Christensen, Jens Hesselberg, Bruce Hewitson, Aristita Busuioc, Anthony Chen, Xuejie Gao, Isaac Held, Richard Jones, Rupa Kumar Kolli, Won-Tae Kwon, René Laprise, Victor Magaña Rueda, Linda Mearns, Claudio Guillermo Menéndez, Jouni Räisänen, Annette Rinke, Abdoulaye Sarr, and Penny Whetton. 2007. "Regional Climate Projections." In *Climate Change 2007: The Physical Science Basis—Contribution of Working Group I to the Fourth Assessment Report of the Intergovernmental Panel on Climate Change*, ed. Susan Solomon, Dahe Qin, Martin Manning, Melinda Marquis, Kristen Averyt, Melinda M. B. Tignor, Henry LeRoy Miller, and Zhenlin Chen, 847–940, Cambridge, U.K.: Cambridge University Press.

CWO (Construction Weekly Online). 2011. "Lebanon Launches Green Building Rating System." *Construction Weekly Online*, June 6. http:// http://www.constructionweekonline.com/article-12670-lebanon-launches-green-building-rating-system/.

Danilenko, Alexander, Eric Dickson, and Michael Jacobsen. 2010. "Climate Change and Urban Water Utilities: Challenges and Opportunities." Water and Sanitation Program, World Bank, Washington, DC.

Dasgupta, Susmita, Benoit Laplante, Siobhan Murray, and David Wheeler. 2009. "Sea-Level Rise and Storm Surges: A Comparative Analysis of Impacts in Developing Countries." Policy Research Working Paper 4901, World Bank, Washington, DC.

Dickson, Eric, Asmita Tiwari, Judy Baker, and Daniel Hoornweg. 2010. *Understanding Urban Risk: An Approach for Assessing Disaster and Climate Risk in Cities.* Washington, DC: World Bank.

Dodman, David, and David Satterthwaite. 2008. "Institutional Capacity, Climate Change Adaptation, and the Urban Poor." *IDS Bulletin* 39 (4): 67–74.

Economist. 2001. "Algeria's Floods: Torrent of Water, Torrent of Rage." *Economist*, November 15. http://www.economist.com/node/863819.

EGBC (Egyptian Green Building Council). 2009. "Egyptian Green Building Council: Formation and Achievements." Housing and Building Research Center, Cairo.

———. 2011. "Current Major Global Rating Tools." EGBC, Cairo. http://www.egypt-gbc.gov.eg/about/global.html.

El-Batran, Manal. 2008. "Urbanization." In *Arab Environment: Future Challenges*, ed. Mostafa K. Tolba and Najib Saab, 32–44. Beirut: Arab Forum for Environment and Development.

El Raey, Mohamed. 1997. "Vulnerability Assessment of the Coastal Zone of the Nile Delta of Egypt to the Impact of Sea Level Rise." *Ocean and Coastal Management* 37 (1): 29–40.

Erian, Wadid, Bassem Katlan, and Ouldbdey Bahbah. 2010. "Drought Vulnerability in the Arab Region: Special Case Study—Syria." Background paper prepared for the *2011 Global Assessment Report on Disaster Risk Reduction*. United Nations Office for Disaster Risk Reduction, Geneva. http://www.preventionweb.net/english/hyogo/gar/2011/en/bgdocs/Erian_Katlan_&_Babah_2010.pdf.

Fernandes, Edesio. 2008. "Informal Settlements in Syria: A General Framework for Understanding and Confronting the Phenomenon." Municipal Administration Modernisation, Damascus. http://www.mam-sy.org/.

Fünfgeld, Hartmut. 2010. "Institutional Challenges to Climate Risk Management in Cities." *Current Opinion in Environmental Sustainability* 2 (3): 156–60.

GAM (Greater Amman Municipality). 2011. *The Amman Plan: Urban Envelope Implementation Plan 2025*. Amman: GAM.

Gelil, Ibrahim Abdel. 2009. "GHG Emissions: Mitigation Efforts in the Arab Countries." In *Arab Environment: Impact of Climate Change on the Arab Countries*, ed. Mostafa K. Tolba and Najib Saab, 13–30. Beirut: Arab Forum for Environment and Development.

Ghoneim, Eman. 2009. "A Remote Sensing Study of Some Impacts of Global Warming on the Arab Region." In *Arab Environment: Impact of Climate Change on the Arab Countries*, ed. Mostafa K. Tolba and Najib Saab, 31–46. Beirut: Arab Forum for Environment and Development.

Government of Syria. 2006. *10th Five Year Plan for Development (2006–2010)*. Damascus: Government of Syria.

Hardoy, Jorgelina, and Gustavo Pandiella. 2009. "Urban Poverty and Vulnerability to Climate Change in Latin America." *Environment and Urbanization* 21 (1): 203–24.

Harris, Cyril M. 2006. *Dictionary of Architecture and Construction*. New York: McGraw-Hill.

Heltberg, Rasmus, Paul Bennett Siegel, and Steen Lau Jorgensen. 2009. "Addressing Human Vulnerability to Climate Change: Toward a 'No Regrets' Approach." *Global Environmental Change* 19(1): 89–99.

Hudes, Karen. 1999. "Groundwater Management in Yemen: Legal and Regulatory Issues." In *Groundwater: Legal and Policy Perspectives—Proceedings of a World Bank Seminar*, ed. Salman M. A. Salman, 133–36. Washington, DC: World Bank.

Huq, Saleemul, Sari Kovats, Hannah Reid, and David Satterthwaite. 2007. "Editorial: Reducing Risks to Cities from Disasters and Climate Change." *Environment and Urbanization* 19 (1): 3–15.

IBE (Institute for Building Efficiency). 2011. *Green Buildings: Measuring Green around the Globe*. IBE. http://www.institutebe.com/Green-Building.

IIED (International Institute for Environment and Development). 2009. *Climate Change and the Urban Poor: Risk and Resilience in 15 of the World's Most Vulnerable Cities*. London: IIED. http://www.iied.org/pubs/pdfs/G02597.pdf.

IPCC (Intergovernmental Panel on Climate Change). 2009. "Working Group 2 Meetings." http://www.ipcc-wg2.gov/meetings/EMs/.

IRIN (Integrated Regional Information Networks). 2005. "Syria: Government Upgrades Disaster Management." *IRIN Humanitarian News and Analysis*. http://www.irinnews.org/report.aspx?reportid=25097.

Jha, Abhas, Jessica Lamond, Robin Bloch, Namrata Bhattacharya, Ana Lopez, Nikolaos Papachristodoulou, Alan Bird, David Proverbs, John Davies, and

Robert Barker. 2011. "Five Feet High and Rising: Cities and Flooding in the 21st Century." Policy Research Working Paper 5648, World Bank, Washington, DC.

KACST (King Abdulaziz City for Science and Technology). 2010. "Launching the First Phase of the National Initiative for Water Desalination Using Solar Energy." Press release, February 12. http:www.kacst.edu.sa/en/about/media/news/Pages/news49.aspx.

Kharoufi, Mostafa. 2011. "Urbanization and Urban Research in the Arab World." Discussion Paper 11, United Nations Educational, Scientific, and Cultural Organization. http://www.unesco.org/most/khareng.htm.

Lange, Jens, Samur Husary, Anne Gunkel, Dirk Bastian, and Tamir Grodek. 2011. "Potentials and Limits of Urban Rainwater Harvesting in the Middle East." *Hydrology and Earth System Sciences* 16: 715–24. doi:10.5194/hess-16-715-2012.

LGBC (Lebanon Green Building Council). 2011. *Green Buildings*. Beirut: LGBC. http://www.lebanon-gbc.org.

McCarthy, Mark P., Martin J. Best, and Richard A. Betts. 2010. "Climate Change in Cities Due to Global Warming and Urban Effects." *Geophysical Research Letters* 37: L09705. doi:10.1029/2010GL042845.

Mearns, Robin, and Andrew Norton, eds. 2010. *Social Dimensions of Climate Change: Equity and Vulnerability in a Warming World*. Washington, DC: World Bank.

Mirkin, Barry. 2010. "Population Levels, Trends, and Policies in the Arab Countries: Challenges and Opportunities." Arab Human Development Report Research Paper, United Nations Development Programme, and Regional Bureau for Arab States, New York.

OAPEC (Organization of Arab Petroleum Exporting Countries). 2010. *OAPEC Annual Statistical Report*. Kuwait City: OAPEC.

Ouroussoff, Nicolai. 2010. "In Arabian Desert, a Sustainable City Rises." *New York Times*, September 25. http://www.nytimes.com/2010/09/26/arts/design/26masdar.html.

Parry, Martin L., Osvaldo F. Canziani, Jean P. Palutikof, Paul J. van der Linden, and Clair E. Hanson, eds. 2007. *Climate Change 2007: Impacts, Adaptation, and Vulnerability—Contribution of Working Group II to the Fourth Assessment Report of the Intergovernmental Panel on Climate Change*. Cambridge, U.K.: Cambridge University Press.

Raleigh, Clionadh, and Lisa Jordan, 2010. "Climate Change and Migration." In *The Social Dimensions of Climate Change: Equity and Vulnerability in a Warming World*, ed. Robin Mearns and Andrew Norton, 103–33. Washington, DC: World Bank.

Sanderson, David. 2000. "Cities, Disasters and Livelihoods." *Environment and Urbanization* 12 (2): 93–102.

Satterthwaite, David, Saleemul Huq, Mark Pelling, Hannah Reid, and Patricia Romero Lankao. 2007. *Adapting to Climate Change in Urban Areas: The Possibilities and Constraints in Low- and Middle-Income Nations*. London: International Institute for Environment and Development.

Sinjab, Lina. 2010. "Syrian Drought Triggers Rural Exodus." BBC, August 29. http://www.bbc.co.uk/news/world-middle-east-11114261.

UNDP (United Nations Development Programme). 2008. *Human Development Report: Fighting Climate Change—Human Solidarity in a Divided World.* New York: United Nations.

———. 2009. *Arab Human Development Report: Challenges to Human Security in the Arab Countries*. New York: United Nations.

UNEP (United Nations Environment Programme) and GEC (Global Environment Centre). 2005. *Water and Wastewater Reuse: An Environmentally Sound Approach for Sustainable Urban Water Management.* Osaka, Japan: UNEP. http://www.unep.or.jp/Ietc/Publications/Water_Sanitation/wastewater_reuse/.

UN ESCWA (United Nations Economic and Social Commission for Western Asia). 2009a. *Status and Prospects of the Arab City: The Reality of the Contradictions and Differences Between Arab Cities—A Critical Vision against the Backdrop of Selected Urban Patterns*. New York: United Nations.

———. 2009b. "Transport for Sustainable Development in the Arab Region: Measures, Progress Achieved, Challenges, and Policy Framework." Presented at the Expert Group Meeting on Transport for Sustainable Development in the Arab Region and Relevant Climate Change Issues, Cairo, September 29.

U.S. Environmental Protection Agency. 2010. "Green Building, Basic Information." Last modified December 22. http://www.epa.gov/greenbuilding/pubs/about.htm.

Verner, Dorte, ed. 2010. *Reducing Poverty, Protecting Livelihoods, and Building Assets in a Changing Climate: Social Implications of Climate Change for Latin America and the Caribbean*. Washington, DC: World Bank.

Wheater, Howard, and Edward Evan. 2009. "Land Use, Water Management, and Future Flood Risk." *Land Use Policy* 26 (Suppl. 1): S251–64.

WHO (World Health Organization). 2005. "A Regional Overview of Wastewater Management and Reuse in the Eastern Mediterranean Region." Regional Office for the Eastern Mediterranean, WHO, Cairo. http://www.emro.who.int/dsaf/dsa759.pdf.

Wilbanks, Thomas J., Patricia Romero Lankao, Manzhu Bao, Frans Berkhout, Sandy Cairncross, Jean-Paul Ceron, Manmohan Kapshe, Robert Muir-Wood, and Ricardo Zapata-Marti. 2007. "Industry, Settlement and Society." In *Climate Change 2007: Impacts, Adaptation, and Vulnerability—Contribution of Working Group II to the Fourth Assessment Report of the Intergovernmental Panel on Climate Change*, ed. Martin L. Parry, Osvaldo F. Canziani, Jean P. Palutikof, Paul J. van der Linden, and Clair E. Hanson, 357–90. Cambridge, U.K.: Cambridge University Press.

World Bank. 2010a. *Climate Change Adaptation and Natural Disasters Preparedness in the Coastal Cities of North Africa*. Washington, DC: World Bank.

———. 2010b. *World Development Report 2010: Development and Climate Change*. Washington, DC: World Bank.

———. 2011a. *Syria Rural Development in a Changing Climate: Increasing Resilience of GDP, Poverty and Well-Being*. Washington, DC: World Bank.

———. 2011b. "Climate Impacts on Energy Systems: Key Issues for Energy Sector Adaptation." World Bank, Washington, DC. http://data.worldbank.org/indicator/SP.URB.TOTL.IN.ZS/countries/1W-1A?display=graph.

———. Forthcoming. *Mayors' Task Force on Urban Poverty and Climate Change*. Washington, DC: World Bank.

CHAPTER 6

Tourism Can Promote Economic Growth and Climate Resilience

Many Arab countries rely heavily on tourism for revenue, employment, and foreign currency. Tourism can also reduce dependence on traditional revenue sources (box 6.1). The average overall industry impact on the economy of Arab countries is 11 percent of gross domestic product (GDP).[1] The contribution of tourism to GDP varies across Arab countries, with eight countries (the Arab Republic of Egypt, Jordan, Lebanon, Morocco, Saudi Arabia, the Syrian Arab Republic, Tunisia, and the United Arab Emirates) receiving 93 percent of tourism revenue.

Climate change will affect tourist attractions and facilities because of coastal erosion and flooding, saline intrusion, temperature increase, and extreme events. The largest and most immediate threat to tourism will be large losses in biodiversity caused by increased temperatures. Adaptive responses can diversify and expand the tourism sector into areas with higher returns and less exposed destinations. Climate change adaptation is increasingly being mainstreamed into national development planning and tourism policies in Arab countries (box 6.2).

Tourism Is Important for Economic Growth, Foreign Exchange, and Employment

The tourism sector provides opportunities for economic growth and diversification and the potential to develop economic resilience to climate change. It also gives people the opportunity to diversify their livelihoods by providing a source of income independent of oil and agriculture.

Photograph by Dorte Verner

BOX 6.1

Promotion of Traditional Culture in Oman

Oman focuses on traditional culture in the development of tourism. The government has made an effort to diversify its economy away from oil and to prioritize tourism by leveraging historic heritage, natural beauty, folklore, and traditional industries. His Majesty Sultan Qaboos initiated the planting of 1 million date palms and 100,000 coconut trees to emphasize their traditional contribution as food, fuel, raw materials for housing and shelter, clothing, and handicrafts. The cityscape has maintained its traditional look with building codes that require maintenance of traditional architecture (Kamoonpuri 2010).

Oman's rich natural landscape provides a varied and fascinating experience with its mountains, deserts, and *wadis*. Developing adventure tourism and ecotourism leverages these assets. Overnight hotel stays grew by 240 percent between 2001 and 2009 because more visitors were attracted and they stayed longer (BMI 2011).

Photograph by Dorte Verner

Sources: Authors, based on BMI 2011; Kamoonpuri 2010.

Travel Has Always Played an Important Role in the Arab Culture

For thousands of years, the Arab region has attracted travelers, and the industry serving these travelers has constantly adapted to take advantage of new opportunities. By 3000 BCE, a strong network of trade and travel

BOX 6.2

Tunisian Strategy for Climate Change Adaptation of the Tourism Industry

The Tunisian National Strategy for Adaptation of the Tourism Industry recognizes the need for effective adaptation to climate change. The strategy recommends immediate adaptation measures to maintain Tunisia's prime position as an attractive tourist destination and to limit the risk of losing the competitive edge to other locations. The primary adaptation proposal focuses on actions that are beneficial regardless of climate change and includes actions to reduce dependence on fossil fuel, improve energy efficiency, conserve water, and limit the exposure of tourist facilities. Climate change should be considered both in the renovation of existing buildings and in the development of new facilities, with strict enforcement of building and zoning codes. One specific recommendation is to reduce Tunisia's reliance on beach-and-sun vacation tourism and to expand tourism into cultural heritage. The report also mentions specific opportunities for development, such as the expansion of natural heritage and biodiversity tourism in the area from Tabarka to Bizerte and in mountainous areas.

Photograph by Dorte Verner

Source: Authors, based on TEC 2010.

routes linked the resource-rich mountains of Afghanistan and Pakistan with the emerging civilizations of the ancient Near East (ArchAtlas 2006; Majidzadeh 1982). The history of the region was shaped by mixing cultures and peoples: the great civilizations of Anatolia, Egypt, Mesopotamia,

FIGURE 6.1

International Tourism by Destination

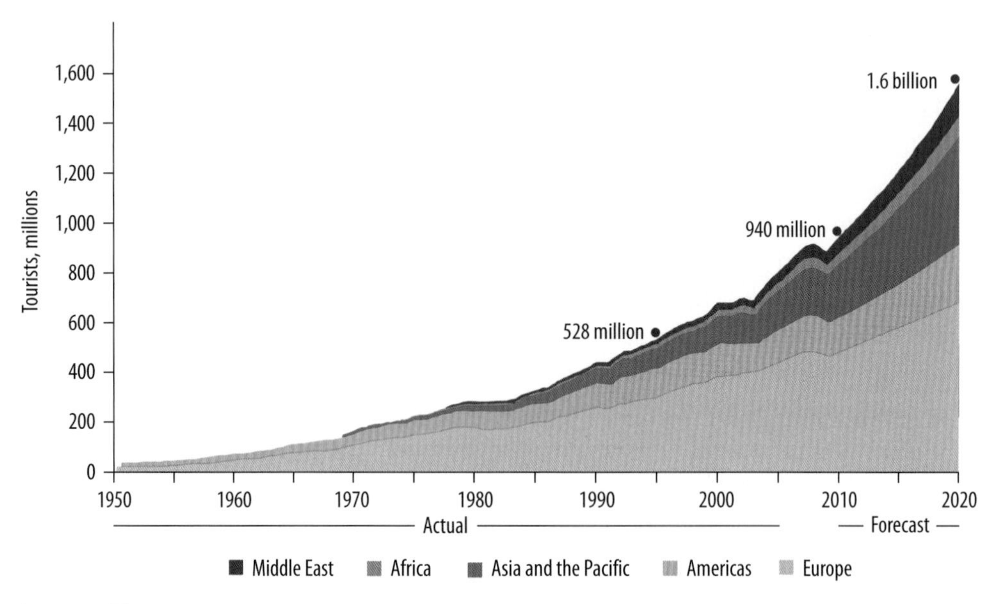

Source: UNWTO 2011a.

and Persia interacted with each other and with Greeks, Minoans, Mycenaeans, Phoenicians, and Romans from the western Mediterranean to create a rich cultural heritage. By the eighth century CE, the tourism industry was flourishing, with taverns, inns, shops, and services catering to visitors (Ansary 2009). Travel narratives and tours by Thomas Cook and Son initiated a European travel craze to the Middle East in the late 19th century. Beginning in the 1970s, modern tourists have increasingly sought out destinations in the Arab region (figure 6.1; see also Waleed 1997).

Tourism Provides Significant Economic Value

Global tourism is a major economic sector with great potential for future growth. The United Nations World Tourist Organization (UNWTO) reports global international tourism receipts of US$919 billion from 940 million tourists in 2010. In the next 20 years, UNWTO (2011b) expects the global tourism industry to grow by 3.3 percent per year, a rate only slightly lower than the average since 2005 (3.9 percent). In 2020, UNWTO forecasts 1.6 billion global international tourist arrivals, with an estimated 96 million to 128 million tourists destined for the Arab region on the basis of its current market share (figure 6.2).

Tourism to the Arab region has been growing at a very high pace over the past two decades (figure 6.1). The number of arrivals to the Middle

FIGURE 6.2

Number of Tourist Arrivals in Selected Arab Countries, 2005–10

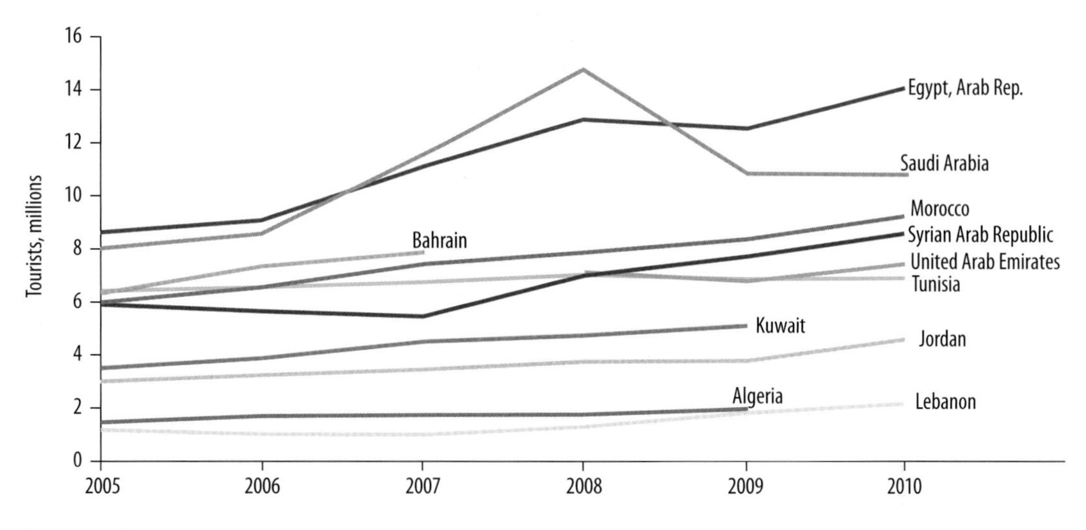

Sources: UNWTO 2011a, 2011b.

East and North Africa grew from 16.6 million to 54.7 million from 1990 to 2010—equivalent to an annualized growth of 6 percent (UNWTO 2011b). The largest growth happened from 2005 to 2008, when arrivals grew by 10 percent per year (UNWTO 2011a, 2011d).

Interim reports for 2011 indicate an 8 percent overall decline in tourism, with higher losses in countries with greater levels of insecurity related to the Arab Spring (UNWTO 2012).

Arab tourists to the region amount to 52 percent of all international arrivals. At 10 percent of all international tourists, nationals residing abroad represent the largest group of tourists. They predominantly visited Algeria and Morocco. The second-largest group originates from Saudi Arabia (9 percent), with half going to Bahrain (figure 6.3).

European tourists constitute 32 percent of all international arrivals, with the majority being French and German. They arrive mainly to explore exotic and historic sites. The majority travel to Egypt, Morocco, and Tunisia, with each of these countries receiving slightly less than one-third of European arrivals.[2]

International tourists to Arab countries provide significant economic benefits (table 6.1). In 2009, 71.5 million tourists generated US$50.2 billion in revenue, constituting a direct contribution of 3 percent to GDP. If all the indirect economic contributions by tourist-related capital investments are included, the impact of tourism climbs to 11 percent of GDP.

With 27.6 million tourist arrivals, the Mashreq receives the most tourists in the Arab region, resulting in a total contribution to GDP of 17 per-

FIGURE 6.3

Origination of Tourists to the Arab Region, 2005–09

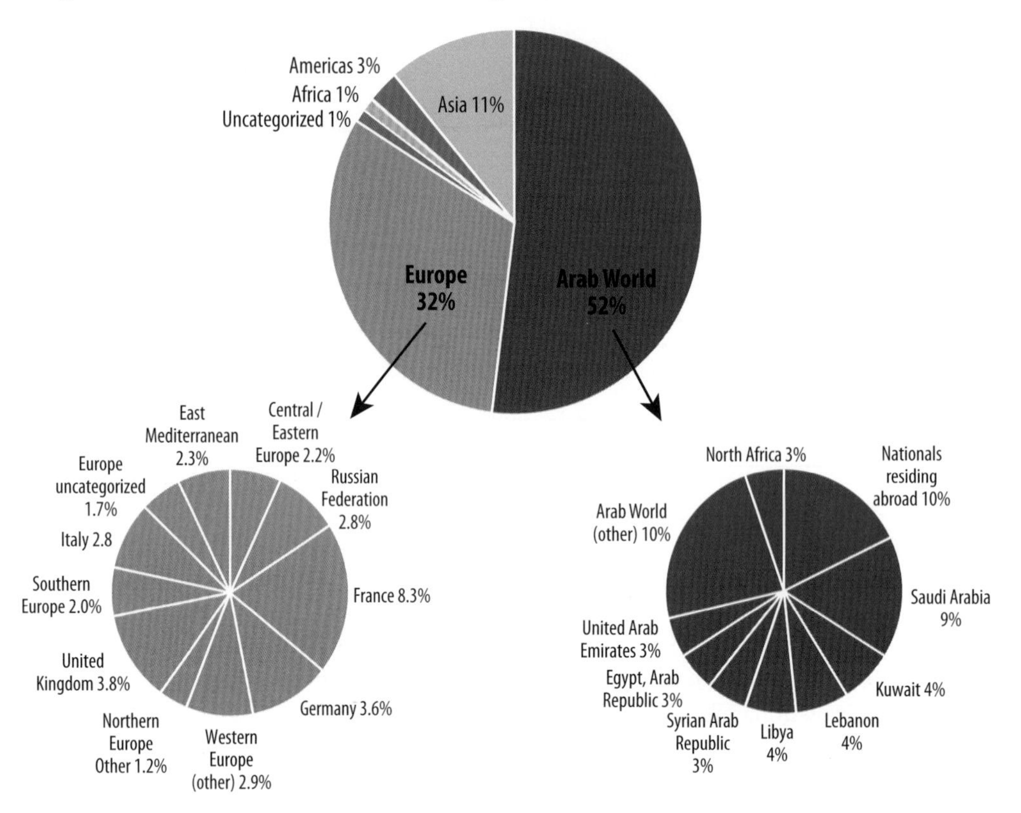

Source: UNWTO 2011b.

cent. With regard to arrivals and tourist spending, Egypt has the largest tourism sector in the Arab region, providing total revenues of US$38 billion, or 20 percent of GDP, but Lebanon is the most dependent on tourism with one-third of its GDP accounted for by total tourist-related revenue. Jordan has a significant tourism sector at 12 percent of GDP. The Maghreb receives a third less than the other regions, but tourism is an important sector for Morocco and Tunisia, earning 7.2 percent and 6.4 percent of GDP, respectively (map 6.1).

The Gulf states receive almost as many international arrivals as the Mashreq, but with the exception of Bahrain, they rely less on tourism. At 20 percent of GDP, however, the total economic contribution of tourism to Bahrain's economy is equal to that of Egypt. In addition, specific areas within countries may depend on tourism; for example, Dubai relies increasingly on tourism revenue (*Economist* 2012; Malik 2012; see also table 6.1).

Tourism contributes a significant and growing share of employment in the Arab region, providing direct employment to about 5 percent of

TABLE 6.1

Tourism Revenue and Direct Contribution to GDP Relative to GDP in 2009

	Tourism receipts (US$ million)	Total GDP contribution (US$ million)	Tourism receipts relative to GDP (%)	Total contribution relative to GDP (%)
Maghreb				
Algeria	267	12,471	0.2	8.9
Libya	50	2,928	0.1	5.0
Morocco	6,557	16,580	7.2	18.2
Tunisia	2,773	7,771	6.4	17.9
Maghreb total	9,647	39,750	2.9	11.9
Mashreq				
Egypt, Arab Rep.	10,755	38,026	5.7	20.2
Iraq	n.a.	n.a.	n.a.	n.a.
Jordan	2,911	5,682	12.2	23.8
Lebanon	6,774	11,626	19.4	33.3
Syrian Arab Republic	3,757	6,980	7.0	12.9
West Bank and Gaza	n.a.	n.a.	n.a.	n.a.
Mashreq total	24,197	62,314	6.6	17.0
LDCs				
Comoros	n.a.	n.a.	n.a.	n.a.
Djibouti	16	n.a.	1.5	n.a.
Mauritania	n.a.	n.a.	n.a.	n.a.
Yemen, Rep.	496	2,498	2.0	9.9
Somalia	n.a.	n.a.	n.a.	n.a.
Sudan	299	n.a.	0.6	n.a.
LDCs total	811	n.a.	1.0	n.a.
Gulf				
Bahrain	1,118	3,948	5.8	20.4
Kuwait	248	9,384	0.2	8.6
Oman	700	4,564	1.5	9.7
Qatar	179	4,761	0.2	4.9
Saudi Arabia	5,995	26,777	1.6	7.1
United Arab Emirates	7,352	38,297	2.7	14.2
Gulf total	15,592	87,731	1.7	9.5
Grand total	*50,247*	*189,795*	*3.0*	*11.2*

Sources: Adopted from IMF 2011; UNWTO 2011a.

Note: LDCs = Least developed countries; n.a. = not applicable. International tourism receipts include all tourism receipts from expenditures made by visitors from abroad. Data are gathered in the framework of the balance of payments (UNWTO 2011b).

the labor force and an additional 6 percent in auxiliary services in 2010.[3] In the past 10 years, employment in the Arab tourist sector grew by 5 percent, with the largest growth taking place in Syria, the United Arab Emirates, and the Republic of Yemen. More mature tourism markets in, Egypt, Lebanon, and Tunisia grew only 2–4 percent (Djernaes, forth-

MAP 6.1

Direct Tourism Revenues as Share of GDP, 2009

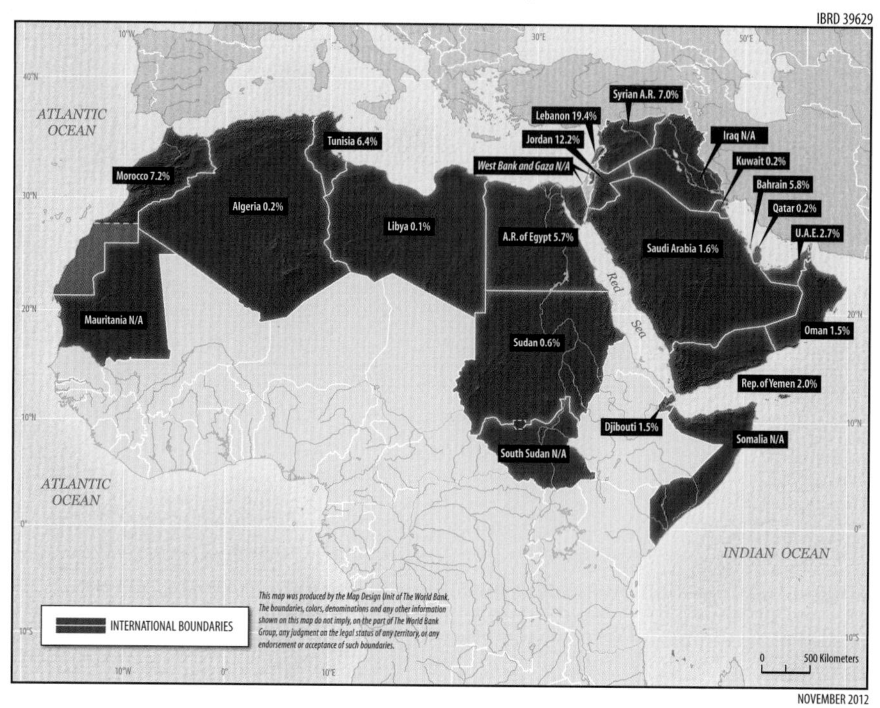

Sources: Authors' representation, based on IMF 2011; UNWTO 2011a, 2011b.

coming). In Bahrain, Egypt, Jordan, Lebanon, Morocco, and Tunisia, tourism contributes more than 15 percent of employment. Lebanon is an outlier with the highest contribution to total employment at 31 percent.

Employment in the tourism sector is directly correlated with the sector's GDP contribution. The economies gain on average just under US$40,000 in increased GDP for each full-time employee in the tourism sector.[4] In general, the highest returns are to be gained from medical tourism (Jordan, Lebanon, and Tunisia). The range of return on recreational and nature tourism is extensive and depends on the degree of luxury provided and on the visitors' purchasing power. Nationals residing abroad provide the lowest return since they often stay with family or friends and have limited interest in tourism offerings (Morocco) (see table 6.6, later in this chapter).

International tourism contributes important foreign currency to several economies: in 2009, tourism receipts for all countries in the Mashreq and Morocco were in excess of 24 percent of the total exports of goods and services, and Tunisia took in 14 percent of its foreign currency from tourism. Tourism is a less important provider of foreign currency for the hydrocarbon-rich Gulf states (World Bank 2011).

Average tourist spending is especially high in Egypt, Lebanon, and the United Arab Emirates. Tourists to those countries spend more than

US$900 on average per stay. Lebanon specifically caters to tourists with a high average spending, which was US$3,660 in 2009. Part of that higher spending may be attributed to Lebanon's medical tourism (Connell 2010).

Climate Change Has a Significant Impact on the Tourism Industry

Climate change will directly affect the tourism industry by rises in temperature, increased aridity, more severe weather, more frequent flooding, increases in water salinity, coastal erosion, and changes in the timing of the seasons. These impacts will affect each tourist segment differently. Climate change will also have indirect impacts on tourism since it may increase migration, cause political instability, create health risks, and spark tariffs on greenhouse gas emissions.

Climate Change Will Directly Affect the Tourism Industry

Climate and customary vacation periods determine tourist seasons. Temperature changes may alter a destination's appeal, rendering it less attractive or even unpleasant. Recent surveys and analyses of tourism patterns suggest that attractiveness peaks at locations with a current monthly average temperature of 21°C (Lise and Tol 2002); under climate change, the northern Mediterranean countries will likely benefit in the long term more than those to the south (Bigano et al. 2006). However, tourism flows are driven by a multitude of factors. For example, surveys show that many tourists select their destination before checking climate and weather details, and beach tourists simply seek out hot destinations, as do those from colder countries (Moreno and Amelung 2009). Climate change will require changes in the details of how facilities are promoted, shifts in peak seasons, and the types of activities; but climate will be only one of many factors prompting shifts in the promotion of tourism. Econometric analyses of domestic and international tourism showed that a trend would occur in which tourists would select higher-latitude and higher-altitude destinations. This finding suggests that many would-be tourists in cooler countries may simply stay home in the new, warmer conditions (Hamilton and Tol 2007b). But in none of the regions studied did climate effects reverse the underlying potential for growth in the tourism sector. A larger cause for concern in the Arab region could be that tourism from within the region might decline if Arab tourists seek cooler destinations; this shift may increase with growing incomes.

Surveys have shown that beach tourists dislike beach erosion and artificial coastline structures (Hamilton and Tol 2007a; Phillips and Jones

FIGURE 6.4

Temperature and Tourist Season in Egypt

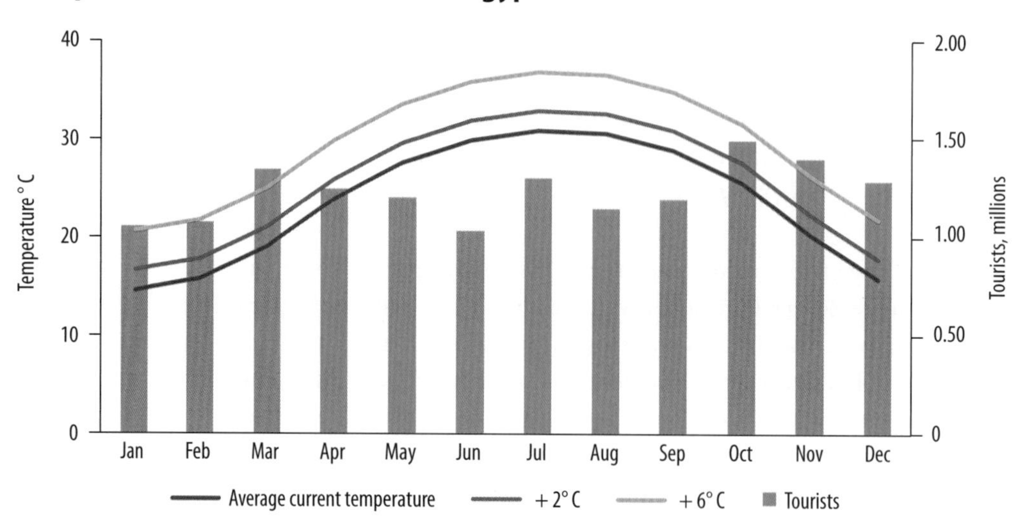

Sources: Ministry of Tourism of Egypt 2010; MSN Weather 2011.

2006), so beach management may be a particular problem, especially along the Nile Delta.

In the region's warmest countries, the height of the tourist season is in the coldest months. Moderate increases in temperatures may effectively increase the attractiveness of these areas in the tourist season. The height of the tourist season in Egypt, for example, is from October through January, with additional customary vacations in March (for Easter) and July. On average, 1.27 million tourists arrive in the coldest months (September through April), and only 1.17 million arrive per month in the warmest period (May through August). Figure 6.4 documents the number of tourists and the temperature by month. Although the country's attractiveness may increase in the colder months as temperatures increase, significant numbers of tourists may choose not to visit in the warmer periods with temperature increases on the order of 4–6°C.

Worldwide demographic changes will cause the average age of tourists to increase. Although older people are more sensitive to heat, they have more flexibility in the timing of their vacations; they may choose to take advantage of the mild winters in the Arab region (Djernaes, forthcoming).

Direct operating expenses in the tourism industry will increase with higher temperatures, increased water scarcity, and a greater frequency or magnitude of floods and saline intrusion. Temperature increases will raise electrical needs for cooling and may require new technology. In Lebanon, energy consumption for cooling is projected to increase by about 2 percent for every degree of warming. Declining precipitation may also reduce hydropower generation and further constrain the energy sector (Republic of Lebanon 2011). Water scarcity results in added costs for transport, infrastructure, and desalination facilities. Higher temperatures, longer arid periods, and salt intrusion may reduce local food production and result in higher food prices (Tolba and Saab 2009). Tourist facilities are likely to experience more extreme conditions and will need to secure backup generators for utilities, manage evacuations, and secure sufficient resources to cover business interruptions and damage to losses in infrastructure (Simpson et al. 2008).[5]

The exposure of the Egyptian tourism industry to climate change is high because a considerable number of Egyptian tourist destinations are in low coastal or river delta areas. Insurance provides measures to spread the financial risks related to disasters and to advance an economy's ability to recover (Geneva Association 2009; Lester 2010). Correctly structured insurance would provide indications of risks, information on adaptation strategies, incentives to avoid or minimize risks, and preparation for catastrophes. However, some properties may not be insurable if their exposure is intolerable. The tourism industry can adapt but only at increasing costs.

Climate Change Presents Specific Challenges to Certain Tourist Segments

Tourists can be categorized on the basis of their motivation for travel: nationals residing abroad, religious observers, Arab vacationers, nature tourists, cultural heritage tourists, and medical tourists. Only 7 percent of international arrivals indicate that they are traveling for business (UNWTO 2011d).[6] Table 6.2 provides a summary of the direct and major indirect impacts of climate change by tourist segment.

One-tenth of all tourists are nationals residing abroad;[7] this segment is very important for Morocco since it amounts to almost half of all Morocco's international arrivals.[8] Ties to family and friends motivate this group to travel; this growing market segment is therefore less vulnerable to climate change. However, climate change impacts could decrease the attractiveness of locations with the Arab region, which could decrease the frequency and length of trips.

Religious tourism is the source of many visitors to Arab countries. The main attraction is Saudi Arabia's holy sites, including Mecca. Saudi Arabia attracts almost as many tourists as Egypt, and almost 90 percent of tourists to Saudi Arabia travel for religious purposes (UNWTO 2011b).[9] In 2010, more than 13 million people visited Mecca, with about 2 million during the hajj (McDermott 2011; Travellerspoint 2011). Religious tourism is among the least vulnerable sectors to climate change. Nevertheless, the pilgrimage includes strenuous walks that can challenge healthy people even in cooler months. Higher temperatures could decrease the number of people able to make the pilgrimage and could lessen interest in repeating it. Over the next decade, the hajj will occur in increasingly warmer months, and then in the hottest months in the 2040s. The difference between the coolest and hottest Hajj months is 10°C. Climate change projections by 2050 indicate a warming of 1–2°C; the ever-improving conditions and management of the pilgrimage should be able to accommodate these increases.

Cultural heritage has been a significant sector for tourism in Arab countries. Although Egypt leads in this category, other players include Jordan, Lebanon, Morocco, Syria, and Tunisia. Climate change could affect cultural heritage tourism by deterring tourists who may be reluctant to visit sites in hotter conditions and by the loss of cultural heritage venues from extreme events or chronic damage. The vulnerability of cultural heritage sites is similar to urban areas: higher temperatures, saline intrusion, flooding, storm effects, and sea-level rise can ruin monuments that are often already in precarious condition. A concern has been raised about high temperatures' worsening of the impact of humidity on wall paintings in ancient tombs, and this concern will have to be factored into better management of tourist traffic (EEAA 2010). Destinations that are already in danger of flooding will be particularly hard hit. Alexandria, one of many low-lying cities in Egypt, attracts more than a million tourists to its beaches and archaeological sites. A 1995 study found that 28 percent of Alexandria's land used for tourism is below sea level, and a half-meter rise in sea level would threaten another 49 percent (El Raey 1995; see also Agrawala et al. 2004). Climate change increases this threat and makes the existing efforts to protect the city against inundation even more important.

Nature tourism relies on highly diverse natural resources, including landscapes, plant and animal species, and ecosystems. The integrity of the natural environment is a significant factor in the long-term success of tourist destinations (Harms 2010). Natural richness alone does not draw the majority of tourists to the region, but enjoyment of natural abundance becomes an integral part of most vacations. Beach tourism is an important part of the nature tourist segment and constitutes 90 percent of Tunisia's tourism revenue (Berriane 1999; UNWTO 2011b; World Bank 2008).[10] Diving is an increasingly large draw for coastal recreation and is completely dependent on the beauty and health of the coral reefs (box 6.3). It constitutes 5 percent of total revenue regionally. Among international tourists enjoying Egyptian coastal recreation, 21 percent come to dive and an additional 37 percent to snorkel. Diving revenue was estimated at US$117 million with an additional US$221 million of indirect revenues

TABLE 6.2

Impacts of Climate Change by Tourist Segment

Segment	Countries most dependent on segment	Increased temperature	Increased aridity	More frequent extreme weather	Increased direct costs	Sea-level rise	Sea salination, acidification
Nationals residing abroad	Algeria, Morocco	• Low effect, given ties to family and friends • Potential for less frequent or shorter visits • Potential positive effect if migration increases	• Medium effect if basic water requirements are met • High effect with lack of water for basic needs	• Low effect, given ties to family and friends • Potential for less frequent or shorter visits	• Low effect, given ties to family and friends • Potential for less frequent or shorter visits • No effect likely on spending per trip	• Medium to high effect of rising sea levels amplified by extreme weather • High effect if seawater intrusion in aquifers	• Low effect
Religion	Saudi Arabia	• Low effect; comfort is not a major concern • Potential high health effects	• Medium effect if basic water requirements are met • Potential high health effects • High effect with severe water shortages	• Potential medium effect if extreme weather occurs during peak pilgrimage season	• Low effect • Potential for less frequent or shorter visits • No effect likely on spending per trip	• Low effect • Potential high effect with seawater intrusion in aquifers	• No or very low effect
Arab vacationers	Bahrain, Lebanon, United Arab Emirates	• Medium effect; the appeal of familiarity of area is more important than weather conditions • Potential positive effect for cooler areas • Potential high effect on niche segments	• Medium effect; this population may be more tolerant to increased aridity than Europeans • Potential positive effect for less arid destinations	• High effect; weather events that hinder recreation would be a deterrent for recreational tourism in general	• Low effect • Potential for less frequent or shorter visits • No effect likely on spending per trip	• High to extremely high effect; dominant tourist destinations are extremely vulnerable to seawater intrusion, coastal erosion, and sea-level rise	• Low to medium effect

(table continues next page)

Impacts of Climate Change by Tourist Segment

Segment	Countries most dependent on segment	Increased temperature	Increased aridity	More frequent extreme weather	Increased direct costs	Sea-level rise	Sea salination, acidification
Cultural heritage	Egypt	• Medium to high effect; travel comfort diminishes with increased temperatures • Potential positive effect for colder destinations • Potential shift of tourism season to colder months	• Medium to high effect; tourists will likely seek alternative destinations if potable water is restricted	• Very high effect; low tolerance for spoiled vacations from extreme weather • Risk of cultural heritage site loss from extreme weather	• Low effect • Potential for less frequent or shorter visits • No effect likely on spending per trip	• High effect if loss of tourism facilities and attractions • High effect if seawater intrusion • Risk of cultural heritage site loss from increased sea levels	• Low to medium effect
Nature	Egypt	• Very high effect; tourists will seek alternative vacation destinations if loss in biodiversity is combined with increasing temperatures	• Very high to extreme effect, increasing with degree of loss of biodiversity and ecosystems	• Very high effect; low tolerance for spoiled vacations from extreme weather	• Low effect • Potential for less frequent or shorter visits • No effect likely on spending per trip	• High effect, increasing with floods • Rising sea levels and seawater intrusion may damage facilities and infrastructure and limit access	• High effect, increasing with loss of biodiversity and ecosystems
Medical reasons	Jordan, Tunisia	• Low effect with adequate air conditioning • Possibility of sector being overwhelmed by increased medical needs	• Low effect; as long as secondary effects remain low	• Low to medium effect, depending on location of medical facility	• Low effect, as long as secondary effects remain low	• Low to medium effect, depending on location of medical facility	• No effect

Source: Djernaes, forthcoming.

Note: Estimates based on available data.

BOX 6.3

Climate Change Impacts on Coral Reefs

Distinct fauna and corals, of which 6 percent are endemic, provide good-quality coral reefs in the Red Sea. It has 13 principal coral communities and a rich biodiversity of approximately 1,000 animal species, of which 14 percent of the fish are endemic. Significantly richer biological composition is present in the south because of differences in reef composition, turbidity, and influxes of nutrient-rich waters from the ocean. Although the Red Sea corals have proved to be more tolerant of high temperatures and siltation than reefs elsewhere, two-thirds are threatened by overfishing, algal blooms, and waste. The Red Sea reefs are considered to be in relatively good condition, but 60 percent are still regarded as threatened. In the Arabian Gulf, 85 percent of corals are considered threatened.

Sources: Burke et al. 2011; Chiffings 1995.

in 2000 (Cesar 2003). However, without better management to reduce the impacts of tourists on the reefs, revenue was projected to peak in 2012 and decline to half by 2050, without considering any additional damage caused by climate change or the associated bleaching of the coral. Nature tourism is highly dependent on biodiversity, and 40 percent of species are seriously threatened with extinction under the projected 2°C rise in temperature (Tolba and Saab 2009). Increasing coastal pressure from storms and elevated sea levels will affect coastlines and lodging facilities. Tourists looking for nature will not endure uncomfortable temperatures if the biodiversity is lost; they will seek alternative vacation destinations.

Arab vacationers predominantly choose Bahrain, Lebanon, and the United Arab Emirates. These countries are less restrictive about dress and alcohol consumption than other Arab countries, and the appeal of a vacation in a familiar but more liberal environment drives the market for Arab recreation. For example, in Bahrain in 2007, 70 percent of tourists were from the Arab region, with 56 percent from Saudi Arabia (Pew Forum on Religion and Public Life 2011; UNWTO 2011b).[11] Although this sector is less climate sensitive than tourism for nature and cultural heritage, industry niches may suffer more than average; the skiing industry in Lebanon and Morocco, for example, may be very hard hit by shortages of snow. Warmer temperatures and reduced precipitation will decrease residence time, intensity, and depth of snow in the mountains (Republic of Lebanon 2011); in Lebanon, a 2°C increase in temperature could shorten the skiing season from about 100 to 45 days. Coastal tourist destinations in Bahrain, Lebanon, and the United Arab Emirates are extremely vulnerable to extreme events, floods, saline intrusion, and loss of coastal areas (Tolba and Saab 2009). A portion of these interregional tourists is driven by shopping, with Dubai as a preferred destination. Their travel decisions are unlikely to be affected by climate change.

Medical tourism includes a wide array of tourists: temporary visitors from abroad, long-term foreign residents, cross-border health care to people from neighboring countries, and outsourced health care for patients sent abroad by health agencies. Other segments that could grow are wellness tourism and travel for discretionary surgery.[12] Morocco is becoming a significant player, and Lebanon has a significant presence in this market. However, solid data on tourism for wellness and discretionary surgery are not easily available and are outside the scope of this chapter.[13]

Medical tourism is a market with great potential for the Arab region. The Middle East attracts 13 percent of European and 2 percent of African and North American medical travelers. Tourists' choices are predominantly determined by quality of service and cost. Jordan is known as a quality medical provider with Anglophone staff trained at premiere Western hospitals (Pigato 2009). As a result, international medical tourism contributed 4 percent of GDP in 2010 (AP 2009). Tunisia has long been a front-runner in medical services to Libya, accounting for 80 percent of Tunisia's health tourism (Pigato 2009). Estimates show that foreign patients generated a quarter of the revenue and half of the employment in the Tunisian health care sector in 2005 (Lautier 2005). Tunisia's tremendous success in attracting Europeans for beach tourism combined with its strong health care sector creates opportunities to expand medical tourism. This tourist segment is not highly vulnerable to climate change. Some patients may be more sensitive to higher temperatures, especially if their treatment requires a longer recovery time, requires outpatient care, or is combined with tourism. The market could prefer cooler locations because of the higher heat sensitivity of an aging population. The segment has large risks of exposure to indirect impacts of climate change, discussed below, but with effective management of those impacts, medical services could become a climate-resilient tourist segment.

Some Results of Climate Change Will Indirectly Affect Tourism

Tariffs on greenhouse gas emissions imposed outside the region may reduce global tourism as the cost of travel increases (Simpson et al. 2008); the farther tourists need to travel to their destination, the higher the impact.

Increased migration as a result of globalization in the past 50 years has strengthened the tourism industry (Mustafa 2010). The desire to visit friends and relatives is a strong motivation for travel. Some destinations have even developed a separate tourism industry for emigrants to return to discover their roots. Climate change may further migration from the Arab region and possibly lead to an expansion of tourism for Arabs returning to their country of origin.

Political stability is essential for tourism. Lack of access to necessities because of climate change could raise the frequency of local and international conflicts. Local conflicts over water rights and access to utilities have already occurred: in August 2011, temperatures near 49°C and an electrical shortage led to public protests in Iraq (Arraf 2011). The risk of conflict will increase as the full impact of climate change materializes over the next century. Political instability could also result in international and national travel restrictions that would hinder tourism and migration. The results of the Arab Spring and its subsequent unrest are a testament to the effects of conflict on the tourism industry: in 2011, Egypt and Tunisia received only two-thirds of the tourists who had visited the prior year. The shortfall left an estimated 100,000 people without work in Tunisia (box 6.4) (Meriemdhaouadi 2011; Salam 2012). However, minor local conflicts have less effect on tourism, since tourists seek alternative destinations in the same nation or region. Medical tourism appears to be very sensitive to levels of conflict.

The health impacts of climate change can affect tourism greatly (see chapter 8 on general health impacts). Tourists prefer destinations with minimum health risks and avoid areas with longer epidemic seasons or elevated levels of tropical and infectious diseases. With an aging tourism population, health impacts may become increasingly important. Epidemics and risks of epidemics have significantly affected tourists' travel patterns in the past. In 2003, the presence of SARS reduced tourism in Thailand by 1.5 million arrivals and resulted in lost revenue of US$3.5 billion (Medical Review.com 2009).[14]

The Tourism Industry Depends on Climate Adaptation

The Arab region is already experiencing climate change impacts and immediate action is necessary, particularly given the long execution periods for investments in buildings and infrastructure for the tourism industry (Simpson et al. 2008). Many adaptations will be part of the general strategy needed for the wider economy: water conservation and protection of water resources are adaptation strategies that benefit the tourism industry,

BOX 6.4

Arab Spring: The 2011 Tourism Effect

Popular uprisings and the Arab Spring reduced tourism across the region. The loss of tourists was a reaction to evacuations in Egypt and Tunisia, civil war in Libya, and Syrian protests with violent government reactions (Smale 2011). Both Egypt and Tunisia lost one-third of tourists relative to 2010 (Salam 2012; TNA 2011; UNWTO 2011c). Egypt had 60 percent fewer tourists in March 2011 compared with 1.3 million in March 2010, and losses in the tourist sector amounted to US$2.2 billion in May 2011 (Middle East Online 2011). In contrast, Oman, Saudi Arabia, and the United Arab Emirates experienced steady growth in 2011 as tourists avoided Egypt and Tunisia (UNWTO 2012). Nevertheless, the tourism industry is confident that tourists will return in 2012 with a new government in Libya and democratic elections in Egypt and Tunisia (Smale 2011).

while protecting nature and increasing food security (boxes 6.5 and 6.6; see also chapters 3 and 4). These efforts would reduce stressors on natural ecosystems and increase their chances of acclimatizing and surviving the impacts of climate change (box 6.2).

Other adaptation measures are specific to the tourism industry: the large investments in fixed immobile capital assets, such as hotels or resorts, make adaptation more difficult and expensive. Table 6.3 provides an overview of the adaptation strategies for different stakeholders in the tourism industry.

Adaptation requires policies and operations and the involvement of all stakeholders. The tourism industry has already had to adapt to climate variation across the world; the hot and arid climate of the Arab region provides added challenges. Table 6.4 presents a summary of the general adaptation strategies required for each climate change impact. Of special importance is the development of stable and predictable government policies and institutional frameworks. Since tourism interacts with many other economic sectors, it will be important to develop integrated policies between government agencies, the private sector, industry and nongovernmental organizations. Success is contingent on the capacity to coordi-

BOX 6.5

Investing in Biodiversity: The Jordanian Dana Biosphere Reserve

The Jordanian Dana Biosphere Reserve uses ecotourism to diversify income and to build the resilience of local ecosystems. The Royal Society for the Conservation of Nature in Jordan works directly with local villages and Bedouin communities to develop tourist ventures. Since 1994, they have combined sustainable development, tourism, and wildlife services in the Jordanian Dana Biosphere Reserve. The reserve provides alternative livelihoods for local people, lessening their dependence on goat grazing and hunting. The Royal Society revived traditional gardens and created new sources of income based on high-value products from handicraft enterprises. In addition, the Dana Biosphere Reserve protects wildlife and scenic areas, while enriching nature by breeding and protecting animals facing extinction (the sand cat and Syrian wolf). The reserve won four international awards for sustainable development and provides significant economic benefits. It has 30,000 visitors a year (refer to spotlight 2 on biodiversity and ecosystem services for more information on biosphere reserves).

Sources: Djernaes, forthcoming; Harms 2010; Namrouqa 2009; RSCN 2011.

BOX 6.6

Unsustainable Water Use in the Tourism Industry

The tourism industry's average water use varies between 100 and 2,000 liters per guest per night. Larger and luxurious accommodations tend to consume considerably more. In comparison, basic household water requirements are estimated at 50 liters per person per day, excluding water for gardens. Indirect consumption for golf courses, irrigated gardens, and swimming pools is the main driver for the high water use in tourism. An 18-hole golf course may require more than 2.3 million liters of water a day.

Source: UNESCO 2006.

TABLE 6.3

Adaptation Strategies for Stakeholders in the Tourism Industry

Target	Tourism operators/ businesses	Tourism industry associations	Governments and communities	Financial sector (investors/ insurance providers)
Technical	• Collect rainwater • Develop systems for water recovery and recycling • Develop buildings and structures to better handle extreme weather events • Use low-impact building structures (green buildings and traditional experience) • Implement new technologies	• Enable access to early-warning equipment (such as radios) to tourism operators • Develop websites with practical information on adaptation measures	• Build water reservoirs • Develop warning systems for weather forecasting and early detection • Build desalination plants • Rationalize water consumption • Develop land-use zoning to consider climate risks • Enforce regulations	• Require advanced building design or material standards for insurance • Provide informational material to customers • Require adherence to regulations
Managerial	• Develop water conservation plans • Close facilities in low season • Diversify products and markets • Diversify regional business operations • Redirect clients away from affected destinations • Integrate environmental management of operations to reduce impacts	• Use short-term seasonal forecasts for planning marketing activities • Develop training programs on climate change adaptation • Encourage environmental management with firms (for example, through certification) • Develop disaster risk management	• Develop disaster management and evacuation plans • Develop impact management plans (such as the Coral Bleaching Response Plan) • Implement convention/ event interruption insurance • Develop risk-reduction strategies • Restrict business licenses to businesses in high-risk locations • Require specific adaptation measures	• Develop insurance that addresses the Arab market and adjust insurance premiums to reflect risks • Restrict lending to high-risk business operations
Policy	• Develop extreme weather interruption guarantees • Stay in compliance with regulation (for example, building codes)	• Coordinate political lobbying and mainstream adaptation efforts • Seek funding to implement adaptation projects	• Develop coastal management plans with setback requirements • Set building design standards (for example, extreme weather)	• Consider climate change in credit risk and project finance assessments
Research	• Assess weather impact of site location • Reduce disaster risk	• Assess awareness of businesses and tourists as well as knowledge gaps	• Monitor environmental risks (for example, predict bleaching risk, beach water quality, dust storms) • Reduce disaster risk	• Understand extreme weather event risk exposure • Understand measures to minimize exposures
Education	• Educate employees and tourists in water conservation • Increase environmental impact awareness	• Develop public education campaigns (such as keep the water clean, benefits of protecting Marine Protected Areas)	• Institute water conservation campaigns • Introduce campaigns on the dangers of UV radiation	• Educate and inform potential and existing customers
Behavioral	• Develop real-time information on weather conditions and forecasts	• Develop water conservation initiatives	• Develop plans for extreme event recovery (including marketing to regain tourists in the event of disaster)	• Develop good in-house practices

Source: Adapted from Simpson et al. 2008.

TABLE 6.4

Adaptation Strategies for Climate Change Impacts on Tourism in General

Temperature increase	Increased aridity	Extreme events	Sea-level rise	Sea salination, acidification
• Implement environmental conservation and preservation • Provide tourists easy access to information related to heat exposure and risk reduction • Implement strict vaccination requirements in advance of travel; may reduce some of the impact of increased range and seasons for diseases • Incorporate traditional building methods and architecture, and new technologies to increase sustainability of facilities • Use sustainable energy for cooling	• Require less watering of surrounding areas, limiting or eliminating grassy golf courses • Increase rainwater collection • Increase recycling and recovery of water • Develop water reservoirs and alternative water supply • Develop public policies on water usage • Increase incorporation of traditional building methods and architecture, and new technologies to increase sustainability of facilities	• Improve weather forecasts and warnings, including local warnings on weather with specific health information (such as ozone levels, dust levels, temperature, storms) • Disseminate information on adaptation measures for extreme weather • Use building design and structure to manage extreme weather exposure • Develop land-use zoning to avoid exposed areas • Develop risk-reduction strategies • Develop disaster risk management and evacuation plans • Develop impact management and contingency plans	• Secure or relocate most exposed infrastructure • Develop coastal reinforcements where necessary, preferably using natural systems and with minimal impact on biodiversity • Implement and uphold zoning restrictions considering disaster risks	• Reduce future emissions and limit the long-term effects

Source: Djernaes, forthcoming.

nate and promote public-private partnerships, disseminate and incentivize best practices, and uphold regulations with transparency (Abaza, Saab, and Zeitoon 2011).

For adaptation of traditional tourist segments, it is important to minimize the growth in direct costs associated with climate change through implementation of adaptation measures (see also tables 6.4 and 6.5):

- Water management (box 6.6)
- Protection of natural environments and cultural heritage
- Zoning of land use for minimal exposure
- Disaster risk management and evacuation plans
- Reinforcement of new and existing buildings to withstand stronger extreme weather
- Coastal erosion control and protection

TABLE 6.5

Adaptation Strategies for Tourism Segments

Segment	Increased temperatures	Increased aridity	Extreme events	Sea-level rise	Sea salination, acidification
Arab vacationers	• Plan events and activities to minimize exposure	• Plan events and activities to minimize exposure	• Protect sites from weather impact, if necessary • Relocate facilities, if necessary Use short-term and seasonal forecasts to plan activities	• Reinforce exposed areas • Secure coastal areas with natural eco-systems to increase biodiversity • Reduce ecosystem stressors to maximize ability to handle climate change stress • Develop alternative tourist destinations away from exposed areas	
Nationals residing abroad	• Implement government policies and programs to incentive solar energy for cooling	• Encourage collection of rainwater and water recycling • Ensure sufficient water for basic needs	• Ensure that communities understand how to minimize exposure • Enforce strict building codes to minimize damage and risks	• Implement and enforce restrictions to ensure that the most exposed areas are not used for settlements	
Religion	• Build additional facilities for shade for pilgrims • Build cooled facilities with energy requirements met through solar energy	• Build additional facilities to ensure water availability	• Enhance warning systems and improve disaster management plans		
Cultural heritage	• Provide cooler facilities at cultural heritage sites supplied by renewable energy • Attract tourists in colder months	• Build additional facilities to ensure water availability at all cultural heritage sites	• Use short-term and seasonal forecasts to plan activities • Protect sites from weather impacts, if necessary	• Secure cultural heritage sites that bring the most tourists • Secure sites with long-term potential to attract tourists	
Nature	• Plan events and activities to minimize exposure	• Plan events and activities to minimize exposure	• Implement incentives for existing facility owners to reduce their exposure to extreme events • Use short-term and seasonal forecasts to plan activities • Relocate facilities, if necessary	• Secure coastal areas with natural eco-systems to increase biodiversity • Reduce stressors on ecosystems to maximize the ability to handle stress associated with climate change • Develop alternative tourist destinations	• Monitor and manage water systems (predict bleaching risk, water quality)
Medical reasons	• Disseminate heat exposure information • Enhance prevention programs • Secure energy availability through local renewable energy	• Ensure sufficient water for basic needs	• Ensure that populations understand how to minimize exposure • Enforce strict building codes for health facilities to minimize damage and risks		

Source: Djernaes, forthcoming.

Risk Reduction Is Necessary for Adequate Adaptation

Large investors in the tourism industry often have facilities in multiple locations. Diversifying income sources reduces the economic risk of a single climate disaster. In contrast, the local population is very dependent on specific tourist locations and has few or no opportunities for developing risk-reduction strategies. Any disaster will be felt more profoundly by these communities, with long-term effects on their livelihoods. National strategies for disaster risk reduction are essential for the social protection of local populations and businesses (see spotlight 1 on disaster risk management).

Shifting to Segments with Higher Returns Will Boost Income and Diversify Livelihoods

Certain sectors of the tourism industry are more productive: visits by nationals residing abroad provide low returns, whereas medical tourism and conference tourism produce some of the highest. In general, employment and economic returns increase with service level, luxury, and tourists' purchasing power. Expanding into these segments can make tourism in the Arab region more productive and more resilient to the effects of climate change.

Medical tourism is a growing segment of the global economy. An expansion into this market relies heavily on capacity building through large capital investments and high levels of education. Such investments could produce substantial secondary effects on the economy as a whole. An expansion of medical facilities for tourism could also contribute to improved overall livelihoods and medical services for the local population.

Business travel and conference participation yield a 10-to-1 return on investment for businesses and present a promising sector for expansion of tourism to Arab countries (see WTTC 2011a).[15] Conference participants are usually influential members of the business world, and positive experiences can generate large benefits in additional business tourists. This segment is developed in Oman and Qatar, but it can expand in other Arab countries. Because business travelers are not primarily seeking recreation, this segment can use existing facilities in nonpeak seasons (for example, the Maghreb in the winter).

The Maghreb countries are well positioned for expansion and diversification in tourism: Algeria and Morocco have very attractive climates and underused tourist attractions. If these countries grew the tourist sector by only 1 percent of GDP, they would create 36,700–47,200 jobs based on their current tourism model. Expanding the tourism industry would increase its economic value by 11 percent in Algeria and 5 percent in Morocco. Another growth opportunity is to develop the tourism model and expand tourist segments with higher returns. For Morocco, the segments to focus on are cultural heritage, nature, and conference. Such an expansion could produce returns resembling those of Tunisia. Morocco can combine these two opportunities: first, by expanding the size of the industry by 5 percent and second, by developing the tourism model. The combined GDP effect would be an estimated US$2 billion or 2 percent growth in GDP (table 6.6). In Algeria, the combined GDP effect would be US$1.5 billion (Djernaes, forthcoming).

Tunisia's strong health care industry has the potential to expand into medical tourism; particularly given Tunisia's successful track record of attracting European tourists for summer vacations. In 2000, the World Trade Organization ranked Tunisia's health system 52nd in the world, ahead of Egypt, India, and Jordan and just behind Thailand (Ahmed and Sanchez-Triana 2008; WHO 2000). Tunisia could also increase returns by expanding its market to cover more business and conference tourism.

Among the Mashreq countries, Egypt could focus on health care and conferences to bring higher returns. Both Jordan and Lebanon have opportunities to expand medical tourism by attracting not only Arab patients but also patients from European and American markets. However, medical tourism is particularly sensitive to political unrest, and the sociopolitical stability of these countries will be a critical factor in further expansion. The perception of instability in Lebanon, for example, appears to have been a major constraint in its ability to tap into the broader international medical tourism market (Connell 2010; Luca 2010).

Least developed countries can in general receive large benefits by developing the tourism sector; however, large up-front capital investments are necessary, and attracting the necessary international investors will be a challenge.

Diversification of Tourist Sectors Provides Opportunities for Geographic Expansion

Tourism has traditionally been confined to limited areas in Arab countries; Tunisia, for example, originally sought to attract the economic benefits of tourism but limited tourist exposures to specific tourist locations (Berriane 1999). The democratization process may provide the opportunity to expand cultural heritage tourism into new regions and away from climate-vulnerable coastal resources. Cultural heritage tourism can further exploit the many impressive ruins left by ancient cultures in the landscape surrounding the Mediterranean. These ruins have great potential as alternative tourist attractions (see boxes 6.1 and 6.7).

TABLE 6.6

Annual GDP Contribution by Tourist Sector per Employee

	GDP contribution per employee in the tourist sector[a] (US$ thousands)	Additional employment if tourism sector grew GDP by 1% (thousands of people employed)
Maghreb		
Algeria	38.1	36.7
Libya	65.2	9.0
Morocco	19.3	47.2
Tunisia	41.0	10.6
Mashreq		
Egypt, Arab Rep.	24.7	76.5
Iraq[b]	n.a.	25.6
Jordan	44.6	5.3
Lebanon	82.2	4.2
Syrian Arab Rep.	28.2	19.1
LDCs		
Comoros[b]	n.a.	0.2
Djibouti[b]	n.a.	0.4
Mauritania[b]	n.a.	1.2
Sudan[b]	n.a.	21.0
Somalia[b]	n.a.	n.a.
Yemen, Rep.	17.4	14.4
Gulf		
Bahrain	117.2	1.6
Kuwait	66.4	16.5
Oman	108.6	4.3
Qatar	503.8	1.9
Saudi Arabia	100.7	37.5
United Arab Emirates	332.2	8.1

Sources: IMF 2011; WTTC 2011b.

Note: LDCs = least developed countries; n.a. = not applicable.

a. GDP per employee is based on data for the period 1988–2009 by country.

b. National historic data are unavailable. Calculations of additional employment are based on an average for countries with GDP per employee of less than US$70,000.

The geographic expansion of tourism provides opportunities to involve different communities and to increase the diversity of their livelihoods. In addition, it will create a tourism industry that is less reliant on only a few natural resources, such as coral reefs, and that will reduce overall industry risks through diversification. These alternative tourist options are unlikely to fully replace lost revenues in the traditional segments. To ensure long-term sustainability, the carrying capacity of the environment must be considered. Some rural areas in the Arab region have a rich cultural tradition that could offer fascinating experiences for tourists; the Bedouin lead vacationers on tours of the desert, offering a unique perspective on sustainable life in a forbidding environment (box 6.7).

BOX 6.7

Cultural and Natural Preservation through Sustainable Tourism: Diversifying Local Livelihoods

Alternative tourism has been successfully implemented on a small scale. These projects develop sustainable tourism that preserves culture and social networks, while promoting cultural heritage and natural preservation.

Tourism in the Sahara is fairly new and has potential as an alternative tourist activity. Saharan tourism is often combined with Tunisian coastal tourism or cultural tourism in Egypt, Morocco, and Tunisia. By involving the Saharan population, tourism can stimulate development, alleviate poverty, and serve as an income-diversifying strategy for nomadic peoples, such as the Tuareg in Algeria. The exploration of fragile natural resources requires a strategy for sustainability to preserve the natural heritage (UNESCO 2003).

The Al-Shouf Cedar Nature Reserve in Lebanon extends from Dahr Al-Baidar in the north to Niha Mountain near Jezzine in the south. The reserve is an example of sustainable tourism aimed at preserving the natural, historical, biological, and environmental heritage while developing the local community. Covered by oak trees, the eastern slopes overlook the beautiful scenery of the Bekaa Valley. At the western slopes' peak, the cedar forest provides a spectacular attraction. The cedar forest is in the process of renewal since the implementation of forest conservation and anti-overgrazing measures. Wolves, hyenas, mountain deer, and ibex reside in the reserve. Local people work as tourist guides and in the production and sale of traditional products to visitors (LAS and UNEP 2007).

The Siwa Oasis project in Egypt is an example of inland sustainable tourism founded on local customs and practices. The project provides more than 200 permanent jobs and 400 auxiliary functions for involvement in the project's design, implementation, management, and traditional handicraft industries. Local capacities and resources were developed while supporting the local social environment. Houses and tourist resorts were built by using 2,500-year-old architectural methods (LAS and UNEP 2007).

The United Nations Educational, Scientific, and Cultural Organization supports tourism that integrates biodiversity conservation with socioeconomic values through its certification of biosphere reserves. The concept provides tourists with experiences that combine educational, cultural, and natural values. The program promotes science, participatory research, education, and environmental monitoring. Tourists are exposed to rural and urban landscapes with opportunities to observe wildlife, interact with traditional communities, and learn about sustainable living.

These programs attract academics, researchers, students, schools, and private sector establishments, in addition to more traditional tourists (UNESCO 2008). The Jordanian Dana Biosphere Reserve is one successful example from the Arab region (box 6.5). It provides a model that can be expanded.

Private and Public Sector Interventions Are Needed for Sustainable Tourism

Tourism is important for economic growth, foreign exchange earnings, and employment:

- *Tourism contributes 3 percent of GDP in the Arab region*, and 93 percent is concentrated in Egypt, Jordan, Lebanon, Morocco, Saudi Arabia, Syria, Tunisia, and the United Arab Emirates.
- *Tourism contributes between 15 and 32 percent of total employment* in Egypt, Jordan, Lebanon, Morocco, and Tunisia.
- *Tourism is important for foreign exchange earnings*; tourism constitutes 24 percent of total exports of goods and services in the Mashreq and Morocco.
- *Tourism is an opportunity to diversify livelihoods and related risks* by providing a source of income that is independent of oil and agriculture.

Tourism is vulnerable to the impacts of climate change, but more important, the tourism industry provides a good opportunity for the Arab region to diversify livelihoods and reduce risk exposure, especially when facing climate change.

- *The largest and most immediate threat to tourism is among tourists seeking nature, culture, and recreation.*
- *Current tourism facilities and important heritage sites may be lost* because of impacts from extreme weather and in the longer term because of higher sea levels.
- *Religious and medical tourists are the least affected by climate change*. Religious tourists are less driven by comfort, and medical tourists are more dependent on indoor climate.

Adaptation

- *Adaptation requires actions with regard to policies and operations with involvement of all stakeholders*. The tourism industry has already had to adapt to climate variations across the world. Nevertheless, climate change adaptation in the Arab region provides added challenges since its climate is already hot and arid.

- *Tourism's ability to provide income and employment for the growing population is important*, and continued sustainable investment in tourism should still be pursued.
- *Adaptation in the tourism industry will require minimizing the growth in direct costs associated with climate change* through implementation of several general measures: (a) water management, (b) protection of natural environments and cultural heritage, (c) land-use zoning for minimal exposure, (d) disaster risk management and evacuation plans, (e) increased efforts to reinforce new and existing buildings to withstand stronger extreme weather impacts, (f) coastal erosion control and protection where necessary, and (g) public health management.
- *Opportunities exist to expand the industry into new areas* (medical, conferences, and inland). Diversification of the tourism industry provides opportunities to involve different populations and to increase the diversity of their livelihoods. In addition, it will create an industry that is less reliant on fragile natural resources, such as coral reefs.

Legislation and Governance

- *Appropriate governance for coordination and cooperation should be developed* across tourist-related sectors for all relevant stakeholders.
- *Governments should foster the design of capacity-building programs for the tourism sector*, focusing on training and creating opportunities to adapt to the effects of climate change in tourism.
- *Legislation and policies need to be developed that seek and create opportunities to increase sustainable business practices* that incorporate climate change adaptation practices and activities.

National Industry Knowledge Development

- *Detailed data should be collected* to calculate the economic contribution of different tourist segments.
- *Data should be generated on the number of tourists* attracted to the countries for the different segments of the industry.
- *Data should be collected on employment* and the contribution to the total employment by segment.

Climate Knowledge Development

- *An information database should be developed, and access improved,* incorporating the analysis and data on climate change modeling and monitoring.
- *Research should be conducted* on the climate change impacts on the sector.
- *Education and campaigns to raise awareness should be carried out* for all stakeholders in the industry.

Notes

1. *Direct spending* includes revenue generated by industries dealing directly with tourists, including visitor exports; domestic spending; and government individual spending. *Total spending* includes capital investment spending by all sectors directly involved in the travel and tourism industry, including investment spending by other industries on specific tourism assets (visitor accommodation, transport equipment, restaurants, and facilities for tourism use).
2. Data are based on UNWTO's statistics of the nationality of arriving nonresident visitors at national borders, except in Oman, Qatar, and the West Bank and Gaza, where records are made at hotels. UNWTO has no data for Djibouti, Mauritania, and Somalia, and no detailed information on tourist origination for the United Arab Emirates. These nations were excluded from the origination analysis.
3. *Direct employment* includes traditional tourist sector services, such as lodging and transportation. *Indirect employment* includes tourism-related investment, public spending, and export of goods by tourists.
4. Based on preliminary findings from analyses of employment and total GDP contribution for the period 1988–2009.
5. The League of Arab States, the United Nations International Strategy for Disaster Reduction, and the Regional Bureau for Arab States brought the relevant Arab governments, civil society organizations, and businesses together in 2010, and the Arab Strategy for disaster risk reduction was adopted by the Council of Arab Ministers Responsible for the Environment in December 2010 and by the Arab Economic and Social Council in September 2011. Currently, work focuses on developing a regional program of implementation: the program's effectiveness has yet to be established (LAS 2011; UNISDR 2011).
6. According to World Trade Organization statistics for the region, only 63 percent of the international tourists indicate a reason for travel.
7. This segment includes Arabs residing outside their country of citizenship who return home for temporary visits. Nationals with changed citizenship are counted as nationals of their acquired homeland; therefore, this group may be underestimated. Also, not all countries report this group of tourists. Historically, Lebanon receives large numbers of tourists who are nationals with changed citizenships.

8. The remainder return to Algeria, Syria, and Jordan (19 percent, 16 percent, and 11 percent, respectively).
9. Of the arrivals to Saudi Arabia, 88 percent originate from nations with Muslim majority populations (based on data from the Pew Forum on Religion and Public Life 2011 and UNWTO 2011d).
10. The Tunisian government supported the development of beach tourism in the early 1960s. The segment experienced substantial growth from 4,000 beds in 1960 to 164,612 beds in 2009 (Berriane 1999).
11. The most recent details available from Bahrain with regard to origination of tourists are for 2007.
12. Wellness tourism consists predominantly of healthy individuals seeking spa vacations for preventative health care and wellness. Discretionary procedures are elective medical procedures that are not required for health reasons.
13. In 2006, the global medical tourism market was approximately US$60 billion and is estimated to grow to US$100 billion in 2012. The largest number of inpatient medical travelers (40 percent) needs advanced technologies that are typically found in the United States. Better care than is available at home drives 32 percent of the market, often from outside the most developed world. Long local wait times on medically necessary procedures drive 15 percent of the market for orthopedics, general surgery, and cardiology. Nine percent of medical travelers are in search of lower costs; and only 4 percent are seeking discretionary procedures, with the United States being the largest consumer (Ehrbech, Guevara, and Mango 2008; Herrick 2007; Pigato 2009).
14. Thailand had four confirmed cases of SARS and two deaths in 2003 (Medical Review.com 2009).
15. The business segment is also much more demanding with regard to direct access by air travel from major business centers, and availability of Internet and technical support.

References

Abaza, Hussein, Najib Saab, and Bashar Zeitoon, eds. 2011. *Arab Environment Green Economy: Sustainable Transition in a Changing Arab World*. Beirut: Arab Forum for Environment and Development.

Agrawala, Shardul, Annett Moehner, Mohamed El Raey, Declan Conway, Maarten van Aalst, Marca Hagenstad, and Joel Smith. 2004. *Development and Climate Change in Egypt: Focus on Coastal Resources and the Nile*. Paris: Organisation for Economic Co-operation and Development. http://www.oecd.org/dataoecd/57/4/33330510.pdf.

Ahmed, Kulsum, and Ernesto Sanchez-Triana, eds. 2008. *Strategic Environmental Assessment for Policies: An Instrument for Good Governance*. Washington, DC: World Bank.

Ansary, Tamim. 2009. *Destiny Disrupted: A History of the World through Islamic Eyes*. New York: Public Affairs.

AP (Associated Press). 2009. "Jordan Launches Medical Tourism Advertising Campaign in U.S." AP, July 13. http://www.haaretz.com/news/jordan-launches-medical-tourism-advertising-campaign-in-u-s-1.279922.

ArchAtlas. 2006. Department of Archaeology, University of Sheffield, U.K. http://www.archatlas.dept.shef.ac.uk/Home.php.

Arraf, Jane. 2011. "Iraq's Government Shuts Down Amid 120-Degree Temps—and No A/C." *Christian Science Monitor*, August 1. http://www.csmonitor.com/World/Middle-East/2011/0801/Iraq-s-government-shuts-down-amid-120-degree-temps-and-no-A-C.

Berriane, Mohammed. 1999. *World Decade for Cultural Development, Tourism, Culture, and Development in the Arab Region*. Paris: United Nations Educational, Scientific, and Cultural Organization.

Bigano, Andrea, Alessandra Goria, Jacqueline Hamilton, and Richard S. J. Tol. 2006. "The Effect of Climate Change and Extreme Weather Events on Tourism." Working Paper 30, Fondazione Eni Enrico Mattei, Milan.

BMI (Business Monitor International). 2011. "Tourism Report: Q4 2011 Oman." London.

Burke, Lauretta, Kathleen Reytar, Mark Spalding, and Allison Perry. 2011. *Reefs at Risk, Revisited*. Washington, DC: World Resources Institute. http://pdf.wri.org/reefs_at_risk_revisited.pdf.

Cesar, Herman. 2003. "Economic Valuation of the Egyptian Red Sea Coral Reef." Monitoring, Verification, and Evaluation Unit of the Egyptian Environmental Policy Program, Cairo. http://earthmind.net/marine/docs/egyptian-red-sea-reefs-valuation.pdf.

Chiffings, Anthony W. 1995. "Marine Region 11: Arabian Seas." In *A Global Representative System of Marine Protected Areas*, vol. 4, ed. Graeme Kelleher, Chris Bleakley, and Susan M. Wells, 40–71. Gland, Switzerland: International Union for Conservation of Nature. http://www.environmentservices.com/projects/programs/RedSeaCD/DATA/PDF/Global_Representative_System_MPA_Arabian.pdf.

Connell, John. 2010. *Health and Medical Tourism*. Wallingford, U.K.: CABI Publishing.

Djernaes, Marie. Forthcoming. "The Impact of Climate Change on Tourism in the Arab World." World Bank, Washington, DC.

Economist. 2012. "A Choice of Models: Theme and Variations—State Capitalism Is Not All the Same." *Economist*, January 21. http://www.economist.com/node/21542924.

EEAA (Egyptian Environmental Affairs Agency). 2010. *Egypt: Second National Communication under the United Nations Framework Convention on Climate Change*. Cairo: EEAA. http://unfccc.int/resource/docs/natc/egync2.pdf.

Ehrbeck, Tilman, Ceani Guevara, and Paul D. Mango. 2008. "Mapping the Market for Medical Travel." *McKinsey Quarterly* (May). https://www.mckinseyquarterly.com/Health_Care/Strategy_Analysis/Mapping_the_market_for_travel_2134.

El Raey, Mohamed. 1995. "Impact of Sea Level Rise on the Arab Region." Arab Climate Resilience Initiative Working Paper, United Nations Development Programme, Regional Bureau for Arab States, Cairo. http://www.arabclimateinitiative.org/Countries/egypt/ElRaey_Impact_of_Sea_Level_Rise_on_the_Arab_Region.pdf.

Geneva Association. 2009. *The Insurance Industry and Climate Change: Contribution to the Global Debate*. Geneva: Geneva Association. http://www.genevaassociation.org/PDF/Geneva_Reports/Geneva_report%5B2%5D.pdf.

Hamilton, Jacqueline M., and Richard S. J. Tol. 2007a. "The Impact of Climate Change on Recreation and Tourism." Working Paper FNU52, Hamburg University Centre for Marine and Climate Research, Hamburg, Germany.

———. 2007b. "The Impact of Climate Change on Tourism in Germany, the U.K., and Ireland: A Simulation Study." *Regional Environmental Change* 7: 161–72.

Harms, Erika. 2010. "Sustainable Tourism: From Nice to Have to Need to Have." In *Trends and Issues in Global Tourism 2010*, ed. Roland Conrady and Martin Buck, 111–17. Berlin and Heidelberg: Springer Verlag.

Herrick, Devon M. 2007. "Medical Tourism: Global Competition in Health Care." Policy Report 304, National Center for Policy Analysis, Dallas. http://www.medretreat.com/templates/UserFiles/Documents/Medical%20Tourism%20-%20NCPA%20Report.pdf.

IMF (International Monetary Fund). 2011. World Economic Outlook Database. http://www.imf.org/external/pubs/ft/weo/2011/02/weodata/index.aspx.

Kamoonpuri, Hasan. 2010. "Royal Orders to Plant Date Palm and Coconut Trees a Boon for Country." *Oman Daily Observer*, October 22. http://omanobserver.om/node/27273.

LAS (League of Arab States). 2011. "The Arab Strategy for Disaster Risk Reduction 2020." UN International Strategy for Disaster Reduction, Geneva. http://www.unisdr.org/we/inform/publications/18903.

LAS and UNEP (League of Arab States and United Nations Environment Programme). 2007. "Guidelines for Sustainable Tourism in the Arab Region." LAS and UNEP.

Lautier, Marc. 2005. "Exportations de Services de Santé des Pays en Développement: Le Cas Tunisien." Notes et Documents 25, Agencé Française de Développement, Paris. http://www.afd.fr/webdav/site/afd/shared/PUBLICATIONS/RECHERCHE/Archives/Notes-et-documents/25-notes-documents.pdf.

Lester, Rodney. 2010. "The Insurance Sector in the Middle East and North Africa: Challenges and Development Agenda." World Bank, Washington, DC. http://siteresources.worldbank.org/INTMNAREGTOPPOVRED/Resources/MENAFlagshipInsurance12_20_10.pdf.

Lise, Wietze, and Richard S. J. Tol. 2002. "Impact of Climate on Tourist Demand." *Climatic Change* 55 (4): 429–49.

Luca, Ana Maria. 2010. "Medical Tourism in Lebanon." *Agenda*, March 10. http://www.nowlebanon.com/NewsArchiveDetails.aspx?ID=152452.

Majidzadeh, Yousef. 1982. "Lapis Lazuli and the Great Khorasan Road." *Paléorient* 8 (1): 59–69. http://www.persee.fr/web/revues/home/prescript/article/paleo_0153-9345_1982_num_8_1_4309.

Malik, Asmaa. 2012. "UAE: Abu Dhabi Achieving Major Developments in Industry, Transport and Tourism." Global Arab Network, January 17. http://www.english.globalarabnetwork.com/2012011712353/Economics/uae-abu-dhabi-achieving-major-developments-in-industry-transport-and-tourism.html.

McDermott, Mat. 2011. "The First Worldwide Green Haji Guide and Eco-Mosques in Qatar: More on Islam and the Environment." TreeHugger Radio, September 13. http://www.treehugger.com/travel/the-first-worldwide-green-hajj-guide-eco-mosques-in-qatar-more-on-islam-the-environment.html.

Medical Review.com. 2009. "Pandemic Threat: Tourism Industry Nightmare?" Medical Review.com, October. http://www.google.com/url?sa=t&rct=j&q=&esrc=s&frm=1&source=web&cd=2&ved=0CC4QFjAB&url=http%3A%2F%2Fwww.tourism-review.com%2Ffm974%2Fmedical-pandemic-threat-tourism-industry-nightmare.pdf&ei=X-5KT7DaI7OmsALpktTqCA&usg=AFQjCNF4fYZzfA_WfhZEALXSqAsXtlZrNQ&sig2=4TvfihxaJVsDtgXCAFaUvg.

Meriemdhaouadi. 2011. "100,000 Jobs Lost in Tunisia Tourism Sector This Year." *Tunisialive*, September 20. http://www.tunisia-live.net/2011/09/20/post-revolution-tunisia-around-100-000-employee-lost-their-jobs-in-tourism-sector/.

Middle East Online. 2011. "Tourism Crisis in Egypt after the Revolution." *Middle East Online*, May 12. http://www.middle-east-online.com/?id=11306.

Ministry of Tourism of Egypt. 2010. "The Statistical Report 2010." Ministry of Tourism, Cairo.

Moreno, Alvaro, and Bas Amelung. 2009. "Climate Change and Coastal and Marine Tourism: Review and Analysis." *Journal of Coastal Research* 56: 1140–44.

MSN Weather. 2011. "Temperature Data." http://weather.uk.msn.com.

Mustafa, Mairna H. 2010. "Tourism and Globalization in the Arab World." *International Journal of Business and Social Science* 1 (1): 37-48. http://www.eis.hu.edu.jo/Deanshipfiles/pub105441268.pdf.

Namrouqa, Hana. 2009. "Jordan Eco-Lodge Ranked High on World List." *Jordan Times*, March 8. http://jordantimes.com/?news=14851&searchFor=GAM%20to%20construct%20exhibition%20centre.

Pew Forum on Religion and Public Life. 2011. "The Future of the Global Muslim Population: Muslim Population by Country." February 26. http://features.pewforum.org/muslim-population/#.

Phillips, Michael R., and Andrew L. Jones. 2006. "Erosion and Tourism Infrastructure in the Coastal Zone: Problems, Consequences and Management." *Tourism Management* 27 (3): 517–24.

Pigato, Miria. 2009. "Challenges and Opportunities in Global Services Trade." *Strengthening China's and India's Trade and Investment Ties to the Middle East and North Africa*. Washington, DC: World Bank. http://siteresources.worldbank.org/INTMENA/Resources/MENA_China_India_Sept08-3.pdf.

Republic of Lebanon. 2011. *Lebanon's Second National Communication to the United Nations Framework, Convention on Climate Change*. Beirut: Ministry of the Environment, Republic of Lebanon.

RSCN (Royal Society for the Conservation of Nature). 2011. Wild Jordan website. http://www.rscn.org.jo/orgsite/Group1/AboutWildJordan/tabid/162/Default.aspx.

Salam, Mohamed Abdel. 2012. "Tourism in Egypt down 27.5 Percent in November, 33 Percent Overall." *Bikyamasr*, January 17. http://bikyamasr.com/54015/tourism-in-egypt-down-27-5-percent-in-november-33-percent-overall/.

Simpson, Murray C., Stefan Gössling, Daniel Scott, C. Michael Hall, and Elizabeth Gladin. 2008. *Climate Change Adaptation and Mitigation in the Tourism Sector: Frameworks, Tools and Practices*. Paris: United Nations Environment Programme, Oxford University, United Nations World Tourism Organization, and World Meteorological Organization. http://sdt.unwto.org/sites/all/files/docpdf/ccoxford.pdf.

Smale, Will. 2011. "Arab Nations Aim to Win Back Tourists." BBC News, November 10. http://www.bbc.co.uk/news/business-15651730.

TEC (Tourisme, Transports, Territoires, Environnement Conseil). 2010. *Stratégie Nationale d'Adaptation au Changement Climatique du Secteur Touristique en Tunisie*. Tunis: Ministère de l'Environnement et du Développement Durable Coopération Technique Allemande.

TNA (Tunisian News Agency). 2011. "Tourist Entries Down 33.3pc (First 10 Months of 2011)." TNA, Tunis, November 11. http://www.tap.info.tn/en/en/economy/7185-tourist-entries-down-333pc-first-10-months-of-2011.html.

Tolba, Mostafa K., and Najib W. Saab, eds. 2009. *Arabic Environment: Climate Change—Impact of Climate Change on Arab Countries*. Beirut: Arab Forum for Environment and Development. http://www.afedonline.org/afedreport09/Full%20English%20Report.pdf.

Travellerspoint. 2011. "Mecca." http://www.travellerspoint.com/guide/Mecca/.

UNESCO (United Nations Educational, Scientific, and Cultural Organization). 2003. *The Sahara of Cultures and People: Towards a Strategy for the Sustainable Development of Tourism in the Sahara, in the Context of Combating Poverty*. Paris: UNESCO.

———. 2006. *Water and Tourism*. UNESCO Water e-Newsletter 155. http://www.unesco.org/water/news/newsletter/155.shtml.

———. 2008. *Friendly Tourism in Arabian Biosphere Reserves: Case Study—Al Reem, Qatar*. Doha: UNESCO Doha Office, 2008. http://unesdoc.unesco.org/images/0018/001803/180332e.pdf.

UNISDR (United Nations International Strategy for Disaster Reduction). 2011. "Strengthening Impact of Cities Campaign." UNISDR, Geneva. http://www.unisdr.org/archive/24170.

UNWTO (United Nations World Tourism Organization). 2011a. *Tourism Highlights*. Madrid: UNWTO.

———. 2011b. *Tourism toward 2030: Global Overview*. Madrid: UNWTO.

———. 2011c. *World Tourism Barometer* 9 (2) and Statistical Annex. http://mkt.unwto.org/en/barometer.

———. 2011d. *Yearbook of Tourism Statistics Data 2005–2009*. Madrid: UNWTO.

———. 2012. "International Tourism to Reach One Billion in 2012." Press release PR12002, UNWTO, Madrid, January 16. http://media.unwto.org/en/press-release/2012-01-16/international-tourism-reach-one-billion-2012.

Waleed, Hazbun. 1997. "The Development of Tourism Industries in the Arab World: Trapped between the Forces of Economic Globalization and Cultural

Commodification." Paper presented at the 30th Annual Convention of the Association of Arab-American University Graduates on the Theme of Arabs, Arab Americans, and the Global Community, Washington, DC, November 1.

WHO (World Health Organization). 2000. "The World Health Report 2000: Health Systems—Improving Performance." Press release, WHO, Geneva. http://www.who.int/whr/2000/en/.

World Bank. 2008. "Tunisia's Global Integration: Second Generation of Reforms to Boost Growth and Employment." Report 40129-TN, World Bank, Washington, DC.

———. 2011. World dataBank website. http://databank.worldbank.org.

WTTC (World Travel and Tourism Council). 2011a. "Executive Summary." *Business Travel: A Catalyst for Economic Performance*. London: WTTC. http://www.wttc.org/site_media/uploads/downloads/WTTC_Business_Travel_2011-_Executive_Summary.pdf.

———. 2011b. "Data Search Tool: Economic Impact Research." http://www.wttc.org/research/.

CHAPTER 7

Gender-Responsive Climate Change Adaptation: Ensuring Effectiveness and Sustainability

The purpose of this chapter is to develop a better understanding of gender-based vulnerabilities and opportunities in climate change adaptation in Arab countries and to propose policy options to address them. The chapter argues that (a) climate change impacts in this region are not gender neutral, (b) specific gender inequalities intensify vulnerability to climate change by increasing sensitivity and reducing adaptive capacity, and (c) to strengthen resilience to climate change, it is essential to build a holistic and gender-responsive approach to adaptation that empowers women as agents of change. This chapter concentrates on rural populations because, generally, they are the most vulnerable to climate change and because the gender-related aspects of climate change are arguably more pronounced. Rural livelihoods are more exposed to climate events, and lower human development, higher poverty rates, and limited access to resources contribute to greater sensitivity and gender disparity. Thus, all references to women and men are to rural populations unless otherwise identified.

This chapter first provides background on vulnerability, the three main arguments listed above, and the global and regional policy background for gender and climate change. The next section develops a profile of gender-based vulnerability to climate change in Arab countries. The third section provides examples of how adaptation can be approached from a gender perspective, with four case studies from the Republic of Yemen, Morocco, the Syrian Arab Republic, and Jordan. Finally, the chapter concludes with policy options.

Photograph by Dorte Verner

Background

Understanding Vulnerability

Climate vulnerability is defined as "a function of the sensitivity of a system to changes in climate, adaptive capacity, and the degree of exposure of the system to climatic hazards" (McCarthy et al. 2001; also see chapter 1 of this volume). Climate change exacerbates the existing vulnerabilities of individuals and households who have limited or insecure access to physical, natural, financial, human, social, political, and cultural assets (Flora and Flora 2008). These assets determine how well people can cope and adapt. Availability of, and access to, assets is socially differentiated, because it is shaped by formal and informal inequalities in many aspects of life, including by gender (Otzelberger 2011; UNFPA and WEDO 2009). Therefore, gender-based vulnerability can be defined as vulnerability caused by inequalities in men's and women's access to the assets, opportunities, and decision-making power that would enable them to adapt successfully to new climate conditions.

Climate Change Impacts Are Not Gender Neutral

Gender is an important variable in the relationship between climate change and its human impacts. Specifically, gender-based inequalities, which persist in different areas of life across Arab countries, result in men, women, boys, and girls facing different vulnerabilities—and potential opportunities—in the face of climate change impacts. However, gender roles and relations are highly context specific and therefore differ among and within the Arab countries. They are also flexible and likely will undergo significant evolution as climate change continues to affect the environment and society (Espey 2011; Resurreccion 2011). These characteristics mean that the gender-differentiated impacts of climate change, and the roles that men and women play in the process of adaptation, must be studied and addressed in local contexts. Despite regional diversity, one can identify some broad trends in patterns of gender roles and relations, which determine poverty and vulnerability among men and women and shape the gender-differentiated impacts of climate change in Arab countries.

The sociocultural gender dynamics and power asymmetries that underpin these patterns tend to make women particularly susceptible to chronic poverty (Espey 2011) and lower adaptive capacity. In particular, the prevailing sociocultural constructions of masculinity and the resulting economic and legal frameworks that privilege men's roles as providers mean that female heads of household and their dependents—whose numbers are increasing—are one of the region's most vulnerable groups (IFAD and FAO 2007).[1] Still, it is important to note that gender inequalities and differences in traditional roles not only affect women, but also can result in men facing specific vulnerabilities (Demetriades and Esplen 2008; Otzelberger 2011; UNFPA and WEDO 2009).

Specific Gender Inequalities Intensify Vulnerability to Climate Change

The drivers of gender-based vulnerability to climate change can be separated into three general areas of inequality: access to resources, opportunity for improving existing livelihoods and developing alternative livelihoods, and participation in decision making. In the rural areas of Arab countries, structural inequalities and sociocultural norms most often disadvantage women, and especially poor women, in these three areas, thus intensifying their exposure and sensitivity to climatic changes. As a result, rural women are more likely to have lower adaptive capacity than men. Their lower adaptive capacity results in exacerbated well-being, impacts on individuals, households, and communities.

Holistic and Gender-Responsive Adaptation Builds Resilience

In looking at gender and climate change adaptation, one must recognize that building resilience requires two types of adaptation actions: first, actions intended to adjust and protect livelihood systems from specific climate change impacts, and second, actions that focus on reducing the underlying drivers of gender-based vulnerabilities to climate change. The first response deals directly with the immediate impacts from a given climate change event; the second is in line with broader sustainable development practices (with a particular focus on factors that drive gender-based vulnerability to climate change, such as unequal access, opportunity, and participation). A commitment to addressing the underlying causes of gender-based vulnerability to climate change—the economic, political, and sociocultural mechanisms that maintain gender-based inequalities—is a prerequisite for building resilience. The two approaches to climate change adaptation are mutually inclusive and beneficial.

Global and Regional Policies Increasingly Recognize Gender as a Factor in Adaptation to Climate Change

A growing number of global and regional polices comprehensively address gender in climate change. These have evolved from simplistic policies that emphasize only women's participation and women's roles toward more holistic approaches with a greater focus on gender equity.

Significant progress has been achieved in integrating climate change into United Nations Framework Convention on Climate Change (UNFCCC) processes. In 2007, the United Nations (UN) and 25 international organizations formed the Global Gender and Climate Alliance (GGCA), which aims to ensure that global climate policies are gender responsive. The Intergovernmental Panel on Climate Change now recognizes gender as one factor that shapes vulnerability to climate change. In 2010, the Cancun Agreements recognized gender equality as integral to adaptation. At the 2011 Conference of Parties (COP-17) in Durban, South Africa, references to gender and women were strengthened in a number of important areas, for instance, in countries' guidelines for programs under the UN National Adaptation Programme of Action (NAPA), in the Nairobi Work Programme,[2] and in the operationalization of the Cancun Agreements, including the Green Climate Fund, the Adaptation Committee, the Standing Committee on Finance, and the Technology Mechanism (WEDO 2011a, 2011b). Overall, however, advocates argue that gender concerns are not yet sufficiently addressed under the UNFCCC (Otzelberger 2011; WEDO 2010a).

Gender has begun to appear on the adaptation agenda in Arab countries as well. Gender-based vulnerabilities and the role of women in adaptation are acknowledged in the Arab Framework Action Plan.[3] At the national level, countries such as Bahrain, the Arab Republic of Egypt, and Jordan have made efforts to mainstream gender into adaptation policy, and several Arab countries have referenced gender in national communications to the UNFCCC.

However, more could be done to mainstream gender into climate-related policies in Arab countries. For example, not all NAPAs specifically incorporate a gender perspective on adaptation (Elasha 2010). This absence is partially due to insufficient collaboration between the institutions concerned and, in many countries, the fact that women's commissions are barely engaged in the activities of ministries of environment or agriculture, and vice versa. The GGCA, the International Union for Conservation of Nature (IUCN), and the Council of Arab Ministers Responsible for the Environment are working to promote instruments for mainstreaming gender into climate action, and encouraging the integration of a national gender perspective into UNFCCC negotiations and COPs.

Gender Roles and Relations Help Shape Responses to Climate Change in Arab Countries

This section looks at the current gender context in Arab countries, and outlines how specific gender roles and inequalities relate to adaptation to changing climatic conditions. Increasingly, the process of adapting to a changing climate will constitute an opportunity for the negotiation of gender roles and relations.

Gender Roles in Rural Livelihoods Create Challenges for Women and Men

Women and men generally have different roles in rural societies in Arab countries. In some countries, women are often responsible for tending the land and livestock during the day and attending to household chores in the evening. Men typically handle the finances, trade in the market, and, if needed, migrate. Because natural resources are directly affected by the climate, women's roles in rural areas are often the most sensitive to climate change. Changes in rural livelihoods across Arab countries represent major challenges, but they also offer potential opportunities for flexibility and long-term change in the economic, political, and sociocultural mechanisms that maintain gender inequalities. Gender roles can be less rigid than they might appear (Obeid 2006; Resurreccion 2011), and climate change may present an opportunity for them to evolve in rural areas.

Rural women play a key role in managing natural resources and sustaining livelihoods

Across Arab countries, the division of labor in poor rural households is such that women undertake a large portion of the labor required to sustain natural resources and rural livelihood systems. Women's roles in agriculture are particularly significant, although their involvement in the sector varies between countries (figure 7.1). They typically work long hours, engaged mainly in nonmechanized, labor-intensive, non-capital-intensive activities (FAO 2005). Women also have primary responsibility for the husbandry of small animals and ruminants, as well as for taking care of large-animal systems, herding, providing feed and water, maintaining stalls, and milking (Elasha 2010; FAO 2005).

The agricultural work traditionally performed by women in Arab countries has long been "invisible," as a crucial but seldom acknowledged contribution to household and national income (FAO 2005). Women work on their own farms and as laborers on other farms, but most of them are not paid for their efforts. About 75 percent of women working in agriculture in the Republic of Yemen are unpaid, as are 66 percent in Syria, 45 percent

FIGURE 7.1

Women and Men Engaged in Agriculture, 2004

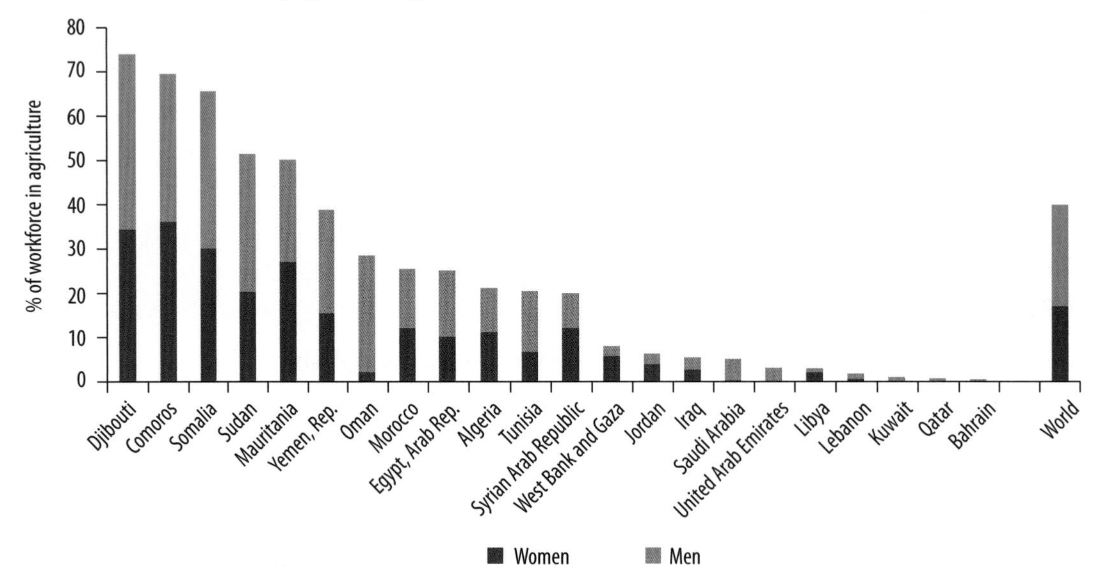

Source: Authors' representation, based on FAOSTAT.

in the West Bank and Gaza (FAO 2005), and 70 percent in Egypt (Egyptian Organization for Development Rights 2011). But the female role in livelihood and natural resource systems is crucial; the work women perform is central to ensuring food security for the family and community and maintaining adequate levels of productivity among the rural labor force (FAO 2005; FAO, IFAD, and ILO 2010; IFAD and FAO 2007).

This traditional role of women as natural resource managers in rural communities in Arab countries makes them especially vulnerable to climate change for two reasons. First, these systems are highly reliant on the climate for their productivity, which makes them highly exposed. Second, because women are often not paid for this work and have less access to and control of land, they have less capacity to adapt. For example, women are the largest group of direct water users, but they do not normally have an equal voice in managing the use or distribution of water.

Women are the primary caregivers in rural families

Women perform vital, but economically unrecognized and unremunerated, activities that contribute to the overall well-being of the household. These activities, which act as a social safety net in the absence of public service provisions, include functioning as caretakers and sometimes as contributors to household income. Caretaking activities are particularly

BOX 7.1

Women's Roles in Buffering against Shocks

"Women in most Arab countries play a key role in adapting households and buffering the family against unexpected climatic shocks. Their knowledge of ecosystems, their skills and abilities, social networks and community organizations help communities mitigate hazardous conditions and events, and respond effectively to disasters."

Source: Elasha 2010.

demanding where birth rates are high, such as in Mauritania, Oman, the West Bank and Gaza, and the Republic of Yemen. This problem is further exacerbated when women suffer poor access to reproductive health services (FAO 2005).

These roles make women highly vulnerable to climate change for two main reasons. First, when climate hazards occur, women play a vital role in maintaining the functioning of the home (see box 7.1). Second, the increased likelihood of illness as a result of climate change is a burden that will fall primarily on women, who are responsible for taking care of sick family members.

Migration decisions are heavily influenced by gender roles[4]

Gender norms, roles, and inequalities play an important role in determining who migrates, when, and where. Migration is not necessarily an equally viable option for all members of the household. For example, those with limited resources or who are responsible for care—often women—are less likely to migrate (Demetriades and Esplen 2008). In Arab countries, men are normally the first to migrate (Elasha 2010). Thus women often remain in the affected community with the burden of assuming the former duties of male household members, in addition to their already heavy workload. At the same time, men face new sets of challenges. For example, when men migrate to the city they are often unskilled and stigmatized. Generally, both women and men face vulnerabilities from migration, as well as potential opportunities.

In Syria, a recent, multi-year drought affected over a million people (Sowers, Vengosh, and Weinthal 2011; and chapters 1, 3, and 6 of this volume), and led to a massive migration from rural areas to the outskirts of nearby urban centers and to Damascus and Aleppo. Estimates for the number of people who have been forced to migrate, either permanently or semipermanently, range from 40,000 to 60,000 families (Solh 2010; UN 2009). In June 2009 it was estimated that 36,000 households (approximately 200,000–300,000 individuals) had migrated from the Al-Hassake governorate alone (ACSAD and UNISDR 2011; Solh 2010; UN 2009). In a region in which men are attributed primary responsibility as income earners, male family members in particular were expected to leave the community to find alternative sources of income (Elasha 2010).

This type of migration has contributed to the increase in the number of female-headed households in many areas.[5] Other contributing factors are the greater numbers of disabled males (caused by conflict), widowhood, and higher divorce rates (FAO 2005; IFAD and FAO 2007; UNDP 2005). In Egypt and Morocco, female-headed households are estimated to be 17 percent of all households, though the true numbers are likely to be higher as a result of male out-migration (FAO 2005).

Although male migration can lead to increased decision-making power for women, it can also contribute to vulnerability. Some female-headed households show improved well-being in some indicators, especially if the woman's situation is out of choice (UNDP 2005). However, most households headed by women are poorer than the small proportion of households headed by unmarried men (UNDP 2005). Among women heads of household in Egypt, roughly 80.5 percent have no landholding, which leaves them without an independent source of income. Dependence on male relatives for access to land and other assets connected to land ownership makes female-headed households particularly vulnerable. Also, the illiteracy rate for rural female heads of household is 73 percent, roughly 10 percent higher than the rate for all rural women (IFAD and FAO 2007). Illiteracy leaves women with less capacity to improve and diversify their livelihoods.

Despite Some Progress, Gender-Based Inequality Persists for Some Indicators

Gender-based inequalities make women and men vulnerable to climate change in different ways. Gender-based vulnerabilities are shaped by the interactions of gender with other factors, because women and men are not able to draw equally on the resources needed for adaptation. The poorest socioeconomic groups are typically the most affected and the least able to adapt. In the rural areas of Arab countries, women make up a large proportion of these groups, which include small farmers and the unemployed (IFAD and FAO 2007).

Countries in the Arab region have achieved great advancements on key gender indicators. Investments in girls and women's education and health,

TABLE 7.1

Selected Gender Indicators for the Arab Region and Subregions, 2009 or Latest Available Data

Subregion	Maternal mortality rate (deaths per 100,000 live births)	Literacy rate (% of population ages 15 and above)		Ratio of female to male primary enrollment (%)	Labor force participation (% of population ages 15–64)		Proportion of seats held by women in national parliaments (%)
		Females	Males		Females	Males	
Least developed countries	613.0	55.0	79.0	84.3	34.9	76.2	11.8
Gulf	19.7	84.3	90.5	100.4	29.0	84.2	5.9
Maghreb	104.1	58.7	78.4	94.3	32.4	81.9	13.3
Mashreq	73.0	66.1	73.0	93.8	57.5	86.3	14.4
All Arab countries	201.7	63.9	81.1	91.7	27.6	78.8	9.2
Low- and middle-income countries, globally	290.0	74.9	86.0	96.0	55.5	83.1	17.6

Source: Authors' calculations based on Genderstat data.

Note: Subregional data are calculated as an average of country data weighted by population except for parliamentary seats held by women, which is calculated as an arithmetic average.

and some advances in women's civil and political participation, have made major inroads in closing the gender gap in these fundamental human rights. These significant improvements in the status of women have resulted in observable progress on important social indicators. Women's life expectancy has increased, and fertility rates and maternal mortality rates have decreased. Arab countries have made significant strides in female literacy and education, largely catching up with other low- and middle-income countries (IFAD and FAO 2007; UN and LAS 2010; UNDP 2005) (see table 7.1). In the Arab Gulf states, more women than men graduate from universities.

But challenges remain, and these improvements have not necessarily translated into gender equality in other domains. Women continue to suffer legal and sociocultural constraints to their agency both within and outside the household (World Bank 2012). Women continue to have less access to land, fewer economic or other livelihood opportunities, and lower civic and political participation rates. In fact, Arab countries have some of the lowest rates for women in the world on these indicators (see table 7.1). However, data on such indicators do not necessarily represent women's economic or social and political roles accurately: much of women's work—especially in rural areas—is invisible in national statistics, and their roles in social and political development are often underestimated (box 7.2).

BOX 7.2

Data: A Fundamental Challenge

Shortcomings in gender-disaggregated data related to the links between gender and adaptation in Arab countries is a fundamental challenge in two ways. First, the availability of socioeconomic data is limited across the region, especially for rural areas. Where data exist, they are typically only available for the most basic indicators. Second, innovative data collection methods that capture gender-relevant information are not as commonly used as they could be. Some of the tools that would improve data collection include qualitative methods such as time-use surveys, focus groups, direct observation, and informant interviews.

Overcoming these shortcomings is crucial for capturing a better picture of women's capabilities and their roles in sustaining livelihoods and human welfare. These roles are significantly undervalued in national accounts across the region, which mostly ignore informal markets and unpaid family labor and production. The indicators on decision-making processes, used in current quantitative methods, do not capture the informal and indirect influence women may have. Women's vital roles in agriculture, as household managers, and as stewards of natural resources and ecosystems are not accounted for in national data collection and statistical tools. This is because these tools derive from models of human production that measure contributions in terms of goods and services exchanged in the market and their cash values, which do not reveal the significance of unremunerated work.

Sources: IFAD and FAO 2007; Jensen et al. 2011; Obeid 2006; UNDP 2005.

Rural women's access to resources is limited

Many people lack secure access to property, land, and resources in rural areas in Arab countries. However, women's access to productive assets, especially fertile agricultural land, is further limited by sociocultural practices that reinforce male control and ownership of those resources, despite the role women often play in their management.

Overall, women make up a small proportion of total landowners:[6] 24 percent in Egypt, 29 percent in Jordan, 14 percent in Morocco, and only 4 percent in Syria. When women do own land, they tend to own smaller plots (IFAD and FAO 2007).

Few formal regulations prohibit women from owning land. Islamic law protects women as independent legal persons entitled to own land, property, and money in their own names, regardless of marital status. However, this protection is sometimes weakened by a combination of the lack of awareness among rural women, in particular of processes of land acquisition and titling and of their rights, customary discrimination, and the persistence of the cultural idea that land is owned by men (Obeid 2006).

Women can lose their formal rights as a result of informal practices such as delaying or controlling women's marriages, giving women cash compensation in place of land, or employing customary laws to ensure that land cannot be divided. These practices exclude women from land ownership to avoid the division of land or to keep land in the family (Obeid 2006). In some countries there is also a lack of land ownership records, and particularly sex-disaggregated data, which compounds the problem and constrains analysis (box 7.2).

Lack of land ownership has serious implications for women's adaptive capacity. These challenges include the following:

- *Difficulty in accessing credit*—Land tenure and property rights are usually required as collateral for loans, which seriously constrains rural women's options for improving agricultural productivity and sustaining their livelihoods in the context of climate change.
- *Insecure access to water*—Because land and water rights are closely related, this lack of water rights also precludes women from membership in water user associations.
- *Limited membership in rural organizations*—Membership is often restricted to heads of households and titled landowners, which can cut women off from decision-making processes, support systems, new technology or techniques, and training.

Gender inequalities constrain rural women's opportunities to improve and diversify their livelihoods

Rural women often face gender-specific challenges to improving or diversifying their livelihoods, thereby increasing their vulnerability to climate change (FAO 2005; IFAD and FAO 2007; UNDP 2005). A limited labor market, restricted mobility, occupational segregation, a mismatch between skills acquired in school and labor market demands, and the strongly gender-based division of labor constrain both men's and women's ability to improve existing livelihoods or to find alternatives. However, rural women's opportunities are particularly limited because they tend to be poorer than men and have lower levels of human development. Poverty lowers a person's capacity to seek, train, or engage in alternative livelihoods, and the low level of human development leads to decreased skills and productivity in existing livelihoods. Rural women face additional challenges to improving existing livelihoods because they have less access to extension services and credit (FAO 2005; IFAD and FAO 2007; Kaisi and Alzoughbi 2007).

The overall rate of female participation in the labor force for the region is only 27.6 percent. This rate is well below the global average for low- and middle-income countries, which is just over 50 percent (see table 7.1). Women represent approximately 50 percent less of the labor force than men in Arab countries. It is not that women do not work; it is that they are often not paid for their labor, and their labor is not always recognized as work (FAO 2005). Where women perform the majority of agricultural labor, as in a number of Arab countries (see figure 7.1), they often have fewer skills to apply to other types of work. In most Arab countries, rural men are expected to maintain their status as the breadwinner for the family, and they will typically seek alternative employment when faced with unsustainable rural livelihoods.

Challenges to human development, especially those related to health and education, are often different for men and women. This disparity has important implications for men's and women's opportunities to improve and diversify their livelihoods. Challenges to human development are exacerbated by continued population growth (IFAD and FAO 2007). Maternal mortality rates are 613.0 per 100,000 births in the least developed country subregion, higher than global developing country averages (see table 7.1). Although data are lacking, clear indicators show that other health problems disproportionately affect rural women and children. This finding is often the result of the inadequate provision of primary health care. Significantly, rural children are 1.7 times more likely to be underweight than urban children in Arab countries, and child mortality rates remain high in the region's least developed countries (IFAD and FAO 2007). Although Arab countries have made progress in education for women, gender gaps also still exist in literacy and education in the Arab world; on average, literacy rates are over 15 percent lower for women. This gender gap in literacy is particularly pronounced among rural populations. This finding is particularly troubling in terms of people's capacity to adapt to climate change, because illiteracy rates are strongly correlated with poverty (IFAD and FAO 2007).

In many cases, the increased productive skills and earning capacity of rural Arab women that result from greater education and human development are not being sufficiently taken advantage of to improve adaptation. For example, the low rate at which women transition from school to the labor market is a major obstacle to development in the region. Simulations using household survey data show that the benefits of enhanced participation of women in the labor force extend to the entire household, raising average household incomes by up to 25 percent (World Bank 2004).

Rural women's limited participation in decision making weakens a country's overall ability to adapt

Low female participation in decision making at household, community, and national levels is a major obstacle to sustainable adaptation. Women's participation in parliament ranges from just below 6 percent to 15 percent in the Arab region, with an average of just over 9 percent (see table 7.1). This finding puts Arab countries significantly below the world average of 18 percent, and behind all other regions. The next lowest regional average is for South Asia, where women hold 14 percent of seats in parliament (World Bank 2009, 2007 data). At the community level, women's agency is also limited, including a limited ability to participate in formal groups, become members of rural organizations, and act independently. Though more research is needed to understand fully why this is the case, certain factors help explain the problem (FAO 2005). These factors include traditional restrictions on women's mobility and autonomy, sociocultural norms that weaken rural women's decision-making power, women's daily labor burdens, and their lack of access and rights to certain livelihood assets.

The importance of women in decision making and social and political development is sometimes underestimated. Gender roles and norms are not impervious to change, and many rural women in Arab countries do play significant roles in decision making at all levels. Within the household, women wield more power in decision making in areas in which they are central to the process of production. For example, in some regions women have significant decision-making power over the management of dairy and poultry production (FAO 2005; IFAD and FAO 2007; Obeid 2006). Recently, decision making related to rural livelihoods has been subject to significant change because of the introduction of new technologies and mechanization. Thus modernization has reduced the need for the time-consuming contributions of women (Chatty 1990, 2006; Obeid 2006) and has been detrimental to their authority. By contrast, women have been instrumental in leading and organizing the popular movements for political change in the region in 2011 and 2012.

Rural women possess valuable knowledge related to adaptation decision making. For example, in the southern region of Syria, women have specific local knowledge of indigenous plants and their uses for food or medical purposes. This experience can help reduce the risk of illness in the wake of exposure to climate change impacts (Kaisi and Alzoughbi 2007). Overall, women's knowledge can contribute to biodiversity protection (and lower sensitivity to climate change impacts), community resilience, and the increased effectiveness of adaptation projects (Demetriades and Esplen 2008).

Finally, it should be noted that 65 percent of the population in Arab countries is below 30 years of age and becoming younger. This demographic change will certainly alter current gender dynamics. The youth are likely to have been raised in smaller families (in households more separated from the extended family than has generally been the norm), where gender disparities are often less pronounced (World Bank 2004).

Regional Case Studies Demonstrate How Gender Consideration Makes a Difference in Adaptation

This section provides contextualized examples of the links between gender, climate change, and adaptation in Arab countries. These examples pertain to water scarcity in the Republic of Yemen, agricultural livelihoods in Morocco, and drought in northeast Syria. A fourth case study discusses Jordan's approach to improving gender responsiveness in adaptation.

Identifying Gender-Related Challenges Is Essential in Managing Water Scarcity in the Republic of Yemen

The Republic of Yemen's water crisis has been particularly detrimental for women and children for several reasons. First, the inadequate access to drinking water has led to the spread of diseases such as malaria, bilharzia, and diarrhea (Assad 2010; World Bank 2011a). Second, water scarcity has negative implications for food security and malnutrition, which often affect women and children disproportionately. Third, water scarcity increases women's workloads as distances to clean water sources increase, making the daily task of collection more time-consuming. In one study, 58.4 percent of women surveyed reported spending time collecting water, compared to just 7.8 percent of men (Koolwal and van de Walle 2010). Because of its impact on women's time, water scarcity has serious implications for many aspects of women's well-being and, by extension, for the well-being of the community as a whole:

- *Female education.* Increased workloads and time burdens for women mean that girls have less time to attend school.[7] An already large gender gap in educational enrollment is widening in some areas because girls are increasingly needed to help collect water (Assad 2010; IRIN 2009; World Bank 2011a, forthcoming b).[8]
- *Women's potential to engage in income-generating activities.* In a study of two rural communities with no piped water or gas stoves, less than 25 percent of women's time was devoted to productive activities. By comparison, in communities with piped water and gas stoves, between 38 and 52 percent of time was devoted to productive activities (World Bank, forthcoming b).
- *Health.* Less time is available for caring for household members, which threatens the health of families. A one-hour reduction per day in the time it takes to collect water improves children's health, especially for girls (Koolwal and van de Walle 2010).

- *Participation in water management and decision making.* Despite women's crucial roles in water and natural resource management, their participation in water-related decision making remains weak (Assad 2010). Increased time burdens mean that women are less available to take part in water organizations and development or adaptation projects.

Box 7.3 gives an example of a water management project in the Republic of Yemen that seeks to address some of these challenges.

BOX 7.3

Community-Based Water Management in the Republic of Yemen: Improving Climate Change Awareness, Water Management, and Child Health, and Empowering Women

A community-based water management project in the Amran district of the Republic of Yemen has a strong focus on gender. The German Agency for Technical Cooperation (GTZ) project emphasizes the following:

- Building women's capacity to participate in water management–related decision making.
- Raising women's awareness of methods to conserve and purify water.

Several project strategies were used to empower women, including the following:

- Literacy classes were taught by young women from the community who received training in teaching adults.
- Thirty-eight village water committees (VWCs) were established and led by community members, with strong female participation.
- Sand filters were distributed to schools, mosques, and nongovernmental organizations. The Yemeni Women's Union played a leading role in raising awareness of the health benefits of sand filters. Women were trained in filter use, cleaning, and maintenance.

Positive results occurred within several areas:

- *Water management.* Female VWC members noted that women's participation in committees meant that they were able to address the real needs of the water sector and to raise awareness about cistern management and use.
- *Health.* Achievements included reductions in waterborne diseases affecting children and decreased expenditure on health.
- *Attitudes toward women.* Women's effective participation, both in the project and in the VWCs, sensitized men and religious and community leaders to gender inequalities in adaptation.
- *Women's empowerment.* Some female committee members noted that women's participation encouraged them to take part in elections for the local council, increasing women's access to decision-making processes.

Sources: Assad 2010.

Women's Engagement in Agribusiness Sustains Rural Livelihoods in Morocco

The small farming communities in the High Atlas region of Morocco, which feed a large proportion of the country's urban population, are largely self-sufficient in terms of food security and productive activities. However, this region is highly exposed to climate change, which includes shifts in temperatures and rainfall patterns (Messouli and Rochdane 2011).

In these communities, women undertake all domestic duties as well as many agricultural tasks—indeed, women carry out more than 50 percent of agricultural labor in these communities. Furthermore, out of economically active rural women across Morocco, 92 percent are engaged in agriculture, around a third of whom are under the age of 19 (Messouli and Rochdane 2011). Women's agricultural tasks include gardening, milking, harvesting, olive collecting, and other work in the fields. Women are also heavily involved in cattle breeding and the cultivation of cereals, legumes, and industrial crops.

Interviews and focus group discussions by Messouli and Rochdane (2011) in these farming communities show that men and women have different priorities. In these discussions, men often spoke of migrating while women talked more about having to take on new activities at home. Men frequently undertake seasonal migration for herding and trading, leaving women to manage natural resources in the increasingly climate-sensitive ecosystem (Messouli and Rochdane 2011). As a result of this dynamic, agriculture has come to be increasingly managed by women in some Arab countries, including Morocco (FAO 2005). Box 7.4 provides a specific example of a women-run initiative in Morocco.

Gender Roles and Relations Inform How Communities Adapt to Drought in Northeast Syria

A recent multi-year drought in northeast Syria has had major impacts on rural women and men (see chapters 1, 3, and 5 of this volume). According to the United Nations, up to 80 percent of those severely affected (mostly women) lived on a diet of only bread and sugared tea. Women, unless pregnant, were often expected to forgo a meal during food shortages. School dropout rates were high, and enrollment in some schools decreased by up to 80 percent, partly because families were migrating and partly because children were being sent to work to supplement household incomes (UN 2009). According to local reports, girls were often the first to be taken out of school.

Migration is a highly gender-related phenomenon. In responding to climate hazards, men and women consider different options, at different

BOX 7.4

Addressing Gender-Related Challenges in Adapting Agricultural Livelihoods in Morocco

In response to growing rural poverty, the Moroccan Ministry of Agriculture's 2020 Rural Development Strategy emphasizes the importance of empowering women farmers and recognizing them "as producers and managers of ecosystems." The government, with international assistance, is helping women improve and diversify their livelihoods. Initiatives to improve women's skill sets include reducing the use of wood for fuel, promoting biogas and solar energy, and digging wells. These initiatives improve agricultural practices and alleviate women's workloads, which enable them to diversify their livelihoods and boost incomes.

In Morocco's semiarid south, around the town of Sidi Ifni, every family has its own plot of land. With backing from the Ministry of Agriculture, young people and women are receiving training and finding employment in the production of new health and cosmetic products from the prickly pear cactus. The previous small-scale production of the plant is being transformed into a significant industry, creating a small economic miracle.

Cactus cooperative members say that the status of women has increased and lives have been transformed. Members stress, "We could never have imagined that we could get such a good income from (cactus products). You don't have to be educated to work in the factories. Our children are feeling the benefits. There is much more money to be made out of cactus and it is women who are earning it."

Source: Messouli and Rochdane 2011.

stages. In response to the Syrian drought, which compelled people to abandon traditional livelihoods (ACSAD and UNISDR 2011; World Bank, forthcoming a, chapter 1), men often travelled south to work as farmers and herders among nomadic tribes near the Jordanian border or to find employment in Jordan and Lebanon. By contrast, women often travelled west to the coastal zones, where greenhouse production of vegetables near Tartous provided employment opportunities. The impact of the drought on rural livelihoods has, in many cases, led to the estrangement of family members (ACSAD and UNISDR 2011). This breach with customary family structures has immense impacts on men, women, and children.

Migration can also provide new opportunities. In 2008, a community of 20 families from the Sba'a tribe emigrated from the Badia rangelands and settled on the outskirts of Palmyra. The children, both girls and boys who previously tended to sheep, are now enrolled in primary school. Migrated families also reported better nutrition and improved access to health care. Among some families it is also acceptable for young women to migrate and take up work. As a result, women may gain greater social and financial independence through new economic opportunities. However, already high levels of unemployment and a lack of marketable skills for urban settings, particularly among young people and women, may mean that migration does not easily yield such opportunities.

In areas where men's skill sets are limited, migration is not necessarily empowering for males. Men may become vulnerable to exploitation, harsh working conditions, and low pay. Such an outcome can be frequent because it is common for young men to work outside Syria, often in Lebanon, and be expected to send home remittances.

Among those remaining in Syria's northeast, there has been a sharp rise in the number of female-headed households. This can be both a blessing and a curse. Some female heads of household, left behind by their migrating husbands, may fall into poverty if they lack the skills to engage in financially productive activities. Others, benefitting from remittances, may gain increased autonomy and authority.

Jordan Has Developed Policy Instruments for Gender-Responsive Adaptation

In May 2011, having recognized that addressing the gender-related dynamics of adaptation is critical to fulfilling the country's development goals, Jordan, in partnership with the IUCN, became the first Arab country to mainstream gender in climate policy.[9] Jordan now has a framework for action (for 2011–16), including practical policy guidelines (referred to in annex 7A, panel a), and is moving toward a more integrated and sustainable approach to adaptation.

The gender mainstreaming process in climate policy and action began in 2008–09. National assessments examined the status of women and gender equality, the nature of climate change impacts, and how these issues are correlated in local contexts. Informed by field visits and research, the program was based on current national priority sectors (water, energy, agriculture and food security, and waste reduction and management). A workshop in 2010 attended by stakeholders from women's organizations; ministries of environment, water and irrigation, agriculture, finance, planning, and health; the United Nations Development Programme;

GIZ (German Agency for International Cooperation);[10] academic institutions; and the National Center for Agricultural Research and Extension also constituted an important part of the process. Following this preparation, the government of Jordan plans to mainstream gender perspectives into its Third National Communication to the UNFCCC. This plan is to be achieved by completing a systematic gender analysis, collecting and using sex-disaggregated data, establishing gender-sensitive indicators and benchmarks, and developing practical tools to support increased attention to gender perspectives in adaptation.

Jordan is a signatory to several key international agreements, which commits it to gender mainstreaming. In 2007 it also ratified the Convention for the Elimination of All Forms of Discrimination against Women.[11] The government recognized that gender equality and women's empowerment are means for promoting development and adaptation, and has adopted a participatory approach in local governance for sustainable and equitable natural resource management. The 2012–16 Jordanian women's strategy, developed by the National Jordanian Commission for Women, incorporates climate change adaptation as an area of concern in its work. This complements and supports the program's emphasis on women as agents of change in adaptation.

Multiple Approaches Are Needed to Incorporate a Gender Perspective in Adaptation

Effective adaptation demands the full potential of entire societies. Thus, adaptation strategies must be inclusive and empowering to those who face barriers to developing their potential, namely women and other vulnerable groups. At the same time, climate change increases the urgency to address the underlying causes of poverty and gender-based inequality and vulnerability, which represent significant challenges in some countries. Persistent gender inequalities are already putting huge strains on political and socioeconomic systems—strains that climate change threatens to exacerbate. To improve the responses to these challenges, it is important to improve gender equality in rural areas, improve the gender responsiveness of rural development and adaptation projects at all levels, and improve the collection and availability of data relevant to gender and climate change in rural areas.

Build Resilience to Climate Change through Increasing Gender Equality

Tackling the specific drivers of gender-based vulnerability to climate change is critically important for adaptation. Holistic development is essential for adaptation and cannot be achieved without tackling gender inequalities. Continued investment in achieving existing development goals, if harmonized within the region's adaptation needs, will reduce sensitivity to climate change and build adaptive capacity.

Arab countries should continue to address gender inequalities by tackling the sociocultural, political, and economic mechanisms that perpetuate them. Priority areas for the region are improving equality in access to resources, livelihood opportunities, and participation in decision making. This goal means continuing the progress toward gender equality, including addressing formal and informal practices that create inequalities in health, education, economic participation, agency, civil rights, autonomy, and participation at all levels. Also important are identifying and addressing drivers of gender-based vulnerability to specific aspects of climate change impacts, such as barriers to women's land ownership, low awareness of climate change, and limited skills for livelihood diversification (see annex 7A, panel c).

Government policies also must address constraints to building adaptive capacity among rural populations in general, including for men. Many Arab countries need to invest in their rural economies and develop social protection frameworks that benefit the most vulnerable (see chapter 4). Such policies should include solutions for poor men and women (da Corta and Magongo 2011).

Increase Gender Responsiveness in Adaptation Strategies and Projects in All Sectors

New frameworks, principles, and capacity-building efforts are needed for developing and implementing gender-responsive adaptation. There is no one-size-fits-all set of policies for addressing all the issues explored in this chapter. Standardized solutions from other contexts—if imported without modification—could lead to the erosion of local methods for increasing resilience (Magnan et al. 2009). The methodological tools developed by international institutions, and in other regions, for building a gender-responsive approach to adaptation are essential, but they should also be adapted to fit local sociocultural contexts. They should ensure the effective participation of women and men and aim to activate the full potential of gender-specific knowledge in adaptation.

National and institutional level

The overarching set of tools for building the capacity of institutions to integrate gender responsiveness in all aspects of adaptation planning and management is encompassed within gender mainstreaming. There is no single way to mainstream gender, and several frameworks exist that can be considered.[12] An institutional framework and guideline of strategies for implementing a gender mainstreaming project at the national government level, based on the Jordanian experience, is outlined in annex 7A, panel a. Since the launch of that project, a number of other Arab countries, including Bahrain and Egypt, have initiated similar projects to develop gender-mainstreaming programs and build the capacity of institutions to address gender-related challenges and opportunities in climate change adaptation.

One key problem with gender mainstreaming, as experienced in other areas of development, is that it can be seen as an end in itself, instead of a tool and a process. For example, the inclusion of female participants, or the requirement to conduct training sessions on gender, has the tendency to become the goal, regardless of the actual impact that taking these steps has. The translation of national-level policies into effective and consistent implementation of gender issues at the project level requires leadership, capacity, accountability, funding, and expertise, both in national governments and among the managers and staff of relevant institutions.[13] Furthermore, although establishing gender-mainstreaming mechanisms is crucial, this effort does not automatically offer solutions to gender-related issues in adaptation. Effective gender-responsive climate change adaptation in Arab countries also requires action on specific points such as those identified in this chapter.

Project level

Adaptation projects in all sectors must incorporate, as a matter of routine, instruments to identify and raise awareness of local gender-adaptation issues, and develop strategies to address them. Essential steps include context-specific gender analyses at the planning stage, monitoring and evaluation during and after implementation, and flexible mechanisms to adjust those steps if they are not working (see annex 7A).

Projects should address both the immediate and practical needs of women and men, as well as the strategic[14] needs of women. Addressing women's strategic needs, by empowering them and building their adaptive capacities, will help tackle the underlying drivers of gender-based vulnerability in the long term. In Arab countries, particular attention should be paid to strengthening—by both quantitative and qualitative measurements—women's participation in decision making and leadership. In addition, project planners should focus on the potential effects of women's lower access to resources and more limited opportunities for alternative livelihoods.

A key concern is to move away from a narrow focus on "vulnerable women" to a more holistic gender analysis that emphasizes the existing gender power relations within society that contribute to other forms of socioeconomic inequality. These gender analyses must identify who holds the power to identify priorities and solutions, shape debates, and make decisions (Demetriades and Esplen 2008). Once these power relations are identified, solutions can be crafted.

Finally, in responding to climate change–related disasters, Arab countries need to incorporate a gender perspective into all stages of planning and implementation of disaster risk management projects. Action is necessary at three stages: (a) predisaster, to build local resilience by identifying sustainable adaptation options to keep affected livelihoods from further deteriorating; (b) during a disaster, by implementing a gender-sensitive response through systematic gender analyses at all stages; and (c) post-disaster, by identifying those most affected and targeting efforts to address their immediate needs and promote new livelihood opportunities.

Improve the Collection and Use of Gender-Disaggregated Data

The systematic, accurate, and sustained collection of data on key indicators and their consistent use in research, policy making, and project design and implementation is crucial for addressing challenges and identifying opportunities related to gender in adaptation in Arab countries. Such data collection makes visible what is otherwise invisible, and tangible what is otherwise abstract. Data collection enables comparisons between communities, regions, and countries; assesses change over time; and measures the effectiveness of policies and projects (Aguilar 2002).

Quantitative as well as qualitative data improve the understanding of the implications of climate change and provide policy makers with the tools necessary for gender-responsive decision making in adaptation. Capturing data to reveal links between gender and adaptation requires research methods based on human welfare models (UNDP 2005). Thus countries must strengthen capacities and promote an enabling environment for qualitative research. Data collection methods could include focus groups, direct observation, and interviews, in addition to quantitative questionnaires and surveys.

Adaptation projects should include local data collection and the development of indicators that target and measure local realities. This approach will help ensure an awareness of underlying gender patterns; accurate assessments of women's and men's different needs for assistance; opportunities for capacity-building and adaptation initiatives; and information that can be used to strengthen the voices of vulnerable groups (box 7.5).

BOX 7.5

Unlocking Women's Potential to Help Drive Adaptation in the Gulf

Adaptation to climate change in Arab countries demands equality of opportunity for men and women. Women are increasingly assuming leadership roles in many areas of life in the Gulf, and the huge increase in women's education and public participation is an important driver of change in the region. However, barriers to the development of women's capabilities persist in the science, technology, and engineering sectors. This situation weakens the potential of Gulf countries to develop creative and innovative climate change adaptation strategies.

Gulf countries are shifting toward building knowledge-based economies based on the sciences, but only a small proportion of students are enrolled in these areas, regardless of sex. More women and men are needed in the sector. Some of the measures Gulf countries can take to improve women's adaptive capacity through education include the following:

- Improving the understanding of how young people make education- and career-related decisions, to identify ways to encourage women to participate in science.
- Strengthening partnerships between higher education institutions, education's governing bodies, and the private sector to provide relevant programs, develop attractive career paths, and reach out to female students.
- Increasing the visibility and accessibility of female role models and building networks for women in science and technology.

Source: Authors' compilation.

Key Messages

- Climate change impacts are not gender neutral.
- Gender inequalities intensify vulnerability to climate change by increasing sensitivity to exposure to climate change impacts and reducing adaptive capacity.
- In Arab countries, women are often among those least able to adapt to the impacts of change because they are more likely to be poor than men, they are often responsible for natural resource and household

management, they lack access to resources and opportunities for improving and diversifying livelihoods, and they have limited participation in decision making.

- Women are key stakeholders in adaptation and important agents of change. Arab countries need to focus on further empowering women to be effective leaders in adaptation.
- Climate change constitutes a threat to development achievements and progress toward the Millennium Development Goals, including gender equality (MDG 3), because it threatens to deepen gender inequalities and worsen poverty.
- To build resilience to climate change, Arab countries need to do the following:
  - Increase gender equality in all domains, particularly in access to resources, opportunities for improving and diversifying livelihoods, and participation in decision-making and political processes.
  - Develop mechanisms to improve the gender responsiveness of adaptation policies and projects at both national and project levels.
  - Improve the collection and use of sex-disaggregated data on relevant indicators.

ANNEX 7A

Matrix of Proposed Policy Options

1. Increase gender-responsiveness in adaptation strategies and projects in all sectors: gender mainstreaming: Proposed policy actions	
National and institutional level (based on the framework for Jordanian gender mainstreaming in climate policy and action)[a]	Ratify international and regional conventions related to gender and adaptation. Establish coordination between all relevant government bodies and other institutions: • Support the development of a network of gender and climate change experts or focal points. • Appoint a permanent gender expert on the national climate change committee or equivalent and establish a consultative support group to work with this expert. Develop understanding of the main gender and adaptation priorities in the local context: • Support research and data collection on the links between gender and adaptation (see panel b). • Disseminate data and results of research. Establish regulations that enforce the incorporation of gender-related criteria in adaptation strategies and programs. Involve women's organizations in adaptation strategies and address adaptation in national gender strategy: • Build awareness of climate change issues among women's organizations to enable them to identify opportunities for their full participation in the relevant processes. Strengthen the capacity of institutions implementing the mainstreaming program: • Develop a specific training protocol to form an integral part of the national mainstreaming program. • Carry out systematic and ongoing gender training workshops and courses on specific issues for all staff, policy advisers, and senior managers—adapted to their specific responsibilities—to build capacity to incorporate gender issues throughout the project cycle and promote gender equality through their roles. • Adopt a learning-by-doing approach to training, to distill lessons from the field, and support bottom-up policy development. Establish and enforce gender-responsive budgeting practices to ensure that adequate resources are available for strengthening gender responsiveness in adaptation. Secure ongoing funding to ensure the continuation of the program, including through bilateral dialogue with international adaptation funding mechanisms. Monitor and evaluate progress regularly and update program priorities, methods, and training materials accordingly: • Establish gender-sensitive reporting, monitoring, and evaluation systems. • Involve gender experts in the preparation of national climate change communications and negotiations. • Consider other measures of progress. For instance, in a key measure in Jordan in May 2011, the Jordanian National Committee for Women's Affairs' National Strategy for Women in Jordan included specific gender- and climate change–related objectives and activities.

(Annex continues next page)

ANNEX 7A

Matrix of Proposed Policy Options (*continued*)

Project level[b]	Allocate adequate financial and human resources for implementing gender mainstreaming. Carry out surveys and analysis of gender roles, norms, power relations, and gender-specific constraints. Ensure equal and effective participation of women and men in project and policy formulation and planning. Identify and address risks and opportunities for men and women. Identify women's and men's needs and preferences. Distinguish practical and strategic gender needs. Identify differing vulnerabilities of men and women along with other differences (such as age, wealth). Integrate results of all of these analyses in project aims and planning. Develop measures to address locally specific constraints, such as to women's participation in the project. Study, document, and build on women's and men's local practices and indigenous knowledge. Ensure equitable sharing of benefits between men and women. Include gender-related criteria in monitoring and evaluation.

2. Improve the collection and use of data disaggregated by sex and by age: Proposed policy actions

There are several steps for improving the availability, accessibility, and analysis of sex- and age-disaggregated data:

- Determine the extent of existing sex- and age-disaggregated data.
- Make available data accessible.
- Strengthen analysis by presenting sex- and age-disaggregated data where available.
- Improve data collection practices to ensure that all relevant data are disaggregated by sex and age where those disaggregated data are not available.

The following are examples of relevant data:

- Impacts of extreme events on health (mortality; diseases see chapter 8).
- Indicators related to the longer-term impacts of climate change (including household-level microeconomic data).
- Participation in rural land and water organizations and committees; sex-disaggregated data on numbers and categories of people participating in rural land and water organizations and committees and training courses.
- Access to or ownership of land.
- Principal source of household income.

Systematically integrate sex- and age-disaggregated data in projects and policy-level development and adaptation interventions via gender mainstreaming frameworks.

Climate change is likely to cause important shifts in, for example, the use of time and division of labor. Data collection must be expanded to encompass these changes, as in the following:

- Study the relationships between gender and climate change adaptation through data collection methods such as surveys on time use and division of labor, focus groups, direct observation, and interviews.

Support the collection of information on local knowledge and practices (for example, on local water management systems, agricultural practices, and biodiversity and ecosystem management) to help base adaptation projects and policies on women's and men's local knowledge and preserve that knowledge.

ANNEX 7A

Matrix of Proposed Policy Options *(continued)*

3. Build resilience to climate change through increasing gender equality: Proposed policy actions

Adaptation means continuing good development. Continue to invest in development and strengthen gains in gender equality in all areas of life: literacy, education, and skills development; health; employment; participation in decision making at all levels; and rights.

Ensure that campaigns to raise awareness of climate change and adaptation options (for example, on sustainable natural resource management and sustainable technologies and practices in agriculture) reach women and children:

- Analyze factors constraining women's access to information and target campaigns accordingly.
- Analyze gender patterns in sources of information to improve targeting of awareness campaigns, particularly where illiteracy rates are high.
- Train female community leaders to raise awareness.
- Link to general literacy and education initiatives.

Promote and invest in innovative new areas of business in rural economies:

- Emphasize improving opportunities for women.
- Conduct gender-sensitive value-chain mapping and foster women-centered value chains.
- Provide business-related training for rural populations.

Increase rural and urban women's skill-development and capacity-building opportunities. Particularly emphasize rural women's skills and opportunities in countries where feminization of rural societies is occurring (owing to out-migration of men) or where there are significant gender gaps in education, nonagricultural skills acquisition, and alternative employment opportunities.

- Develop training in community and political participation skills.
- Develop business-related training.
- Link to general literacy and education initiatives.

Promote inclusive extension services, addressing gender-specific barriers to access.

Train women extension agents.

Tailor infrastructure development to reduce women's domestic burden (increase service delivery for all):

- Improve roads.
- Increase access to fuel for heating and cooking.
- Improve water access.

Implement targeted social protection, including insurance schemes, rural pensions, access to credit, and cash transfer programs:

- Assess the needs of women-headed households in particular.

Reform regulations for accessing credit:

- Analyze gender differences in access to credit.
- Remove the criterion of being a named landowner.
- Design credit schemes specifically for women.

Reform property rights law and practices related to land and property:

- Allow joint spousal titling of land and property, to guarantee equal rights to property acquired during marriage.
- Reduce significance of marital status for legal status and land or property ownership.
- Review and reform laws and practices related to inheritance.
- Simplify and disseminate knowledge on land laws.
- Create mechanisms to improve the enforcement of existing land ownership laws.
- Improve existing land access programs, especially by increasing emphasis on gender issues in access to land.
- Support women's collective schemes for securing land access rights.

Reform membership practices of rural, land, and water organizations:

- Analyze gender differences in participation.
- Remove the common criterion of having to be a named landowner.

Source: Authors' compilation.

a. Based on the IUCN (Regional Office for West Asia) Program for Mainstreaming Gender into Climate Change Initiatives in Jordan 2010, http://www.iucn.org/ROWA.

b. Adapted from Otzelberger 2011; FAO 2010.

Notes

1. This sociocultural framework has been supported by macroeconomic factors that have limited women's participation in the workforce. The kind of social contract that governments in the region have adhered to since the mid-20th century, which has been underpinned by generous but costly welfare states, large public sectors, and generous subsidies, has often reinforced women's roles as homemakers (World Bank 2012).
2. The Nairobi Work Programme helps developing countries "improve their understanding and assessment of impacts, vulnerability and adaptation to climate change" and "make informed decisions on practical adaptation actions and measures" (see http://unfccc.int/adaptation/nairobi_work_programme/items/3633.php).
3. An Arab regional framework plan for climate change being developed by the League of Arab States includes specific programs on adaptation (including water, land, and biodiversity; agriculture and forestry; industry, construction, and building; tourism; population and human settlements; health; and marine and coastal zones), which will be implemented over a period of 10 years.
4. Although the relationship between climate change impacts and migration cannot be reduced to simplistic causality, the impact of climate change will likely be an important determinant of migratory behavior in the future.
5. Female-headed households are those headed by women in the absence—whether temporary or permanent—of adult males who otherwise supply the main source of income for the household.
6. Landowners may not live on or work the land they own. Landholders live on and work the land they own. Many middle-class female landowners—who often inherit land—rent their land to others, mainly men (IFAD and FAO 2007).
7. Existing constraints include the need for segregated classrooms and sometimes resistance from male family members. The female youth literacy rate is 24 percent lower than for male youth (72 percent versus 96 percent), and the ratio of female to male primary enrollment is 80 percent.
8. Some research has found that boys and men may assist girls and women in the collection of fuelwood in the Republic of Yemen when they have to travel long distances or it is dangerous. Energy poverty can therefore affect male community members also (El-Katiri and Fattouh 2011). This may also be the case where water is scarce.
9. The program was launched in November 2010. See http://www.iucn.org/ROWA.
10. At the time of the workshop, this entity was the German Agency for Technical Cooperation (GTZ), prior to its merger with two other agencies to form the German Agency for International Cooperation (GIZ).
11. Article 2(a).
12. Refer to World Bank (2011b) for one example.
13. Refer to World Bank (2011b) for more specifics on how to mainstream gender.
14. Practical gender needs are the immediate needs of individuals to ensure their survival, within existing social structures (typically concerning living conditions, health, nutrition, water, and sanitation). Strategic gender needs (of women) are needs whose fulfillment requires strategies to challenge male dominance and privilege, through addressing gender-based inequalities. Often practical and strategic gender needs overlap (Reeves and Baden 2000), but this distinction can be helpful in developing gender responsiveness in adaptation.

References

ACSAD (Arab Center for the Study of Arid Zones and Dry Lands) and UNISDR (United Nations International Strategy for Disaster Reduction). 2011. *Drought Vulnerability in the Arab Region: Case Study—Drought in Syria: Ten Years of Scarce Water (2000–2010)*. Damascus: ACSAD; Geneva: UNISDR.

Aguilar, Lorena. 2002. "Gender Indicators: Gender Makes the Difference." Fact Sheet, Gland, Switzerland, International Union for Conservation of Nature.

Assad, Ruby. 2010. Field notes on the gender-related aspects of a GTZ community-based water management project in the Republic of Yemen. GTZ (German Agency for Technical Cooperation), Eschborn, Germany.

Chatty, Dawn. 1990. "Tradition and Change among the Pastoral Harasiis in Oman." In *Anthropology and Development in North Africa and the Middle East*, ed. Muneera Salem-Murdock, Michael M. Horowitz, with Monica Sella, 336–49. Oxford, U.K.: Westview Press.

———. 2006. "Adapting to Biodiversity Conservation: The Mobile Pastoral Harasiis Tribe of Oman." In *Modern Oman: Studies on Politics, Economy, Environment, and Culture of the Sultanate*, ed. Andrzej Kapiszewski, Abdulrahman Al-Salimi, and Andrzej Pikulski. Wydawca: Ksiñegarnia Akademicka.

da Corta, Lucia, and Joanita Magongo. 2011. "Evolution of Gender and Poverty Dynamics in Tanzania." Working Paper 203, Chronic Poverty Research Centre, Manchester, U.K.

Demetriades, Justin, and Emily Esplen 2008. "The Gender Dimensions of Poverty and Climate Change Adaptation." *IDS Bulletin* 39(4)4.

Egyptian Organization for Development Rights. 2011. *Report on the Economic Deprivation and Legal Exclusion of Rural Women in Egypt*. Cairo: Egyptian Organization for Development Rights.

Elasha, Balgis Osman. 2010. "Mapping Climate Change Threats and Human Development Impacts in the Arab Region." Arab Human Development Report Research Paper, United Nations Development Programme, Regional Bureau for Arab States, New York.

El-Katiri, Laura, and Bassam Fattouh. 2011. *Energy Poverty in the Arab World: The Case of Yemen*. Oxford, U.K.: Oxford Institute for Energy Studies.

Espey, Jessica. 2011. "Women Exiting Chronic Poverty: Empowerment through Equitable Control of Households' Natural Resources." Working Paper 74, Chronic Poverty Research Centre, Overseas Development Institute, Manchester, U.K.

FAO (Food and Agriculture Organization of the United Nations). 2005. *Breaking Ground: Present and Future Perspectives for Women in Agriculture*. Rome: FAO.

———. 2010. "UN Joint Programmes: Integrating Gender Issues in Food Security, Agriculture, and Rural Development." Rome: FAO. http://www.fao.org/docrep/013/i1914e/i1914e00.pdf.

FAO (Food and Agriculture Organization of the United Nations), IFAD (International Fund for Agricultural Development), and ILO (International Labour Organization). 2010. *Gender Dimensions of Agricultural and Rural Employment: Differentiated Pathways Out of Poverty—Status, Trends, and Gaps*. Rome: FAO, IFAD, and ILO.

FAOSTAT. http://faostat.fao.org

Flora, Cornelia Butler, and Jan L. Flora. 2008. *Rural Communities: Legacy and Change*, 3rd ed. Boulder, CO: Westview Press.

IFAD (International Fund for Agricultural Development) and FAO (Food and Agriculture Organization of the United Nations). 2007. *The Status of Rural Poverty in the Near East and North Africa*. Rome: IFAD and FAO.

IRIN (Integrated Regional Information Networks). 2009. "Yemen: Clambering Up Mountains to Find Water." *IRIN Humanitarian News and Analysis*. http://www.irinnews.org/Report.aspx?ReportID=86847.

Jensen, Jillian, Sandra Russo, Kathy Colverson, Malika A-Martini, Alessandra Galie, and Bezaiet Dessalegn. 2011. "Integrating Gender Approaches into Research at the WLI Benchmark Sites: Synopsis of the WLI Gender Training Short Course, March 2011." Water and Livelihoods Initiative, University of Florida, Gainesville, and International Center for Agriculture Research in the Dry Areas, Aleppo, Syrian Arab Republic.

Kaisi, Ali, and Samira Alzoughbi. 2007. "Women's Contributions to Agricultural Production and Food Security: Current Status and Perspectives in Syria." Paper presented at the GEWAMED Workshop on Integrating Gender Dimensions in Water Resource Management and Food Security, Larnaca, Cyprus, March 12–14.

Koolwal, Gayatri, and Dominique van de Walle. 2010. "Access to Water, Women's Work, and Child Outcomes." Policy Research Working Paper 5302, World Bank, Washington, DC.

Magnan, Alexandre, Benjamin Garnaud, Billé Raphaël, François Gemenne, and Stéphane Hallegatte. 2009. *The Future of the Mediterranean: From Impacts of Climate Change to Adaptation Issues*. Paris: Institut du Développement Durable et des Relations Internationals.

McCarthy, James J., Osvaldo F. Canziani, Neil A. Leary, David J. Dokken, and Kasey S. White, eds. 2001. *Climate Change 2001: Impacts, Adaptation, and Vulnerability—Contribution of Working Group II to the Third Assessment Report of the Intergovernmental Panel on Climate Change*. Cambridge, U.K.: Cambridge University Press.

Messouli, Mohamed, and Saloua Rochdane. 2011. Field notes related to a project called "Vulnerability Assessment and Risk Level of Ecosystem Services for Climate Change Impacts and Adaptation in Morocco." Department of Environmental Sciences, University Cadi Ayyad, Marrakech, Morocco.

Obeid, Michelle. 2006. *Women's Access and Rights to Land: Gender Relations in Tenure—Jordan, Yemen, and Morocco*. Ottawa: International Development Research Centre.

Otzelberger, Agnes. 2011. *Gender-Responsive Strategies on Climate Change: Recent Progress and Ways Forward for Donors*. Brighton, U.K.: BRIDGE/Institute of Development Studies.

Reeves, Hazel, and Sally Baden. 2000. "Gender and Development: Concepts and Definitions." Report 55, BRIDGE/Institute of Development Studies, Brighton, U.K.

Resurreccion, Bernadette. 2011. "The Gender and Climate Debate: More of the Same or New Pathways of Thinking and Doing?" Asia Security Initiative Policy Paper 10, Centre for Non-traditional Security Studies, Rajaratnam School of International Studies, Singapore.

Solh, Mahmoud. 2010. "Tackling the Drought in Syria." *Nature Middle East*, September 27. http://www.nature.com/nmiddleeast/2010/100927/full/nmiddleeast.2010.206.html.

Sowers, Jeannie, Avner Vengosh, and Erika Weinthal. 2011. "Climate Change, Water Resources, and the Politics of Adaptation in the Middle East and North Africa." *Climatic Change* 104 (3–4): 599–627.

UN (United Nations). 2009. *Syria Drought Response Plan*. New York: UN. http://reliefweb.int/sites/reliefweb.int/files/resources/2A1DC3EA365E87FB8525760F0051E91A-Full_Report.pdf.

UN (United Nations) and LAS (League of Arab States). 2010. *The Third Arab Report on the Millennium Development Goals 2010 and the Impact of the Global Economic Crises*. New York: UN.

UNDP (United Nations Development Programme). 2005. *Arab Human Development Report: Towards the Rise of Women in the Arab World*. New York: UNDP.

UNFPA (United Nations Population Fund) and WEDO (Women's Environment and Development Organization). 2009. *Climate Change Connections*. New York: UNFPA and WEDO.

WEDO (Women's Environment and Development Organization). 2010a. "Cancun Climate Agreements: Taking Great Strides for Women's Rights and Gender Equality." WEDO, New York. http://www.wedo.org/news/cancun-climate-agreements-taking-great-strides-for-women%E2%80%99s-rights-and-gender-equality.

———. 2011a. "Gender Equality Language in Durban Outcomes: Outcome of the Work of the Ad Hoc Working Group on Long-Term Cooperative Action." Draft Decision [-/CP.17]. WEDO, New York. http://www.wedo.org/wp-content/uploads/Gender-Equality-Language-in-Durban-Outcomes.pdf.

———. 2011b. "The Outcomes of Durban COP17: Turning Words into Action." Press release, December 16. http://www.wedo.org/themes/sustainable-development-themes/climatechange/the-outcomes-of-durban-cop-17-turning-words-into-action.

World Bank. 2004. *Gender and Development in the Middle East and North Africa: Women in the Public Sphere*. Washington, DC: World Bank.

———. 2009. *The Little Data Book on Gender*. Washington, DC: World Bank.

———. 2011a. *Costing Adaptation through Local Institutions: Synthesis Report*. Washington, DC: World Bank.

———. 2011b. *Field Guide: Integrating Gender into Climate Change Adaptation and Rural Development in Bolivia*. Washington, DC: World Bank.

———. 2012. "Capabilities, Opportunities, and Participation: Gender Equality and Development in the Middle East and North Africa Region—A Companion Report to the WDR 2012." World Bank, Washington, DC.

———. Forthcoming a. "Syria Rural Development in a Changing Climate: Increasing Resilience of Income, Well-Being, and Vulnerable Communities." Report 60765–SY, World Bank, Washington, DC.

———. Forthcoming b. "The Women of Rural Yemen, Egypt, and Djibouti: Trade-off and Complementarities of Productive versus Reproductive Activities." World Bank, Washington, DC.

CHAPTER 8

Human Health and Well-Being Are Threatened by Climate Change

For thousands of years, climate change has affected human life, civilization, and the fundamental requirements for health: clean air, safe drinking water, food, and shelter. Historical evidence demonstrates that the most serious threat to human health has been from impaired food yields, mostly attributable to droughts. In many cases, this has resulted in famines and deaths, as well as in infectious disease epidemics (McMichael 2012). This is not peculiar to the Arab region, which has for millennia experienced frequent episodes of extreme and variable climate, often with serious implications for human health.

During the Younger Dryas, nearly 12,000 years ago, the Natufians, in today's northern Syria and the settlements along the Nile Valley, were hard hit by this climatic shock, and only a few survived as a result of its reduction in crop yields. With the start of the fourth millennium BCE, changing climatic conditions in southern Mesopotamia (Sumeria), which encompassed the lower Tigris and Euphrates River floodplains, increased food insecurity, hunger, and malnutrition. Despite efforts to cope with these changes, starvation spread widely, and with it the authority of the state diminished, ultimately leading to its being conquered by the upstream Akkadian empire (northern Mesopotamia).

This situation, however, did not last for long, and the empire collapsed around 2200 BCE because of the extension of droughts in the north, which caused more widespread starvation and malnutrition (McMichael 2012). The region is also historically no stranger to a number of climate-sensitive vector-borne diseases. Decipherable descriptions of malaria were recorded in Egyptian and Mesopotamian texts as early as 5,000 years ago. Although eradicated in a number of countries in the

Photograph by Dorte Verner

region since then, in recent years, cases have started to reemerge in previously nonendemic countries. In contrast, water-, food-, rodent-, and other vector-borne diseases recorded in ancient scripts from the region remain to this day and in some cases are increasing because of a warmer regional environment.

Climate change poses a significant threat to public health at a global level, which includes the Arab countries (see, for example, box 8.1). The Intergovernmental Panel on Climate Change (IPCC) in its *Fourth Assessment Report* declared that climate change "contributes to the global burden of disease and premature death" (Parry et al. 2007). Addressing the impacts of climate change on human health is challenging because of the wide spectrum of determinants that influence health. These determinants include the physical and social environment, old and new technologies, and changing political landscapes that reshape social and economic conditions. Health is directly and indirectly related to the impacts of climate change. Direct impacts of climate change on human health include mortality and morbidity from extreme weather events (floods, heat waves, droughts, and hurricanes); indirect impacts include longer-term climatic changes that affect the range and reproductive rates of disease vectors, extend transmission seasons, increase the incidence of food- and water-borne diseases, and lead to poor air quality and food insecurity (Parry et al. 2007). Figure 8.1 provides a conceptual framework depicting the links between climate change and human health.

In a recent report, the World Health Organization (WHO) attributed 0.2 percent of annual global mortality to climate change (WHO 2009b). The same report attributes about 1.2 million annual deaths to urban air pollution, 2.2 million to diarrhea, 3.5 million to malnutrition, and 60,000 to natural disasters, all of which are climate-sensitive outcomes prone to increase with a warmer or more variable climate. The United Nations Development Programme (UNDP) *Human Development Report 2007/2008* pointed out that major deadly diseases could expand their coverage, especially in developing countries; for example, an additional 220 million to 400 million people could be exposed to malaria—a disease that already claims about 1 million lives per year (UNDP 2007). Meanwhile, in certain geographic locales, climate change may positively affect health outcomes: milder winter seasons in some areas may lead to reductions in cold-related morbidity and mortality, while increasing temperatures in some settings may limit the expansion of a number of disease-transmitting vectors, positively impacting the spread and endemicity of related illnesses (WHO, WMO, and UNEP 2003). Despite the multiplicity of possible impacts, the threat that climate change is likely to pose to human health and security should not be underestimated (WHO EMRO 2008); the interactions between environment and health are com-

BOX 8.1

Recent Health Impacts Associated with Weather-Related Extreme Events in Selected Arab Countries

Least Developed Countries

- Sudan has experienced serious droughts in the past few years that have left nearly 1.5 million Sudanese near starvation (Wakabi 2009). Many Sudanese grapple with water scarcity, live in unhygienic environments, and suffer from malnutrition (WHO EMRO 2008).
- The Republic of Yemen experienced one of its worst floods in October 2008, which left 58 dead and 20,000 without shelter (CNN 2008). Another flood occurred in May 2010 that killed at least seven people in a shantytown around Sana'a (Reuters 2010).
- In 2011, Somali droughts and conflict left an estimated 2.4 million people, or 32 percent of the country's population, dependent on foreign aid (UN News Centre 2011).

The Gulf

- In 2009, heavy rains in Saudi Arabia left 150 people dead and substantially damaged shantytowns on the outskirts of Jeddah (BBC 2009). Recently, Saudi Arabia experienced a similar disaster that stranded people for days, killed 10, and damaged local buildings (Byron 2011).
- Oman was hit by powerful Cyclone Phet, which killed at least 44 people in 2010 (chapter 1).

The Maghreb

- Algeria experienced flash floods in 2008 that killed 29 people, destroyed around 600 homes, and left hundreds homeless (AP 2008).
- In Morocco, at least 30 people were killed in November 2010 after heavy rain and floods. The floods affected the livelihoods of nearly 75,000 people (EM-DAT 2011).

The Mashreq

- The Arab Republic of Egypt suffered one of its worst storms in memory in 2010, which destroyed more than 40 homes and 57 electrical towers and killed 15 people (Hassan 2010).
- Jordan suffered one of its worst droughts in 50 years in 1999, which affected the livelihoods of nearly 180,000 people and had the largest impact on food production. The country was also hit by another drought the following year that affected another 150,000 people (EM-DAT 2011).

Source: Authors' compilation.

plex, highly dependent on local conditions, and dynamically shaped by national, regional, and international societal developments.

Although a framework for action to protect health from climate change—through an effective and coordinated mechanism to strengthen institutional capacity—was endorsed in the 55th session of the Regional Committee for the Eastern Mediterranean in 2008 (WHO EMRO 2005),

FIGURE 8.1

Links between Climate Change and Human Health

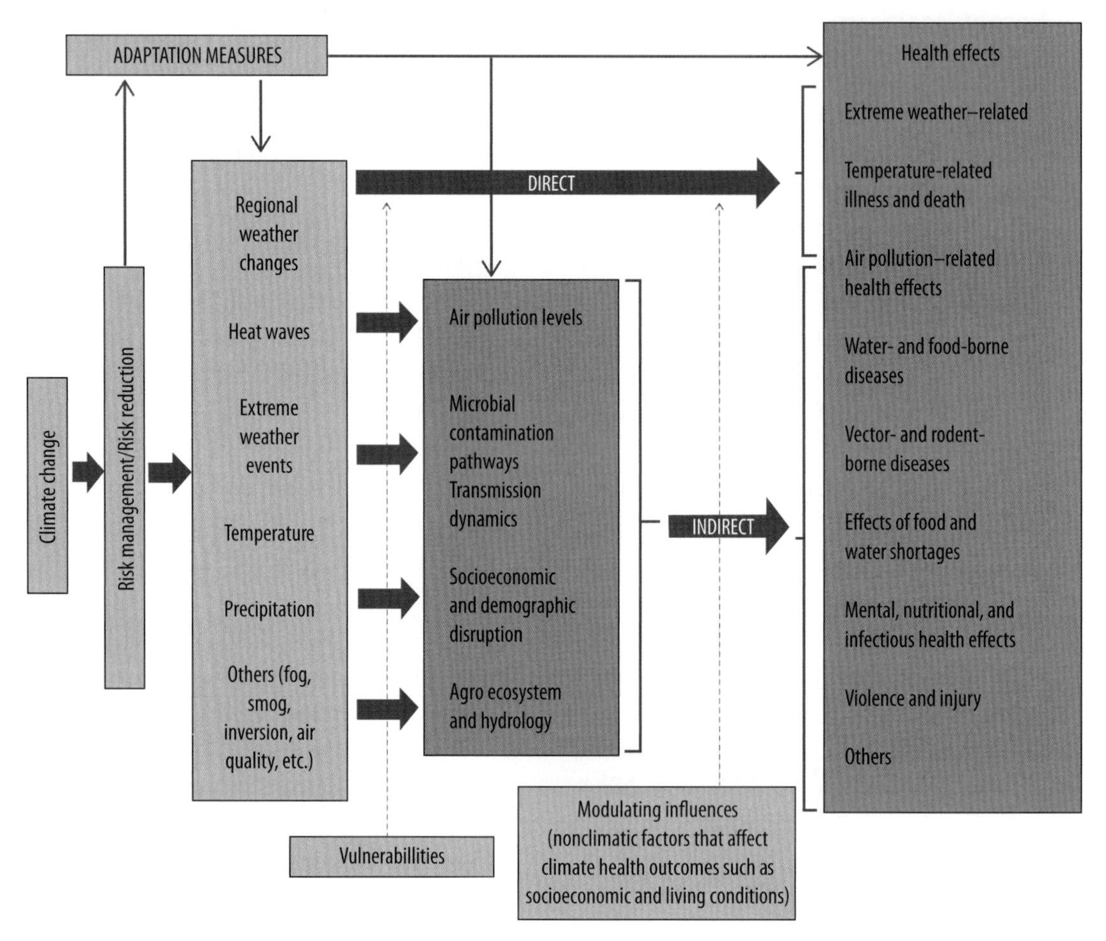

Source: Modified from Patz et al. 2000.

technical and research-based reports addressing this issue specific to the Arab context and the adaptive capacities of health sectors are still scarce (Habib, El Zein, and Ghanawi 2010).

This chapter examines the climate change–related public health challenges that the Arab region faces. On the basis of this assessment, a course of action is recommended to help to reduce and manage the health impacts of climate change.

Climate Change Threatens Human Health in the Region

Extreme Weather Events Pose Direct Health Threats

Extreme weather events, such as floods, droughts, storms, fires, and heat waves, impact morbidity and mortality in Arab countries. Countries in all four subregions have experienced these extreme weather events, which include floods, droughts, cyclones, or landslides resulting in death and illness (Musani and Shaikh 2008). Between 1990 and 2011, more than 200 natural disasters (droughts, extreme temperatures, floods, storms, and wildfires) resulted in more than 5,800 deaths in the Arab region (box 8.1) (EM-DAT 2011 and spotlight 1 of this volume). These extreme weather events are expected to increase and intensify as a result of global climate change (Parry et al. 2007). Victims of these disasters often find shelter in inadequate housing that lacks basic services, which increases the risk of water-, air-, and vector-borne illnesses and the rates of hunger and malnutrition (WHO EMRO 2008).

Flood-related diseases

Floods disrupt basic sanitation systems and increase probabilities of diarrheal disease outbreaks. Pathogens from human or animal sewage often contaminate drinking-water sources during floods, which leads to increased risks of waterborne diseases, such as dysentery, hepatitis A, cholera, typhoid fever, or leptospirosis. Floods may also indirectly lead to an increase in vector-borne diseases through the expansion in the number and range of vector habitats. For example, increased rates of diarrhea and malaria were reported following a 1988 flood in Sudan (Woodruff et al. 1990).

Heat wave–related mortality and morbidity

The region experienced seven extreme temperatures events from 1990 to 2011 with more than 100 deaths (EM-DAT 2011). Heat waves can cause death (a) directly through heat-related illnesses or (b) by aggravating preexisting heat-sensitive medical conditions. Increased heat-related morbidity can occur, including mortality, through direct heat illness or through aggravation of preexisting diseases. Either of these outcomes may require primary care or hospitalization, and a proportion of this morbidity may eventually result in mortality. Additional risk factors for heat-related mortality arise from pressures on water and electrical systems, risk-associated behavioral responses, and worsening environmental conditions, especially air quality.

Although heat-related deaths are often documented as such, heat illnesses often may go unreported in low- and middle-income countries, calling for the use of proxies for baseline data when planning and designing public health interventions. Heat waves increase the risk of hyperthermia (excessive body temperature), which poses serious health risks including heat exhaustion, cramps, heatstroke, and death. People with preexisting illnesses such as cardiovascular and respiratory diseases are more prone to heat cramps, heat syncope, heat exhaustion, and heatstroke as a result of elevated temperatures (McGeehin and Mirabelli

2001). Research indicates that extreme weather events also are associated with numerous stress-related health problems: myocardial infarction, sudden cardiac death, and cardiomyopathy (Suzuki et al. 1997; Watanabe et al. 2005).

Psychological impacts

Extreme weather events have been associated with a variety of psychological impacts attributable to loss, social disruption, displacement, and repeated exposure to natural disasters. The Centers for Disease Control and Prevention projects that 200 million people will be displaced by climate change–related factors worldwide by 2050 (CDC 2010e). In the Arab Republic of Egypt alone, 2 million to 4 million Egyptians may be displaced by a 0.5-meter rise in sea level (Rekacewic 2008).

Natural disasters, geographic displacement, loss or damage of property, and death or injury of loved ones will exacerbate mental health problems and stress-related disorders, from posttraumatic stress disorders to depression, anxiety, insomnia, and possible drug or alcohol abuse (Silove and Steel 2006). These events impact livelihoods and socioeconomic situations, which in turn, could result in increased violence and injuries.

Mental health disorders and psychiatric diagnoses are more frequent in countries that are politically volatile or violent (Al-Krenawi 2005), and this includes Arab countries. The effects of extreme weather attributable to climate change will exacerbate already prevalent mental health and stress disorders (Fritze et al. 2008; IWGCCH 2010).

Climate Change Is Associated with Many Indirect Health Consequences

Climatic changes, such as increased temperatures and more variable precipitation, have indirectly affected health by influencing the endemicity of vector-, water-, and food-borne diseases. These changes also interfere with agricultural systems and affect crop yields, which could create food and water shortages leading to malnutrition. The relationship between increased temperatures and escalating morbidity and mortality has been supported by a number of studies from the region (El-Zein, Tewtel-Salem, and Nehme 2004; Husain and Chaudhary 2008). WHO estimated that in 2004, the world faced a total of 141,277 deaths and a disease burden of 5.4 million disability-adjusted life years (DALYs) as a result of global climate change; the Eastern Mediterranean Region (EMR)[1] had its share of 20,000 fatalities and a disease burden of 755,870 DALYs (WHO 2004). Actual figures may be much higher, given that the primary outcomes measured were diarrhea, lower respiratory infections, and malaria.

Vector-borne diseases

Climate-related environmental changes, such as increased temperatures, variable humidity, and rainfall trends, may affect the density of vector populations, their transmission patterns, and infection rates (WHO, WMO, and UNEP 2003). The stress posed on water supplies by rapid human development has led to new dam and irrigation canal construction, spurring changes in mosquito populations. Vector-borne illnesses—closely associated with temperature and humidity conditions—such as malaria, dengue fever, Rift Valley fever, and West Nile virus may intensify, reemerge in previously endemic areas, or emerge in areas and countries previously unaffected. See box 8.2 for a description of vector-borne diseases.

Water-related illnesses

Water scarcity is a major concern for many Arab countries (see chapter 3).[2] Inadequate water supplies, in quantity or quality, may lead to increased risk of waterborne illnesses such as diarrhea, typhoid, hepatitis, dysentery, giardiasis, bilharziasis, and cholera. Studies in Sudan and Lebanon indicate that poor water quality, inadequate access to water, warm weather, poor sanitation and hygiene, and poverty are contributing factors to diarrheal diseases (El Azar et al. 2009; Musa et al. 1999). Outbreaks of cholera in the past three decades have been affected by seasonal patterns, especially in countries near the equator (Emch et al. 2008). Cholera epidemics are likely to emerge in areas experiencing warmer weather and water scarcity (Emch et al. 2008; Huq and Colwell 1996), such as tropical zones and the Arabian Peninsula. Cases of cholera were reported in Djibouti, Iraq, Oman, and Somalia between 1995 and 2005 (WHO EMRO 2005). Cholera is endemic to Somalia with seasonal outbreaks in urban and rural settlements resulting in many deaths.

The potential of water scarcity in the future threatens environmental sustainability in the region. The Jordan and Yarmouk Rivers' waters may substantially recede, affecting Jordan and nearby territories (European Council 2008). The per capita renewable water resources in the region decreased from 4,000 cubic meters per year in 1950 to a current 1,100 cubic meters and are expected to decline to 550 cubic meters by 2050 (World Bank 2007). This decline in water availability may ultimately increase the incidence of cholera (Huq and Colwell 1996), although cholera epidemics have been mostly absent since 2000.

As discussed in chapters 3 and 4, pressures on freshwater resources have increased the use of wastewater for irrigation. If improperly treated, wastewater can pose health risks to farmers, their families, and consumers (WHO EMRO 2008). A study of the waterways and canals in the Nile Delta carried out by the State Ministry for Environmental Affairs in 2009

BOX 8.2

Vector-Borne Diseases in Arab Countries

Malaria is currently endemic in the Comoros, Djibouti, Somalia, Sudan, and the Republic of Yemen (WHO EMRO 2009). Cases have also been reported in the Arab Republic of Egypt, Morocco, Saudi Arabia, and the United Arab Emirates, although not endemic in these countries (Al-Mansoob and Al-Mazzah 2005; Al-Taiar et al. 2006; Bassiouny 2001; Hamad et al. 2002; Hassan et al. 2003; Himeidan et al. 2007; Malik et al. 1998; Noor et al. 2008; WHO EMRO 2009).

Leishmaniasis is a zoonotic disease caused by protozoan parasites (genus *Leishmania*) and transmitted to humans through carrier-phlebotomine sand flies. Increased incidence of leishmaniasis has been associated with increased temperatures (Oshaghi et al. 2009), yearly rainfall and continentality in Tunisia (Ben-Ahmed et al. 2009), and heavy rainfall and drought in Sudan (Neouimine 1996). Leishmaniasis has recently reemerged in Saudi Arabia, Sudan, the Syrian Arab Republic, and Tunisia (Rathor 2000; Sturrock et al. 2009).

Lymphatic filariasis, dengue fever, and Rift Valley fever (transmitted by the vector *Culex pipiens* mosquito) are similarly affected by climate change as temperature increases and high rainfall play an important role in propagating the vectors that transmit them (Parry et al. 2007). Lymphatic filariasis is mainly endemic in the Nile Delta, exposing an estimated 2.7 million people (Epstein 2002). Filariasis is endemic to Egypt, Sudan, and the Republic of Yemen (Government of Egypt 2010) and affects around 0.4 million people in the three countries (El Setouhy and Ramzy 2003; Sturrock et al. 2009). Djibouti and Somalia have seen outbreaks of dengue fever in the past decade. In 2009, dengue fever was reported in Saudi Arabia, Sudan, and the Republic of Yemen (WHO EMRO 2009). In fact, in October of 2009, the city of Taiz in the Republic of Yemen detected at least 350 confirmed cases of dengue. According to local authorities, dengue is not an epidemic but rather a recurring disease in the governorate, which appears every two to three years.

Schistosomiasis (bilharziasis) is a visceral parasitic disease caused by blood flukes and transmitted to humans through freshwater snails. Egypt has had problems in the past with schistosomiasis and there are reports of it reemerging in the region (Government of Egypt 2010; Wiwanitkit 2005). Climate change and higher temperatures may affect the life cycle of snails leading to an increase in the incidence of schistosomiasis. Increased cases of schistosomiasis infection have been reported during hot weather in Egypt and Morocco (Khallaayoune and Laamrani 1992; Malone et al. 1997; Yousif et al. 1996), and increased incidence was reported during rainfall in Saudi Arabia (al-Madani 1991).

Trachoma is an infectious disease of the eye and is transmitted in human feces. Incidences of trachoma are associated with hot, dry climates and in living environments that are overcrowded and have poor sanitation and limited access to clean water. Cases of trachoma were reported but not confirmed in Iraq, Libya, and Somalia, whereas the illness is endemic in Djibouti, Egypt, Mauritania, Morocco, Oman, Sudan, and the Republic of Yemen (Polack et al. 2005).

Source: Authors' compilation.

found that organic matter concentrations and *E. coli* bacterial counts exceeded the permissible levels in canal waters (EEAA 2010). In addition to the increased release of untreated sewage, the poor control of the use of sewage water for irrigation may possibly be a factor in some locations.

Food-borne diseases and malnutrition

Nearly 800 million people suffer from malnutrition, which causes 3.5 million deaths per year (WHO 2009a; WHO EMRO 2008). Currently, 32 million of these malnourished people come from 16 Arab countries: 8 million in Sudan and 6 million in the Republic of Yemen (FAO 2010). It is quite possible that a majority of the Eastern Mediterranean region will suffer from food shortages by 2025 (Al-Salem 2001). Malnutrition is a main contributor to child mortality, causing about half of the deaths for children younger than five years (WHO EMRO 2009). In general, analyses show that increased health expenditures per capita drastically reduce child mortality (figure 8.2). Hence, addressing health expenditures and increasing those for children can improve child health and reduce mortality in the most affected countries.

Studies have indicated an association between increased temperatures and food poisoning (Fleury et al. 2006; Kovats, Hajat, and Wilkinson 2004; Lake et al. 2009). Food-borne illnesses result from the ingestion of spoiled or contaminated food, such as seafood contaminated with metals

FIGURE 8.2

The Relationship between Health Expenditures Per Capita and Child Mortality

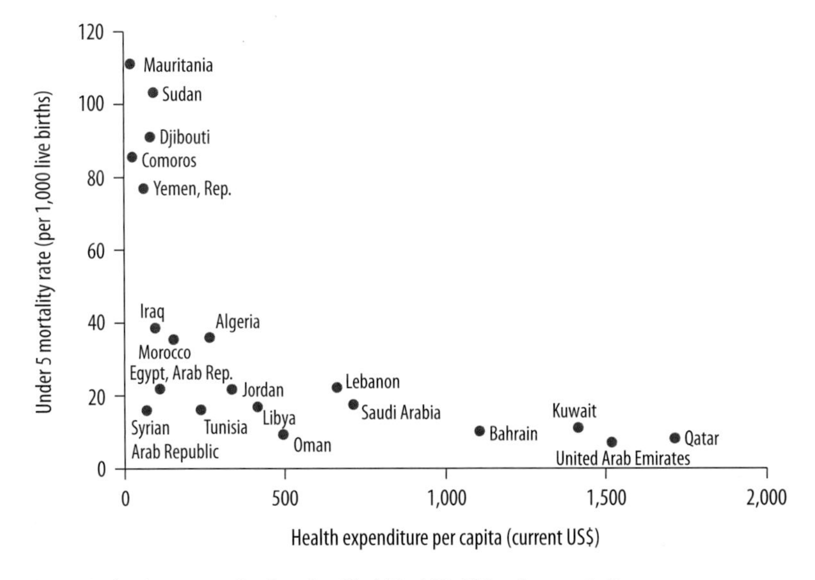

Source: Authors' representation, based on World Bank World Development Indicators.

or crops with pesticide residues or microbes (IWGCCH 2010). Extreme weather events such as droughts encourage the proliferation of crop pests and the spread of mold that may be harmful to humans. Changes in climate may also affect environmental ocean parameters that lead to the proliferation of existing or new pathogens that are harmful to human health (CDC 2010c). Harmful algal blooms produce toxins that, when ingested through shellfish, also cause human diseases (Parry et al. 2007). The increase in pests and weeds could also lead to a wider use of pesticides and a higher risk of pesticide exposure (CDC 2010c).

Other health effects

Weather conditions, including temperature, humidity, and wind, also affect ambient air quality, which is largely determined by anthropogenic sources of pollutants (CDC 2010a; Cizao 2007; IWGCCH 2010). Climate-related environmental factors, including dust storms, rainfall, and increases in temperature, raise the ambient concentrations of aeroallergens (including pollen and dust), ground-level ozone, and suspended particulate matter, which exacerbate respiratory illnesses (IWGCCH 2010). In Egypt, inhabitants suffer serious health effects from air pollution, resulting annually in 15,000 bronchitis cases, 329,000 respiratory infections, and 3,400 deaths in Cairo alone (Anwar 2003; UNEP 2007).

Increased human exposure to toxic substances as a result of climate change–related factors may be linked to cancers in humans (IWGCCH 2010). For example, volatile and semivolatile carcinogens are transferred from water and wastewater into the atmosphere as a result of higher ambient temperatures, and toxic pollutants are washed out by heavy rains and floods, which contaminate runoffs and ultimately water resources (Macdonald et al. 2003). However, limited evidence has been established on these transfers and their impact on people's exposure to carcinogens, and ultimately, their impact on cancer outcomes (IWGCCH 2010). There are no conclusive studies exploring linkages between climate or climate-mediated carcinogens and cancers in Arab countries (Habib, El Zein, and Ghanawi 2010), although limited government oversight of heavy industry heightens the probability that Arab populations are exposed to these substances.

Cardiovascular illnesses have been linked to climate change–related variables such as average daily temperatures (Basu and Samet 2002; Braga, Zanobetti, and Schwartz 2002; Ebi et al. 2004). Physiological adjustments to cold and warmth are associated with changes in blood pressure, blood viscosity, and heart rate, all important determinants of mortality related to cardiovascular diseases and strokes (Martens 1998). Studies in Egypt, Kuwait, Lebanon, the Syrian Arab Republic, and the United Arab Emirates show a significant association between climate change, specifically temperature and humidity increases, and cardiovascular disease (Douglas et al. 1991; El-Zein and Tewtel-Salem 2005; El-Zein, Tewtel-Salem, and Nehme 2004; Shanks and Papworth 2001; Zawahri 2004). For some Gulf countries—Bahrain, Oman, Qatar, and the United Arab Emirates—an assessment of climate change–related human health risks projected that between 2070 and 2090 rising mortality rates from cardiovascular diseases, respiratory illnesses, and thermal stress will be experienced (Husain and Chaudhary 2008).

Studies have shown a steady increase in some forms of birth defects (Correa-Villaseñor et al. 2003). This could suggest a possible link to climate-related environmental changes such as increased temperatures and changes in rainfall that may increase human exposure to toxic substances such as pesticides (IWGCCH 2010). Human development stages are most vulnerable at preconception, the fetal period, and early childhood (Wadhwa et al. 2009). During these periods, exposures may alter developmental changes and cause functional deficits through mechanisms such as genetic mutations and epigenetic changes (Wadhwa et al. 2009). The estimated total annual birth defects in Arab countries is 622,000, constituting 7 percent of the estimated total annual birth defects in the world (8 million) (Christianson et al. 2006).

Climate change may increase the incidence of neurological disorders in humans. Malnutrition, exposure to pesticides, hazardous chemicals, biotoxins, and metals in the air, food, and water increase the risk of neurological disorders, and all are expected to be worsened by climate change (CDC 2010a; Handal et al. 2007; Kar, Rao, and Chandramouli 2008). Studies have shown that the onset of Alzheimer's and Parkinson's disease are occurring at an earlier age and with more severe symptoms because of changes in environmental conditions (Bronstein et al. 2009; Mayeux 2004). It is believed that changes in climatic conditions may also be partly responsible for the increasing number of children affected by learning disabilities (Bronstein et al. 2009; Mayeux 2004). The effects of climate change will also result in higher risks of neurological illnesses from ingestion of or exposure to neurotoxins from harmful algal blooms in water and seafood (S. Moore et al. 2008; Sandifer et al. 2007). Although there have been very few studies on the occurrence of neurological disease in the Arab region (Benamer and Shakir 2009), a study in Libya reports 11,908 neurological patients in 2006 and a total of 6,892 new neurological cases per year (Benamer 2007). A study in Syria relates the incidence of ischemic cerebral lesions leading to stroke to climatic changes of high summer temperatures and unstable humidity (Zawahri 2004). Neurological disorders generally place a huge financial burden on national health care systems and on families and take a substantial toll on the quality of life of both the patient and the caregiver.

Some Population Groups Are Particularly Vulnerable

Certain groups or individuals may be particularly vulnerable to the health impacts of climate change. Vulnerability depends on various factors such as population density, socioeconomic well-being, nutrition, environmental determinants, preexisting health status, and the quality of and accessibility to affordable health care (WHO, WMO, and UNEP 2003). Some groups may be directly affected by their physical location: inhabitants of regions geographically located at the margins of areas where malaria and dengue fever are endemic will be more susceptible to these diseases if climate change results in the growth and geographic expansion of the vectors carrying these illnesses (WHO, WMO, and UNEP 2003). But geographic regions are also linked by ongoing sociocultural processes and events such as urbanization, rural development, immigration, and so forth, all of which reverberate through the lives of area residents and produce distinct health outcomes.

Mass Gatherings Are More Hazardous in Hot Weather

The annual *hajj* in Saudi Arabia—a mass gathering of Muslim pilgrims—often takes place during the hot summer (see also chapter 6 on tourism). This gathering exposes visitors to high temperatures and associated heat-related morbidity and mortality. In addition, infectious diseases are of particular concern among pilgrims because of the potential rapid spread of the disease. For instance, in 1987, a meningococcal disease spread among the pilgrims in Mecca and was carried elsewhere in Saudi Arabia, to other Gulf states, and to Pakistan, the United States, and the United Kingdom (P. Moore et al. 1989; Wilson 1995).

Residents of Conflict Areas Will Face Even More Dire Conditions

During conflicts, the social fabric and structural integrity of societies break down: housing and infrastructure are often destroyed, as was the case in Gaza during the 2009 bombardment (Abu-Rmeileh, Hammoudeh, and Giacaman 2010); state institutions may stop functioning; and many social resources are disrupted. Environmental events arising during conflict may aggravate existing health catastrophes or precipitate new ones. In addition, ecological processes associated with climate change may be a partial antecedent of conflict, as was the case in Darfur, Sudan.[3] A number of Arab countries have been maligned by armed conflict in recent years, including Iraq, Lebanon, Libya, Somalia, the West Bank and Gaza, and the Republic of Yemen. Nations may have to deal with the concurrent burden of conflict and of acute or prolonged natural disasters.

Institutions responsible for securing people's health may be ill-prepared for the overwhelming burden of multiple catastrophic events.

Residents of Low-Lying Areas Are Threatened by Rising Sea Levels

Rising sea levels cause flooding and the intrusion of saltwater into the groundwater, which can lead to soil salinization and the deterioration of drinking water (El-Raey 2008 and chapter 6 of this volume). Increased salinity of surface water and groundwater also reduce crop and livestock yields and thereby further threaten food security and nutrition status (CDC 2010d). Harmful pathogens may also proliferate as a result of high water salinity (IWGCCH 2010).

Residents of Camps, Shantytowns, Slums, and Rural Areas May Be Unable to Cope

Residents of poor neighborhoods and slums are most likely to have inadequate housing and limited access to water, electricity, food, and medical resources (Habib et al. 2011). Residents of these communities often deal with a number of social and environmental problems and, as a result, are more sensitive to infectious and chronic diseases and mental illnesses (Habib et al. 2011; *Yemen Times* 2011). Because of their precarious living environment, these residents may be more exposed to the negative impacts of climate change; and without adequate financial resources or social capital, they may be unable to adapt to environmental changes precipitated by climate change (Government of Jordan 2009).

In the Arab region, rural areas have higher rates of poverty than do urban areas. Rural areas are also generally underserved by health care facilities and emergency response services, which are often located far from people in need. In the event of extreme weather conditions, such as floods, hurricanes, heat waves, and dust storms, people living in remote locations are more vulnerable to environmental problems and less able to deal with the consequences of such events. For example, underserved populations in Lebanon are mostly concentrated in the rural north and east of the country, although poverty and inadequate housing and infrastructure are widespread, even in the larger urban centers (Government of Lebanon 2011).

Internally Displaced Populations and Refugees Are Extremely Vulnerable

Populations are frequently displaced during extreme weather events, such as droughts or floods, and during conflict. There are many refugees and internally displaced people (IDPs) in the Arab region. They include Iraqis, Lebanese, Libyans, Somalis, Sudanese, Syrians, people in the West Bank and Gaza, and Yemenis (Habib 2010; Habib et al. 2011; IDMC 2011). Many of the displaced are extremely vulnerable to disease because of their disadvantaged situation and their often precarious living conditions (St. Louis and Hess 2008). For example, in Lebanon, many refugees from Iraq and the West Bank and Gaza live in camps and shantytowns with poor infrastructure, low-quality housing, and an inadequate water supply, all of which increase the risk of disease (Government of Lebanon 2011). In the 2006 war in Lebanon, approximately 1 million IDPs took refuge in schools, gardens, or garages. During the conflict period, there were frequent reports of increased diarrhea, typhoid fever, hygienic illnesses (rash and lice), and an increase in incidence of chronic illnesses (Government of Lebanon 2011). When many IDPs returned to their villages to find their homes destroyed, they occupied temporary tents that were inundated by water during the first heavy rain of winter.

Outdoor Workers Face Harsh Weather

Outdoor occupations in such sectors as agriculture, farming, fishing, and construction are particularly vulnerable to heat illnesses, especially in hot climates and places with high humidity (Shanks and Papworth 2001). Strenuous and physically taxing work coupled with extreme outdoor conditions may lead to heat exhaustion, hyperthermia, dehydration, and heatstroke (McDonald, Shanks, and Fragu 2008). Construction workers, for example—who in Arab countries are usually unskilled and poorly paid—labor in high temperatures and humidity for long hours, which causes poor health and reduced productivity (Kjellstrom 2009). Heat stress can lead to serious physical injury, fainting, organ damage, and even death. Health centers in the Gulf region have dealt with heat stress among workers in outdoor occupations and are familiar with the problem (Deleu et al. 2005; Shanks and Papworth 2001). Recently, the government of Qatar—a country where about 24 percent of workers are employed in the construction industry (Qatar Statistics Authority 2007)—restricted the work hours of certain occupations during the hottest parts of the day (Yeo 2004).

Older Adults, Women, and Children Are among the Most Vulnerable

Older adults, children, and pregnant women may be particularly vulnerable to environmental changes and temperature increases. The elderly have higher risk of heat-related illnesses such as heatstroke, cardiovascular disease , respiratory disease, and heat-related mortality (CDC 2010b).

Pregnant women experience physiological changes that make them more susceptible to thermal stress. A large proportion of Arab women who are not generally involved in livelihood-related decisions are more vulnerable to poor environmental conditions resulting from climate change (UNDP 2009). Growing children are also more vulnerable to climate-induced environmental changes (Bartlett 2008). Children have rapid metabolism, immature organs and nervous systems, developing cognition, limited experience, and various behavioral characteristics and are at increased risk of heatstrokes, heat exhaustion and dehydration, injury, and infectious disease outbreaks (Bartlett 2008).

Data and Research Are Needed to Fill Gaps

To effectively address climate change and its impacts in the health sector, information is needed in Arab countries. The Arab region suffers from a dearth of climate change research and evidence-based policies mainly because of deficient national and regional health surveillance systems (Habib, El Zein, and Ghanawi 2010). Many Arab countries lack the appropriate institutions and infrastructure to carry out the WHO approach of surveillance. The majority of the surveillance systems engage in collection of data related to communicable and noncommunicable diseases, without correlation to environmental indicators, with the exception of Kuwait, Oman, and the West Bank and Gaza, which do report such indicators (WHO EMRO 2009).

The surveillance systems of the region vary widely between fully functional and productive in a few countries to rudimentary and almost nonexistent in others (Hallaj 1996). This is partly because of the lack of resources and human capacity, as well as other factors, such as political instability and conflicts.

Data surveillance systems should ideally be linked to major health indicators related to climate change including food- and vector-borne illnesses, respiratory illnesses, malnutrition and hunger, population displacement, land loss, physical hazards, and others. For example, the coverage and quality of data on malaria in high-risk countries is still very limited and unreliable because of the lack of confirmatory facilities (WHO EMRO 2009). Only two countries (Sudan and the Republic of Yemen) have carried out national malaria prevalence assessments to measure past exposure in areas with low transmission (WHO EMRO 2009). In countries where such systems are less developed, researchers may be forced to elicit information from key stakeholders to explain how communities adapt to climate change, perceive risk, and measure their vulnerability.

Collection of data related to specific diseases (HIV, malaria, maternal health) is often guided by WHO Eastern Mediterranean Regional Office (EMRO) initiatives. Toward that end, the WHO EMRO conducted an in-depth review of core capacities of the surveillance system in several countries, and technical guidelines and information were provided to the countries of the region (WHO EMRO 2009). Many Arab countries had included a plan for strengthening their surveillance systems in action plans for adaptation to climate changes presented at the Regional Seminar on Climate Change and Health held in Cairo in June 2008.

Scope of Research Is Limited

A comprehensive review of literature published on climate change and health in the EMR between 1990 and 2010 found a lack of evidence-based research directly exploring the relationship between climate change and health (Habib, El Zein, and Ghanawi 2010). The available studies were geographically limited to Egypt, Lebanon, Sudan, and countries in the Gulf (Habib, El Zein, and Ghanawi 2010). Despite the absence of a clearly established relationship, there is evidence of potential health impacts of climate change in the region. Several studies from the Arab region have demonstrated the relationship between environmental causes of illness (heat, natural disasters, air quality, and infectious agents) health consequences, and environmental conditions that may be affected by climate change (Habib, El Zein, and Ghanawi 2010). The scientific literature has inadequately addressed two types of climate change knowledge: (a) analytical epidemiological research that studies specific infectious diseases and chronic illnesses particularly impacted by climate change (Habib, El Zein, and Ghanawi 2010) and (b) cost estimation studies that estimate the health burden and costs of climate change (Ebi 2008). Governments that have submitted "National Communications" have provided basic cost assessments of health and climate change, but these are probably underestimated because of the absence of many activities and diseases in these estimates (Kovats 2009).

Climate change research is particularly difficult because it requires both a multidisciplinary approach and an awareness of multilayered social, environmental, and health processes. In the Arab region, the ability to research climate change issues is further complicated by the absence of effective and coordinated surveillance systems, limited resources and institutional capacity, low level of government investment in research and development, and a lack of political will (Habib, El Zein, and Ghanawi 2010).

There are several key gaps in regional research. Longitudinal studies over extended periods of time serve an integral role in investigating the relationship between climate and health but have not been carried out in Arab countries. Disease mapping can be used in countries with limited research capabilities and where vector-borne diseases pose a major health risk. Disease mapping is considered less data intensive and allows for the assessment of disease in time and space and for analysis of patterns of presence and absence of disease (Bertollini and Martuzzi 1999). In addition, specific models should be used to project the health impacts on vulnerable populations, given various socioeconomic conditions (Parry et al. 2007). These models may be adapted from international studies and tailored to regional and local settings. Areas of interest include but are not limited to aeroallergens, malaria transmission (especially in highly endemic areas), and water quality in coastal cities and areas already dealing with water scarcity. Research at the regional and national levels must further distinguish the relationship between climate, geography and environment, and population health.

Adaptive Strategies Exist to Improve Human Health and Strengthen the Health Sector

Roles and Responsibilities

To meet the challenges of climate change, Arab countries should take steps toward national preparedness and adaptation.[4] Because of the cross-cutting nature of climate change issues, implementing initiatives requires horizontal and vertical coordination between multiple levels of regional and international organizations, government institutions, nongovernmental organizations (NGOs), the business sector, and stakeholders from civil society and the public (see chapter 9).

Although there is substantial progress in disaster risk reduction and strengthening of disaster preparedness plans in the region, this progress is still uneven across countries and the planning action for preparedness and adaptation is still underdeveloped (UNISDR 2009). Greater efforts and investments should go toward improving disaster prevention, especially given its economic efficiency. Every US$1.00 invested in preventative activities is estimated to save between US$2.50 and US$13.00 in disaster aid (DFID 2005). Broad logistical constraints are a result of (a) the inadequacy of the legislative systems of many countries in the region and (b) lack of resources and funding to address the challenges posed by disasters (OCHA 2010).

The IPCC (Parry et al. 2007) has flagged the need for both anticipatory and responsive strategies in which the roles of stakeholders are clearly identified. Responsive strategies aim at reducing current vulnerabilities to climate change that has already occurred and at preparing for possible extreme weather events. Anticipatory strategies aim at addressing health outcomes associated with future projections of climate extremes and changes in average temperature and precipitation. Both types of strategies would be implemented by a variety of stakeholders that include international and regional organizations, national governments, health institutions and professionals, the general public, and academia. In addition, every opportunity must be exploited to involve civil society organizations and the private sector—both of which can be involved in service provision and outreach—as well as community-driven climate preparedness initiatives at the household level.

The selection, cost-benefit or cost-effectiveness analyses, and evaluation of adaptive strategy alternatives are generally left to planners and decision makers according to strategic priorities and baseline conditions.

International and regional initiatives are already under way

The role of international and regional organizations is critical to provide support toward regional preparedness by establishing a skills and knowledge base among health professionals and policy makers in Arab countries so they can support future developments in climate change research, advocacy, and intervention. Further, these organizations bring a wealth of knowledge, experience, and lessons learned across the region and on a global level, allowing the cross-fertilization of knowledge to benefit the entire region.

To date, much has been accomplished within the regional and international spheres pertaining to climate change. WHO is the parent to two organizations involved in programs related to climate change in the region: the Center for Environmental Health Activities and WHO EMRO. These organizations have worked closely with Arab governments, health professionals, and civil society to develop national frameworks and systems for pressing environmental and health issues by offering technical assistance, research grants, access to information, and other forms of support. For instance, international organizations and government ministries in Jordan have joined efforts in establishing programs on water quality, health protection, and food security to enhance Jordan's capacity to adapt to climate change; these programs have linked the work of the UNDP with the Jordanian Ministry of Environment, WHO with the Ministry of Health, the Food and Agriculture Organization with the Ministry of Agriculture, and the United Nations Educational, Scientific, and Cultural Organization with the Ministry of Education (UNDP Jordan 2011).

The Ministers of Health endorsed a new resolution during the 52nd session of the Regional Committee for the Eastern Mediterranean in 2005 encouraging member states to "further strengthen national emergency preparedness and response programs through legislative, technical, financial and logistical measures" (WHO EMRO 2005). In practical implementation of this resolution, the Arab health ministers are called to apply the basic pillars of health disaster preparedness as outlined in the World Health Assembly (Resolution 58.1) issued in 2005 by seeking WHO resources to (a) develop and implement their country-specific health-related emergency preparedness plans, (b) ensure adequate institutional response to critical health needs during crises, (c) benefit from WHO health expertise for response operations, and (d) design, plan, and implement a transition and recovery program.

National governments need to play an assertive role in adaptation

The role of national governments focuses on establishing a framework for action and ensures its implementation by the various stakeholders and institutions involved. National governments would be primarily responsible for prioritizing, designing, and overseeing implementation of adaptive policies as well as for emergency preparedness programs. Further, it is imperative that national governments allocate resources to cover such efforts and ensure both technical and allocative efficiency in their use. From another perspective, national governments would need to set up regulatory frameworks to ensure that policies are adhered to and programs implemented.

The public health sector needs to play a central role in providing for and protecting the health of local populations against the adverse affects of climate change. Health care systems in the Arab region are typically unable to provide sufficient care for all people. Many national systems suffer from inadequate funding, poor governance, and weak institutional capacity and coordination, as well as a sizable preexisting disease burden (Khogali 2005). Climate change, coupled with rapid population expansion and continued environmental degradation, will increase the disease burden and pressure health care systems.

Providing a strong public health sector is a first important step toward preparedness and adaptation. The availability of health resources in the region differs substantially from country to country, depending on political, social, and economic conditions. The lack of funding and basic infrastructure in some states limits the adaptive capacity of their health sectors.

Fundamental climate change adaptation measures for many Arab countries are complementary to the primary national health and infrastructural priorities, such as developing healthy environmental infrastructure, containing the spread of infectious diseases, establishing extensive primary health services, and developing secondary and tertiary care capacity to address emergent health crises. Despite a clear need for basic adaptive measures (as expressed in national communications), there is little evidence that governments have made headway toward implementing such measures (ESCWA and LAS 2006).

Efforts to address the health aspects of climate change fall primarily on the health sector, and in particular, national health ministries. These government agencies would benefit by employing and empowering climate change specialists to help ensure the sustainability of governmental efforts to adopt climate change adaptation strategies. These task forces should also be responsible for involving other government offices as partners in climate change adaptation. The magnitude and cross-cutting nature of climate change problems require inter- and intragovernmental collaboration, commitment, and action.

In sum, public health policies need to focus on adaptive capacity to reduce vulnerability, enhance public health services, strengthen disaster and emergency preparedness and response, and improve clinical management. The health sector needs to shift from being reactive to proactive, while recognizing that managing climate-related health impacts is a "no-regrets" approach.

Health institutions must build their capacities

One of the major responsibilities of health institutions is to build their capacities in terms of human capital and training. Health agencies and institutions should be mandated to design, implement, or enhance specific programs and interventions in disease surveillance; epidemic forecasting; prevention; outpatient outreach; vector control; water, sanitation, and hygiene; and nutrition.

The public must be involved

The public must understand the impacts of climate change on human health. Climate change and health need to be presented in lay terms to make the problem more personally relevant, significant, and comprehensible to all members of society. Communities have an important perspective to share on climate change. Facilitating input from the public and civil society will strengthen adaptation policy and will spur local community action to reduce the health impacts of climate change.

Civil society organizations and the private sector are yet to be engaged in adapting to climate change

Although there are many environmental NGOs registered in the Arab region, few have been actively engaged in climate change issues (Bluhm

2008). NGOs can play a strong role in mobilizing popular support through media campaigns, networking with key stakeholders from different sectors, and lobbying policy makers in government. In 2007, a network of NGOs formed the Arab Climate Alliance, which advocates for just environmental policy and regional action on climate change. The alliance includes member organizations from Bahrain, the Comoros, Egypt, Jordan, Morocco, Syria, Tunisia, the United Arab Emirates, the West Bank and Gaza, and the Republic of Yemen. A similar effort was initiated by the Arab Network for Environment and Development (RAED), which includes more than 200 NGOs that focus on environment, sustainable development, and climate change–related issues in the region.[5]

The private sector also plays an important role in forwarding environmental goals through business ventures and "green" investment. Green architecture, construction, and landscaping companies have found a niche in several Arab countries. Focused on building and landscaping that is environmentally friendly, green development companies may be able to establish residential projects in areas most affected by climate change. For example, these ventures may be able to invest in low-cost and innovative construction methods that protect residents from extreme weather by using new heating and cooling techniques. Many of these projects may require joint partnerships between the public and private sectors in that there may be limited consumer demand for residential green development, especially among low-income, migrant, or displaced populations. The private sector is also involved in consulting governmental and nongovernmental institutions on climate change issues. The private sector may be able to attract additional expertise on climate change issues that can inform health policy making and initiatives as well as play an important role in adaptation.

Academic and research institutions are vital for producing evidence-based knowledge and training capacities

The region has recently witnessed a huge surge in the number of private universities and research institutions. There are a number of established academic institutions with at least masters programs in environmental health or science (for example, the Graduate Environmental Science program at the American University of Beirut and the graduate program of the Institute for Environmental Studies and Research at Ain Shams University in Cairo). The role of academic and research institutions is to help coordinate the surveillance and monitoring of health and environmental indicators, initiate research projects around climate change issues, and train professionals who will inevitably take leading roles in government and civil society. Academic institutions have a vital role in research and

BOX 8.3

Current Research Needs in the Arab Region

Research on climate change and health should draw on the expertise of scientists and researchers from Arab countries and include subregional, regional, and national priorities. Some priorities for this research include the following topics:

Heat-related morbidity and mortality: Multi-disciplinary research on the adaptive capacity of vulnerable populations should explore socioeconomic, environmental, physical, and health parameters. In particular, the many countries in the Arab region that experience long hot summers would benefit from this research.

Respiratory diseases: Techniques that monitor air quality and climate-sensitive exposures linked to respiratory diseases should be developed, and epidemiological studies on the connection between changing climate variables and the onset of respiratory diseases should be pursued. In the Arab region, increased temperatures, reduced precipitation, and the tendency to desertification can increase air pollution and the factors affecting respiratory diseases.

Infectious diseases: Research and monitoring techniques are needed to address the growing threat of vector-borne diseases and the effects of ecological disruption; surveillance and early warning systems would enhance communication and prevention strategies.

Research on water quality and water-related illnesses: Research is needed on the vulnerability of water systems to flooding or sewer overflow during extreme weather, and the resultant proliferation of pathogens, toxins, and chemicals. A focus on detecting and monitoring the incidence of health threats related to sea-level rise and climate change is also warranted.

Capacity building: Policy makers, scientists, and health professionals should be trained in disaster management and data collection and monitoring; in how to respond to health emergencies; and in how to raise awareness among citizens regarding climate-related health issues.

providing support at a national and regional level. Specialties in epidemiology, biostatistics, communicable diseases, environmental health, occupational health, health services administration, and others are vital for providing the support for academic service coordination to deliver high-quality public health services. See box 8.3 for current research needs in the Arab region.

Cross-Sectoral Collaboration

In light of the cross-sectoral natural of climate change and health impacts, climate health strategies will require cross-sectoral collaboration. The preparedness of the energy sector to increase power surge capacity for ventilation in the event of heat waves is as important as the availability of rapid response medical teams for the medical emergencies that occur during heat waves. Another example is the cooperation between the health sector and other departments, such as water, planning, public works, natural resources, meteorology, and environment, that will be required for conducting environmental risk assessments to monitor environmental health threats during floods.

Certainly, enhancement of adaptive capacity in general, and particularly for human health, should be regarded as a component of broader sustainable development initiatives and programs. Underdevelopment greatly constrains adaptive capacity in every relevant dimension of vulnerability. Improving adaptive capacity can strongly support multisectoral development processes. Vulnerability assessments, data collection, and dialogue among stakeholders conducted to enhance adaptation are likely to generate useful information to aid the design of development programs.

Although planning for adaptation will almost invariably be complicated by uncertainties and competing priorities and interests, management of climate risks can oftentimes benefit from existing decision support tools and regulatory frameworks. Further, planned adaptations to climate risks are more likely to be implemented when developed as integral parts or modifications of existing programs and strategies (Parry et al. 2007).

Recommendations for Enhanced Adaptive Capacity in the Health Sector

Health systems often lack the capacity to effectively implement climate change adaptation strategies. Many Arab countries have a substantial "adaptation deficit" and would benefit from exploring the following policy recommendations:

Establish or strengthen information systems linking health and climate change–related outcomes

Countries in the Arab region are encouraged to adopt the following strategies:

- Develop climate-sensitive surveillance systems and evaluation techniques for health. This involves collecting information on frequency and magnitude of climate change–related health outcomes and linking them to environmental and meteorological indicators. Governments should invest in monitoring stations for ambient temperatures, humidity, and precipitation, and for ambient air and water quality. These indicators are important for early warning systems. Consequently, this results in operationalizing an effective integrated health-environment management information system (see below).

- Strengthen health-environment management information systems to enable evidence-based decision making for planning, designing, financing, and implementing adaptation programs to address the climate change–related burden of disease.
- Collect and analyze information on groups vulnerable to climate change. This includes identifying their specific vulnerabilities and characterizing risk exposures, describing their geographical locations and social and economic status, and evaluating their access to social protection services.

Build capacity for climate resilience

Building capacity for climate resilience requires that Arab countries establish a foundation for strengthening expertise in climate change and public health–related disciplines and for providing services to support adaptation efforts in the health sector. Arab governments need to develop or strengthen a comprehensive framework for human resources for health, both technical and managerial, that integrates climate change into its core strategic mission. This effort can be achieved by incentivizing academic institutions to invest in climate change and health research and by building or expanding graduate training programs in climate change and health sciences. In addition, academic and research institutes can provide technical assistance to health ministries, while civil society has a vested interest in raising public awareness about climate change, health, and household and community adaptive measures.

Protect the poor and vulnerable through social services

Countries in the Arab region are encouraged to adopt the following strategies:

- Develop an emergency preparedness response plan. The plan would incorporate transition and recovery programs to deal with emergent crises, assist the poor and vulnerable to relocate away from high-risk areas, and provide basic needs such as adequate shelter, access to food, clothing, and medication for the poor and most vulnerable.
- Strengthen health care service delivery by adopting the following: upgrading primary health care and emergency and ambulatory services

to cope with emergent health crises, ensuring equitable access (both physical and financial), and improving the quality of care.

- Protect populations against catastrophic expenditure and health shocks by ramping up efforts for social protection in health, especially for the poor and vulnerable. This should avoid regressive redistribution policies whereby the rich might benefit more than the poor in accessing health care services and in benefiting from reduced health care costs. Moreover, the expansion of health insurance, social assistance, and safety net programs are invaluable to protecting the poor and vulnerable.

Create an institutional framework for decision making and formulating supportive policies

Countries in the Arab region could reach sound health policies by implementing the following steps:

- Develop a situational analysis to outline the current strategies adopted by the health care sector in dealing with emergency preparedness and in mapping health outcomes.
- Develop a SWOT (strengths, weaknesses/limitations, opportunities, and threats) analysis to identify the strengths, weaknesses, opportunities, and threats in the health care sector.
- Develop a stakeholder analysis that defines roles and responsibilities of each party (government, private sector, civil society organizations, the public, as well as others).
- Develop or update climate change–specific clinical practice guidelines and standard operating procedures covering various levels of health care: primary, secondary, tertiary, and specialized care.
- Establish a national climate change steering committee consisting of focal points representing concerned ministries and other stakeholders to initiate dialogue on climate change policies. The committee reports to the cabinet of ministers and coordinates with regional and international organizations such as UNDP, WHO, United Nations Environment Programme, the World Bank, IPCC, and others.

The above recommendations will serve to establish an effective institutional framework as a prerequisite for achieving the following core outcomes:

- National health policies that address climate change from various aspects, in line with results of the situational and SWOT analyses, to reduce the burden of disease and deal with health outbreaks, impacts of tourism, extreme weather events, and natural disasters, to name a few.

- A regional platform for dealing with cross-boundary climate change–related issues within the health sector with a threefold objective: (a) containing epidemics and infectious disease outbreaks, (b) facilitating technical and operational cooperation, and (c) supporting a public health research and development agenda (activities launching a public health research forum; establishing an environment conducive to multidisciplinary research).

Secure financial resources to fund potential opportunities to alleviate the burden of climate-sensitive diseases

Countries in the Arab region can employ a number of tools to analyze the gaps and deficiencies that may exist in the health care system, namely:

- Health sector–specific public expenditure review to account for revenues and expenditures to inform budgetary decisions and sector-specific budget allocation with a climate change adaptation perspective.
- Health system analysis focusing on arrangements for governance, organization, financing, and delivery of health services, including both micro- and macroeconomic evaluation of climate change–related health interventions and services. This would inform dialogue on health sector reform.

Upgrading of Public Health Systems Is Necessary

Although climate health research is still nascent in the Arab region, preliminary evidence indicates that a number of health crises—ranging from the expansion of malaria, to growing malnutrition caused by food insecurity, to an increasing frequency and intensity of natural disasters that jeopardize health—may be precipitated by climate change. These events will have a cumulative impact on the burden of disease in Arab countries. The impact will be especially acute in the Arab region, because of its largely arid landscape, limited water resources, large burden of disease, wide base of low- and no-income people, weak government institutions, and lack of other institutional resources or capacity to pursue adaptation measures.

But climate change adaptation programs are particularly important because they correspond with the national development priorities of most Arab states and are in the interest of people in Arab countries, especially population groups most vulnerable to environmental and health crises.

Implementing the robust programs required to combat the impacts of climate change and secure the health of the people of Arab countries requires coordinated and principled planning and action from all sectors of society. Adaptation is a necessary pursuit for Arab governments and should be a priority issue that is tackled in partnership with international, regional, and national stakeholders. Evidence-based policy, both at the national and regional level, would provide the required tools to implement adaptation strategies in Arab countries. Short-term action plans such as developing environmental health services related to water and air quality, data monitoring and surveillance, and strengthening health care systems are essential components of the adaptation framework. Long-term action plans include climate change and health-related research jointly planned and implemented by all Arab countries.

Key Messages

1. According to WHO, climate change–related illnesses are increasing in the Arab region, specifically malaria, cardiovascular disease, and malnutrition- and waterborne-related illnesses (for example, diarrheal diseases).

2. Vulnerability to climate change–related illnesses in the Arab region is evident among a wide segment of the population. The most vulnerable are the internally displaced; those with low socioeconomic status; residents of low-lying areas, camps and slums; and those who work in specific occupations, such as outdoor workers.

3. The health problems that are exacerbated by climate-related environmental changes are not properly mapped because of the dearth of climate change and health research and the lack of evidence-based public health policies. Consequently, health care systems in most Arab countries are currently unable to provide for the health needs of the Arab populations.

4. Climate change will necessitate the upgrading of the health care systems (with resources and expertise) to provide health care services to the most vulnerable populations.

5. Sound adaptation strategies that include core elements, such as the proper assessment of public health vulnerabilities, the enhancement of health monitoring and surveillance systems, and the development of early warning systems and emergency preparedness plans to deal with health crises, should be implemented.

6. Adaptation strategies should involve all key stakeholders (ministries, researchers in government and academic institutions, local communities, NGOs, development partners, and the private sector) in a strong regional cooperation to prevent and deal with cross-boundary health crises.
7. The promotion and funding of interdisciplinary research on climate-related priority health outcomes and the strengthening of environmental health services related to water and air quality are essential tools in combating the health effects of climate change.

Notes

1. According to the WHO classification, the Eastern Mediterranean Region (EMR) comprises 22 economies (19 Arab economies and 3 non-Arab economies). The Arab economies in EMR include Bahrain, Djibouti, Egypt, Iraq, Jordan, Kuwait, Lebanon, Libya, Morocco, Oman, Qatar, Saudi Arabia, Somalia, Sudan, Syria, Tunisia, the United Arab Emirates, the West Bank and Gaza, and the Republic of Yemen—excluding Algeria, the Comoros, and Mauritania. The three non-Arab economies are Afghanistan, the Islamic Republic of Iran, and Pakistan.
2. Countries with water scarcity concerns include Algeria, Bahrain, the Comoros, Djibouti, Egypt, Jordan, Morocco, Sudan, Syria, Tunisia, and the United Arab Emirates.
3. Ban Ki-Moon, Secretary General of the United Nations, interpreted the Darfur conflict through this lens when he described how the expansion of desert lands in western Sudan and extended periods of drought led to competition over scarce land resources, which in turn, erupted into a years-long conflict (Moon 2007).
4. Adaptation to climate change refers to the actions taken to address the consequences of climate change, while preparedness refers to institutional readiness to deal with its effects.
5. For more information about RAED, visit the network's website at http://www.aoye.org/Raed/raed1.html.

References

Abu-Rmeileh, Niveen M. E., Weeam Hammoudeh, and Rita Giacaman. 2010. "Humanitarian Crisis and Social Suffering in Gaza Strip: An Initial Analysis of Aftermath of Latest Israeli War." *Lancet*–Palestinian Health Alliance abstract. http://download.thelancet.com/flatcontentassets/pdfs/palestine/S014067361060846X.pdf.

Al-Krenawi, Alean. 2005. "Mental Health Practice in Arab Countries." *Current Opinion in Psychiatry* 18 (5): 560–64.

al-Madani, Abdulkarim Ali. 1991. "Problems in the Control of Schistosomiasis in Asir Province, Saudi Arabia." *Journal of Community Health* 16 (3): 143–49.

Al-Mansoob, M. A. K., and M. M. Al-Mazzah. 2005. "The Role of Climate on Malaria Incidence Rate in Four Governorates of Yemen." *Medical Journal of Malaysia* 60 (3): 349–57.

Al-Salem, Saqer S. 2001. "Overview of the Water and Wastewater Reuse Crisis in the Eastern Mediterranean Region." *Eastern Mediterranean Health Journal* 7 (6): 1056–60.

al-Shaibany, Saleh. 2010. "Powerful Cyclone Phet Barrels Toward Oman." Reuters, June 2. http://www.reuters.com/article/2010/06/02/us-oman-hurricane-idUSTRE6513C420100602.

Al-Taiar, Abdullah, Shabbar Jaffar, Ali Assabri, Molham Al-Habori, Ahmed Azazy, Nagiba Al-Mahdi, Khaled Ameen, Brian M. Greenwood, and Christopher J. M. Whitty. 2006. "Severe Malaria in Children in Yemen: Two Site Observational Study." *British Medical Journal* 333 (7573): 827.

Anwar, Wagida A. 2003. "Environmental Health in Egypt." *International Journal of Hygiene and Environmental Health* 206 (4–5): 339–50.

AP (Associated Press). 2008. "29 Die, Hundreds Rescued in Algeria Flash Floods." AP, October 2. http://www.msnbc.msn.com/id/26990849/ns/weather.

Bartlett, Sheridan. 2008. "Climate Change and Urban Children: Impacts and Implications for Adaptation in Low- and Middle-Income Countries." Institute for Environment and Development, London.

Bassiouny, Hassan K. 2001. "Bioenvironmental and Meteorological Factors Related to the Persistence of Malaria in Fayoum Governorate: A Retrospective Study." *Eastern Mediterranean Health Journal* 7 (6): 895–906.

Basu, Rupa, and Jonathan M. Samet. 2002. "Relation between Elevated Ambient Temperature and Mortality: A Review of the Epidemiologic Evidence." *Epidemiologic Reviews* 24 (2): 190–202.

BBC. 2009. "Saudi Arabia Floods Leave 77 Dead." *BBC News*, November 26. http://news.bbc.co.uk/2/hi/middle_east/8380501.stm.

Ben-Ahmed, Kais, Karim Aoun, Fakhri Jeddi, Jamila Ghrab, Mhamed-Ali El-Aroui, and Aïda Bouratbine. 2009. "Visceral Leishmaniasis in Tunisia: Spatial Distribution and Association with Climatic Factors." *American Journal of Tropical Medicine and Hygiene* 81 (1): 40–45.

Benamer, Hani T. S. 2007. "Neurological Disorders in Libya: An Overview." *Neuroepidemiology* 29 (3–4): 143–49.

Benamer, Hani T. S., and Raad A. Shakir. 2009. "The Neurology Map of the Arab World." *Journal of the Neurological Sciences* 285(1–2): 10–12.

Bertollini, Roberto, and Marci Martuzzi. 1999. "Disease Mapping and Public Health Decision-Making: Report of a WHO Meeting." *American Journal of Public Health* 89 (5): 780.

Bluhm, Michael. 2008. "Group of NGOs Will Push Arab League to Adopt Uniform Policy on Climate Change." *Daily Star* [Beirut], February 12.

Braga, Alfésio L., Antonella Zanobetti, and Joel Schwartz. 2002. "The Effect of Weather on Respiratory and Cardiovascular Deaths in 12 U.S. Cities." *Environmental Health Perspectives* 110 (9): 859–63.

Bronstein, Jeff, Paul Carvey, Honglei Chen, Deborah Cory-Slechta, Donato Di-Monte, John Duda, Paul English, Samuel Goldman, Stephen Grate, Johnni

Hansen, Jane Hoppin, Sarah Jewell, Freya Kamel, Walter Koroshetz, James W. Langston, Giancarlo Logroscino, Lorene Nelson, Bernard Ravina, Walter Rocca, George W. Ross, Ted Schettler, Michael Schwarzschild, Bill Scott, Richard Seegal, Andrew Singleton, Kyle Steenland, Caroline M. Tanner, Stephen Van Den Eeden, and Marc Weisskopf. 2009. "Meeting Report: Consensus Statement—Parkinson's Disease and the Environment; Collaborative on Health and the Environment and Parkinson's Action Network (CHE PAN) Conference 26–28 June 2007." *Environmental Health Perspectives* 117 (1): 117–21.

Byron, Katy. 2011. "Flooding in Saudi Arabia Kills 10." CNN, January 29. http:/articles.cnn.com/2011-01-29/world/saudia.arabia.flooding_1_jeddah-rain-water-rescue-operations?_s=PM:WORLD.

CDC (Centers for Disease Control and Prevention). 2010a. "Climate Change: Health and Environmental Effects." CDC, Atlanta.

———. 2010b. "Climate Change and Public Health: Cardiovascular Disease and Stroke." CDC, Atlanta.

———. 2010c. "Climate Change and Public Health: Foodborne Diseases and Nutrition." CDC, Atlanta.

———. 2010d. "Coastal Zones and Sea Level Rise." CDC, Atlanta.

———. 2010e. "Mental Health and Stress-Related Disorders: Impacts on Risk." CDC, Atlanta.

Christianson, Arnold, Christopher P. Howson, and Bernadette Modell. 2006. *Global Report on Birth Defects: The Hidden Toll of Dying and Disabled Children.* White Plains, NY: March of Dimes.

Cizao, Ren. 2007. "Evaluation of Interactive Effects Between Temperature and Air Pollution on Health Outcomes." PhD dissertation, Faculty of Health, Queensland University of Technology, Brisbane, Australia.

CNN. 2008. "Yemen Floods Leave 58 Dead, 20,000 without Shelter." CNN, October 28. http://articles.cnn.com/2008-10-25/world/ yemen.flooding_1_heavy-rains-emergency-shelter-floods?_s=PM:WORLD.

Correa-Villaseñor, Adolfo, Janet Cragan, James Kucik, Leslie O'Leary, Csaba Siffel, and Laura Williams. 2003. "The Metropolitan Atlanta Congenital Defects Program: 35 Years of Birth Defects Surveillance at the Centers for Disease Control and Prevention." *Birth Defects Research Part A: Clinical and Molecular Teratology* 67 (9): 617–24.

Deleu, Dirk, Abbas El Siddig, Saadat Kamran, Ahmed A. Kamha, Ibrahim Y. M. Al Omary, Hisham A. Zalabany. 2005. "Downbeat Nystagmus Following Classical Heat Stroke." *Clinical Neurology and Neurosurgery* 108 (1): 102–4.

DFID (U.K. Department of International Development). 2005. "Natural Disaster and Disaster Risk Reduction Measures: A Desk Review of Costs and Benefits." DFID, London.

Douglas, Alexander S., H. al-Sayer, John M. Rawles, and T. M. Allan. 1991. "Seasonality of Disease in Kuwait." *Lancet* 337 (8754): 1393–97.

Ebi, Kristie L. 2008. "Adaptation Costs for Climate Change–Related Cases of Diarrhoeal Disease, Malnutrition, and Malaria in 2030." *Globalization and Health* 4: 9–18.

Ebi, Kristie L., K. Alex Exuzides, Edmund Lau, Michael A. Kelsh, and Anthony Barnston. 2004. "Weather Changes Associated with Hospitalizations for Cardiovascular Diseases and Stroke in California, 1983–1998." *International Journal of Biometeorology* 49 (1): 48–58.

EEAA (Egyptian Environmental Affairs Agency). 2010. *Egypt State of Environment Report 2009*. Ministry of State for Environmental Affairs, Cairo.

El Azar, Grace E., Rima R. Habib, Ziyad Mahfoud, Mutassem El-Fadel, Rami Zurayk, Mey Jurdi, and Iman Nuwayhid. 2009. "Effect of Women's Perceptions and Household Practices on Children's Waterborne Illness in a Low Income Community." *Ecohealth* 6 (2): 169–79.

El-Raey, Mohamed. 2008. *Impact of Sea Level Rise on the Arab Region*. University of Alexandria and Regional Center for Disaster Risk Reduction, Arab Academy of Science, Technology and Maritime Transport, Alexandria, Egypt.

El Setouhy, Maged, and Reda M. Ramzy. 2003. "Lymphatic Filariasis in the Eastern Mediterranean Region: Current Status and Prospects for Elimination." *Eastern Mediterranean Health Journal* 9 (4): 534–41.

El-Zein, Abbas, and Mylene Tewtel-Salem. 2005. On the Association between High Temperature and Mortality in Warm Climates. *Science of the Total Environment* 343 (1–3): 273–75.

El-Zein, Abbas, Mylene Tewtel-Salem, and Gebran Nehme. 2004. "A Time-Series Analysis of Mortality and Air Temperature in Greater Beirut." *Science of the Total Environment* 330 (1–3): 71–80.

EM-DAT. 2011. International Disaster Database. Centre for Research on the Epidemiology of Disasters. http://www.emdat.be/.

Emch, Michael, Caryl Feldacker, M. Sirajul Islam, and Mohammed Ali. 2008. "Seasonality of Cholera from 1974 to 2005: A Review of Global Patterns." *International Journal of Health Geographics* 7 (31): 1–33.

Epstein, Paul R. 2002. "Climate Change and Infectious Disease: Stormy Weather Ahead?" *Epidemiology* 13 (4): 373–75.

ESCWA and LAS (Economic and Social Commission for Western Asia and League of Arab States). 2006. "Arab Region State of Implementation on Climate Change." ESCWA and LAS, Cairo.

European Council. 2008. "Climate Change and International Security." European Council, Brussels.

FAO (Food and Agricultural Organization of the United Nations). 2010. *Statistics: Food Security*. Rome: FAO.

Fleury, Manon, Dominique F. Charron, John D. Holt, Brian Allen, and Abdel R. Maarouf. 2006. "A Time Series Analysis of the Relationship of Ambient Temperature and Common Bacterial Enteric Infections in Two Canadian Provinces." *International Journal of Biometeorology* 50 (6): 385–91.

Fritze, Jessica G., Grant A. Blashki, Susie Burke, and John Wiseman. 2008. "Hope, Despair, and Transformation: Climate Change and the Promotion of Mental Health and Wellbeing." *International Journal of Mental Health Systems* 2 (1): 13–23.

Government of Egypt. 2010. "Second National Communication under the United Nations Framework Convention on Climate Change." Ministry of State for Environmental Affairs, Cairo.

Government of Jordan. 2009. "Second National Communication under the United Nations Framework Convention on Climate Change." Ministry of Environment, Amman.

Government of Lebanon. 2011. "Second National Communication under the United Nations Framework Convention on Climate Change." Ministry of Environment, Beirut.

Habib, Rima R. 2010. "Policy and Governance in Palestinian Refugee Camps: Addressing Poor Living Conditions to Improve Health in Palestinian Camps in Lebanon." Research and Policy Memo 4, Issam Fares Institute for Public Policy and International Affairs, Beirut.

Habib, Rima R., Kareen El Zein, and Joly Ghanawi. 2010. "Climate Change and Health Research in the Eastern Mediterranean Region." *EcoHealth* 7 (2): 156–75.

Habib, Rima R., Nasser Yassin, Joly Ghanawi, Pascale Haddad, and Ziyad Mahfoud. 2011. "Double Jeopardy: Assessing the Association between Internal Displacement, Housing Quality, and Chronic Illness in a Low-Income Neighborhood." *Journal of Public Health* 19 (2): 171–82.

Hallaj, Zuhair. 1996. "Constraints Facing Surveillance in the Eastern Mediterranean Region." *Eastern Mediterranean Health Journal* 2 (1): 141–44.

Hamad, Amel A., Abd El Hamid D. Nugud, David E. Arnot, Haider A. Giha, Abdel Muhsin A. Abdel-Muhsin, Gwiria M. H. Satti, Thor G. Theander, Alison M. Creasey, Hamza A. Babiker, and Dia Eldin A. Elnaiem. 2002. "A Marked Seasonality of Malaria Transmission in Two Rural Sites in Eastern Sudan." *Acta Tropica* 83 (1): 71–82.

Handal, Alexis J., Betsy Lozoff, Jaime Breilh, and Siobán D. Harlow. 2007. "Neurobehavioral Development in Children with Potential Exposure to Pesticides." *Epidemiology* 18 (3): 312–20.

Hassan, Amro. 2010. "Egypt: Heavy Rains, Flooding Kill 15." *Los Angeles Times*, January 19. http://latimesblogs.latimes.com/babylonbeyond/2010/01/egypt-fifteen-dead-because-of-heavy-floods.html.

Hassan, A. N., M. A. Kenawy, H. Kamal, A. A. Abdel Sattar, and M. M. Sowilem. 2003. "GIS-Based Prediction of Malaria Risk in Egypt." *Eastern Mediterranean Health Journal* 9 (4): 548–58.

Himeidan, Yousif E., E. E. Hamid, Lukman Thalib, Mustafa I. Elbashir, and Ishag Adam. 2007. "Climatic Variables and Transmission of Falciparum Malaria in New Halfa, Eastern Sudan." *Eastern Mediterranean Health Journal* 13 (1): 17–24.

Huq, Anwar, and Rita R. Colwell. 1996. "Environmental Factors Associated with Emergence of Disease with Special Reference to Cholera." *Eastern Mediterranean Health Journal* 2 (1): 37–45.

Husain, Tahir, and Junaid Rafi Chaudhary. 2008. "Human Health Risk Assessment Due to Global Warming: A Case Study of the Gulf Countries." *International Journal of Environmental Research and Public Health* 5 (4): 204–12.

IDMC (Internal Displacement Monitoring Centre). 2011. *Global Statistics*. Geneva: IMDC.

IWGCCH (Interagency Working Group on Climate Change and Health). 2010. "A Human Health Perspective on Climate Change." Environmental Health Perspectives and the National Institute of Environmental Health Sciences, Research Triangle Park, NC.

Kar, Bhoomika R., Shobini L. Rao, and B. A. Chandramouli. 2008. "Cognitive Development in Children with Chronic Protein Energy Malnutrition." *Behavioral and Brain Functions* 4: 31.

Khallaayoune, Khalid, and Hammou Laamrani. 1992. "Seasonal Patterns in the Transmission of Schistosoma Haematobium in Attaouia, Morocco." *Journal of Helminthology* 66 (2): 89–95.

Khogali, Mustafa. 2005. "Health and Disease in a Changing Arab World 2000/2025/2050: Global, Environmental, and Climate Change and Emerging Diseases." *Ethnicity and Disease* 15 (1 Suppl. 1): S1-74–75.

Kjellstrom, Tord. 2009. "Climate Change, Direct Heat Exposure, Health, and Well-Being in Low and Middle-Income Countries." *Global Health Action* 2. doi:10.3402/gha.v2i0.1958.

Kovats, R. Sari. 2009. *Adaptation Costs for Human Health*. London: Imperial College.

Kovats, R. Sari, Shakoor Hajat, and Paul Wilkinson. 2004. "Contrasting Patterns of Mortality and Hospital Admissions during Hot Weather and Heat Waves in Greater London, U.K." *Occupational and Environmental Medicine* 61 (11): 893–98.

Lake, Iain R., Iain A Gillespie, Graham Bentham, Gordon L. Nichols, Chris Lane, G. K. Adak, and E. John Threlfall. 2009. "A Re-evaluation of the Impact of Temperature and Climate Change on Foodborne Illness." *Epidemiology and Infection* 137 (11): 1538–47.

Macdonald, Robie W., Donald Mackay, Yi-Fan Li, and Brendan Hickie. 2003. "How Will Global Climate Change Affect Risks from Long-Range Transport of Persistent Organic Pollutants?" *Human and Ecological Risk Assessment* 9 (3): 643–60.

Malik, Gaafar M., Osheik Seidi, Abdelmageed El-Taher, and Abdin Shiekh Mohammed. 1998. "Clinical Aspects of Malaria in the Asir Region, Saudi Arabia." *Annals of Saudi Medicine* 18 (1): 15–17.

Malone, John B., Mohamed S. Abdel-Rahman, Mohamed M. El Bahy, Oscar K. Huh, M. Shafik, and Maria Bavia. 1997. "Geographic Information Systems and the Distribution of Schistosoma Mansoni in the Nile Delta." *Parasitology Today* 13 (3): 112–19.

Martens, Willem J. 1998. "Climate Change, Thermal Stress, and Mortality Changes." *Social Science and Medicine* 46 (3): 331–44.

Mayeux, Richard. 2004. "Dissecting the Relative Influences of Genes and the Environment in Alzheimer's Disease." *Annals of Neurology* 55 (2): 156–58.

McDonald, Oliver F., Nigel J. Shanks, and Laurent Fragu. 2008. "Heat Stress: Improving Safety in the Arabian Gulf Oil and Gas Industry." *Professional Safety* 53 (8): 31–38.

McGeehin, Michael A., and Maria Mirabelli. 2001. "The Potential Impacts of Climate Variability and Change on Temperature-Related Morbidity and Mortality in the United States." *Environmental Health Perspectives* 109 (Suppl. 2): 185–89.

McMichael, Anthony J. 2012. "Insights from Past Millennia into Climatic Impacts on Human Health and Survival." *Proceedings of the National Academy of Sciences of the United States of America*. http://www.pnas.org/cgi/doi/10.1073/pnas.1120177109.

Moon, Ban Ki. 2007. "A Climate Culprit in Darfur." *Washington Post*, June 16, A15.

Moore, Patrick S., Benjamin Schwartz, Michael W. Reeves, Bruce G. Gellin, and Claire V. Broome. 1989. "Intercontinental Spread of an Epidemic Group A Neisseria Meningitidis Strain." *Lancet* 2 (8657): 260–63.

Moore, Stephanie K., Vera L. Trainer, Nathan J. Mantua, Micaela S. Parker, Edward A. Laws, Lorraine C. Backer, and Lora E. Fleming. 2008. "Impacts of Climate Variability and Future Climate Change on Harmful Algal Blooms and Human Health." *Environmental Health* 7 (Suppl 2): S4.

Musa, Hassan A., P. Shears, Shamsoun Kafi, and S. K. Elsabag. 1999. "Water Quality and Public Health in Northern Sudan: A Study of Rural and Peri-urban Communities." *Journal of Applied Microbiology* 87 (5): 676–82.

Musani, Altaf, and Irshad A. Shaikh. 2008. "The Humanitarian Consequences and Actions in the Eastern Mediterranean Region over the Last 60 Years—A Health Perspective." *Eastern Mediterranean Health Journal* (Suppl. 14): S150–56.

Neouimine, Nikolai I. 1996. "Leishmaniasis in the Eastern Mediterranean Region." *Eastern Mediterranean Health Journal* 2 (1): 94–101.

Noor, Abdisalan M., Archie C. A. Clements, Peter W. Gething, Grainne Moloney, Mohammed Borle, Tanya Shewchuk, Simon I. Hay, and Robert W. Snow. 2008. "Spatial Prediction of Plasmodium Falciparum Prevalence in Somalia." *Malaria Journal* 7: 159. doi:10.1186/1475-2875-7-159.

OCHA (United Nations Office for the Coordination of Humanitarian Affairs). 2010. "Regional Office for the Middle East, North Africa, and Central Asia: Fast Facts." OCHA, Cairo. http://www.unocha.org/where-we-work/regional-office-middle-east-north-africa-and-central-asia-romenaca.

Oshaghi, Mohammed Ali, N. Maleki Ravasan, Ezzatodin Javadian, Yavar Rassi, Javid Sadraei, Ahmadali A. Enayati, Hassan Vatandoost, Z. Zare, and Sara N. Emami. 2009. "Application of Predictive Degree Day Model for Field Development of Sandfly Vectors of Visceral Leishmaniasis in Northwest of Iran." *Journal of Vector Borne Diseases* 46 (4): 247–55.

Parry, Martin L., Osvaldo F. Canziani, Jean Palutikof, Paul van der Linden, and Clair E. Hanson, eds. 2007. *Climate Change 2007: Impacts, Adaptation, and Vulnerability—Contribution of Working Group II to the Fourth Assessment Report of the Intergovernmental Panel on Climate Change*. Cambridge, U.K.: Cambridge University Press.

Patz, Jonathan A., Michael A. McGeehin, Susan M. Bernard, Kristie L. Ebi, Paul R. Epstein, Anne Grambsch, Duane J. Gubler, Paul Reiter, Isabelle Romieu, Joan B. Rose, Jonathan M. Samet, and Juli Trtanj. 2000. "The Potential

Health Impacts of Climate Variability and Change for the United States: Executive Summary of the Report of the Health Sector of the U.S. National Assessment." *Environmental Health Perspectives* 108 (4): 367–76.

Polack, Sarah, Simon Brooker, Hannah Kuper, Silvio Mariotti, David Mabey, and Allen Foster. 2005. "Mapping the Global Distribution of Trachoma." *Bulletin of the World Health Organization* 83 (12): 913–19.

Qatar Statistics Authority. 2007. *Population Statistics: Labor Force Sample Survey*. Doha: Qatar Statistics Authority.

Rathor, Hamayun R. 2000. "The Role of Vectors in Emerging and Re-emerging Diseases in the Eastern Mediterranean Region." *Dengue Bulletin* 24: 103–9. http://www.searo.who.int/en/Section10/Section332/Section522_2535.htm.

Rekacewic, Philippe. 2008. Map drawn from a UNEP GRID-Arendal using data from the Food and Agriculture Organization of the United Nations and Aquastat. UNEP GRID-Arendal, Arendal, Netherlands.

Reuters. 2010. "Seven Die in Worst Floods in Yemen Capital in Decade." http://www.reuters.com/article/2010/05/06/us-yemen-floods-idUSTRE6451OP20100506.

Sandifer, Paul A., Carolyn Sotka, David Garrison, and Virginia Fay. 2007. "Interagency Oceans and Human Health Research Implementation Plan: A Prescription for the Future." Interagency Working Group on Harmful Algal Blooms, Hypoxia, and Human Health of the Joint Subcommittee on Ocean Science and Technology, Washington, DC.

Shanks, Nigel J., and Greg Papworth. 2001. "Environmental Factors and Heatstroke." *Occupational Medicine* 51 (1): 45–49.

Silove, Derrick, and Zachary Steel. 2006. "Understanding Community Psychosocial Needs After Disasters: Implications for Mental Health Services." *Journal of Postgraduate Medicine* 52 (2): 121–25.

St. Louis, Michael E., and Jeremy J. Hess. 2008. "Climate Change: Impacts on and Implications for Global Health." *American Journal of Preventive Medicine* 35 (5): 527–38.

Sturrock, Hugh J. W., Diana Picon, Anthony Sabasio, David Oguttu, Emily Robinson, Mounir Lado, John Rumunu, Simon Brooker, and Jan H. Kolaczinski. 2009. "Integrated Mapping of Neglected Tropical Diseases: Epidemiological Findings and Control Implications for Northern Bahr-el-Ghazal State, Southern Sudan." *PLoS Neglected Tropical Diseases* 3 (10): e537.

Suzuki, Shunji, Susumu Sakamoto, Masanobu Koide, Hideki Fujita, Hiroya Sakuramoto, Tatsumi Kuroda, Kintaka Taigo, and Takefumi Matsuo. 1997. "Hanshin-Awaji Earthquake as a Trigger for Acute Myocardial Infarction." *American Heart Journal* 134 (5): 974–77.

UNDP (United Nations Development Programme). 2007. *Human Development Report 2007/2008: Fighting Climate Change—Human Solidarity in a Divided World*. New York: UNDP.

———. 2009. *Resource Guide on Gender and Climate Change*. New York.

UNDP Jordan. 2011. "Adaptation to Climate Change to Sustain Jordan's MDG Achievements." UNDP Jordan, Amman. http://www.undp-jordan.org/index.php?page_type=projects&project_id=16&cat=3.

UNEP (United Nations Environment Programme). 2007. "Air Quality and Atmospheric Pollution in the Arab Region." Regional Office for West Asia, UNEP, Manama.

UNISDR (United Nations International Strategy for Disaster Reduction). 2009. *Global Assessment Report on Disaster Risk Reduction 2009: Risk and Poverty in a Changing Climate*. Geneva: United Nations

UN (United Nations) News Centre. 2011. "Drought-Hit Somalia on Brink of Humanitarian Disaster, Warns UN Expert." Press release, March 2. http://www.un.org/apps/news/story.asp?NewsID=37660.

Wadhwa, Pathik D., Claudia Buss, Sonja Entringer, and James M. Swanson. 2009. "Developmental Origins of Health and Disease: Brief History of the Approach and Current Focus on Epigenetic Mechanisms." *Seminars in Reproductive Medicine* 27 (5): 358–68.

Wakabi, Wairagala. 2009. "Fighting and Drought Worsen Somalia's Humanitarian Crisis." *Lancet* 374 (9695): 1051–52.

Watanabe, Hiroshi, Makoto Kodama, Yuji Okura, Yoshifusa Aizawa, Naohito Tanabe, Masaomi Chinushi, Yuichi Nakamura, Tsuneo Nagai, Masahito Sato, and Masaaki Okabe. 2005. "Impact of Earthquakes on Takotsubo Cardiomyopathy." *Journal of the American Medical Association* 294 (3): 305–7.

WHO (World Health Organization). 2004. "Health Statistics and Health Information Systems: Risk Factors Estimates for 2004." WHO, Geneva.

———. 2009a. "Global Health Risks: Mortality and Burden of Disease Attributable to Selected Major Risks." WHO, Geneva.

———. 2009b. "Protecting Health from Climate Change: Connecting Science, Policy and People." WHO, Geneva.

WHO (World Health Organization), WMO (World Meteorological Organization), and UNEP (United Nations Environment Programme). 2003. "Climate Change and Human Health: Risks and Responses." WHO, Geneva.

WHO EMRO (World Health Organization Eastern Mediterranean Regional Office). 2005. "Report of the Regional Committee for the Eastern Mediterranean: Fifty-Second Session." WHO EMRO, Amman.

———. 2008. *Technical Discussion on Climate Change and Health Security*. EM/RC55/Tech.Disc.1. WHO EMRO, Amman.

———. 2009. "Annual Report of the Regional Director," January 1–December 31, 2009." WHO EMRO, Amman.

Wilson, Mary E. 1995. "Infectious Diseases: An Ecological Perspective." *British Medical Journal* 311 (7021): 1681–84.

Wiwanitkit, Viroj. 2005. "Overview of Clinical Reports on Urinary Schistosomiasis in the Tropical Asia. *Pakistan Journal of Medical Sciences* 21 (4): 499–501.

Woodruff, Bradley A., Michael J. Toole, Daniel C. Rodrigue, Edward W. Brink, El Sadig Mahgoub, Magda Mohamed Ahmed, and Adam Babikar. 1990. "Disease Surveillance and Control after a Flood: Khartoum, Sudan, 1988." *Disasters* 14 (2): 151–63.

World Bank. 2007. *Making the Most of Scarcity: Accountability for Better Water Management Results in the Middle East and North Africa*. Washington, DC: World Bank.

Yemen Times. 2011. "Psychological Health in Yemen." *Yemen Times*, March 3. http://www.yementimes.com/DefaultDET.aspx?SUB_ID=35692.

Yeo, Theresa Pluth. 2004. "Heat Stroke: A Comprehensive Review." *AACN Clinical Issues* 15 (2): 280–93.

Yousif, Fouad, Mamdouh Roushdy, Ahmed Ibrahim, Karem el Hommossany, and Clive Shiff. 1996. "Cercariometry in the Study of Schistosoma Mansoni Transmission in Egypt." *Journal of the Egyptian Society of Parasitology* 26 (2): 353–65.

Zawahri, Mohamed Z. 2004. "Stroke and the Weather." *Neurosciences* 9 (1): 60–61.

PLAN DE SITUATION DES CENTRES DE VIE - COMMUNE DE IN AMGUEL -
WILAYA D'ADRAR
Algerie
2010

CHAPTER 9

Implement Policy Responses to Increase Climate Resilience

Climate change affects people in Arab countries and across the world. Least developed countries (LDCs) and poor people and communities are the most vulnerable because they often live in exposed locations, depend on natural resources, and have limited adaptive capacity. As this report demonstrates, increasing climate resilience requires a diverse set of policy actions aimed at different time horizons and at different actors, including all levels of government, the private sector, civil society, and households.

A national climate change strategy, a National Adaptation Programme of Action (NAPA), and national communication are not enough. To be effective, adaptation strategies need to be supported by strong domestic policies, legislation, and action plans. It is also essential to mainstream climate change adaptation plans into existing public financial management systems and national policies, plans, and programs—in particular those related to climate-sensitive economic sectors such as agriculture, health, tourism, and water. Most actions aimed at increasing climate resilience will also have broader local development benefits by, for example, contributing to improved environmental governance and facilitating social inclusion and sustainable growth. Even in the absence of extreme events and other climate variability, these actions are likely to present wins for Arab leaders.

Taking an integrated approach to climate change adaptation requires strong political leadership. Although Arab leaders are already responding to the social and economic impacts of climate change, continued and additional action is essential. Climate change has considerable momentum, and without sustained commitment from Arab leaders, economic, human, and social development will be adversely affected for decades to come.

Photograph by Dorte Verner

This report is intended as a resource to begin to assess climate risks, opportunities, and actions at a regional level. The information highlighted here explains the potential impacts of climate change in key sectors, as well as in urban and rural settings, and then goes on to discuss possible policy options to reduce climate risk and better adapt to climate variability and change. This chapter attempts to provide guidance to policy makers in Arab countries on how best to move forward on this agenda at a country level. It does this in two ways. First, it provides a framework for moving forward by revisiting the Framework for Action on Climate Change Adaptation (Adaptation Pyramid), presented in chapter 1. Second, it puts forward a typology of policy approaches that are relevant to the region to help decision makers formulate effective policy responses. These approaches include (a) the provision of reliable and accessible data, (b) the provision of human and technical resources and services, (c) the provision of social protection for the poor and most vulnerable, (d) the development of a supportive policy and institutional framework, and (e) the development of the capacity to generate and manage revenue and to analyze financial needs and opportunities associated with adaptation. Finally, a policy matrix outlines key policy recommendations covered in each of the chapters.

Adaptation Is an Integrated Part of Public Sector Management for Sustainable Development

The prospect of climate change adds another element to be integrated into national planning. Government, with assistance from the private sector and civil society, can ensure that a country's development policies, strategies, and action plans build resilience to a changing climate.

Adaptation is a long-term, dynamic, and iterative process that takes place over decades. Decisions will have to be made despite uncertainty about how both society and climate will change, and adaptation strategies and activities must be revised as new information becomes available. Although many standard decision-making methodologies are appropriate, alternative, robust methods for selecting priorities within an adaptive management framework will be more effective.

The Adaptation Pyramid provides a framework to assist stakeholders in Arab countries to integrate risks and opportunities into development activities (see chapter 1, figure 1.4). It is based on an adaptive management approach but also highlights the importance of leadership and political commitment, without which adaptation efforts are unlikely to achieve what is necessary to minimize the impacts of climate change.

Assess Climate Risk Impacts and Opportunities

In this first step, a wide range of analyses could be used.[1] All rely on access to climate and socioeconomic data to provide information on climate impacts, including on vulnerable groups, regions, and sectors. To help understand the risks and impacts, data are needed on current climate variability and change as well as projections and uncertainty about the future climate. Similarly, information on past adaptation actions and on coping strategies needs to be gathered and evaluated in light of the changing climate. These analyses will have to be undertaken at national and local levels and consider different contexts, including direct impacts and the indirect effects, for example, of climate-induced rural-to-urban migration (box 9.1).

Prioritize Options

The second step is to identify and prioritize adaptation options within the context of national, regional, and local priorities and goals, particularly with financial and capacity constraints in mind. Expectations of climate

BOX 9.1

Impacts of Climate-Induced Rural-to-Urban Migration on Water Supply

Accelerating rural-to-urban migration, due in part to the difficulty in maintaining viable agriculture in a changing climate, is increasing water stress in some regions and is posing challenges to water supplies and the provision of water services. For example, in Beirut, where half the Lebanese population lives, water shortages are very frequent because local supplies are incapable of meeting the rising demand. Lacking access to adequate water services, people often illegally tap and deplete shallow aquifers, resulting in seawater intrusion. In an attempt to reduce pressure on heavily populated Cairo, the government has encouraged urban development in desert areas, which has presented serious challenges to the population in procuring water supplies over large distances. In Jordan, the population is increasingly concentrated in the highlands, several hundred meters above most prospective water resources.

Source: Authors' compilation.
Note: See chapter 3.

change make it more important to consider longer-term consequences of decisions, because short-term responses may miss more efficient adaptation options or even lead to maladaptive outcomes, such as the further development of highly vulnerable locations. The prioritization of adaptation responses requires an understanding of the links among projected climate impacts, associated socioeconomic impacts, and adaptation responses. One technique for such prioritization could include landscape mapping. For example, it may be possible to map where increasing rainfall or aridity will affect current cropping and land-use systems and to similarly map where options exist to move to more drought-hardy, tree-based oil, fruit, and forage systems when increasing aridity makes traditional cereal crops no longer viable. Another possible approach to prioritizing options is robust decision making (RDM) (box 9.2), which seeks to identify choices that provide acceptable outcomes under many future scenarios.

Implement Responses in Sectors and Regions

Adaptive responses will often be somewhat at odds with immediate local priorities, and thus the third step of implementing responses needs cooperation and understanding at the national, sectoral, and regional or local levels (often jointly). At the national level, adaptation needs to be integrated into national policies, plans, and programs and financial management systems. This integration includes five-year plans prepared in a number of Arab countries. Moreover, it includes sustainable development and poverty reduction strategies and plans; policies, regulations, and legislation; investment programs; and the budget. In addition, national adaptation strategies can help mainstream adaptation into other national policies as well as implement it at the sectoral and local levels. This implementation could involve the formation of an interministerial committee at various levels with the participation of the private sector, academia, and civil society.

Monitor Outcomes

The fourth step is to monitor outcomes to ensure that adaptation-related strategies and activities have the intended adaptation outcomes and benefits. Comprehensive qualitative and quantitative indicators can help project proponents recognize strengths and weaknesses of various initiatives and adjust activities to best meet current and future needs. The monitoring framework should explicitly consider the effects of future climate change, particularly for projects with a long time horizon.

This process is iterative—hence the next step will be to reassess activities while taking into account new and available information, such as future climate change or the effectiveness of previously applied solutions.

BOX 9.2

Robust Decision Making and the Transport Sector

Decision making throughout both the public and private sectors has been dominated by the "predict and act" approach. Many approaches and methods exist, but the essential elements are to develop a model to predict future conditions, apply some form of optimization technique to select the "best" option, and then act on that decision. However, in many situations, such a wide range of possible futures exists that the "predict and act" approach is not effective.

Analyzing options for transport and related infrastructure is one such case. The effectiveness of a particular design depends on many other decisions outside the control of the planners. For example, land-use and settlement changes are decided on by the millions of users of the transport system as they cope with its strengths and weaknesses. Climate change adds yet another level of uncertainty to transport planning with its widely different scenarios of the effectiveness of mitigation and the uncertainties of the climate projections. Transportation and infrastructure cannot adapt incrementally to climatic changes as the investments are capital intensive and require long lead times. To extract the best value out of the lifetime of an investment, planners should choose a robust strategy that will not require expensive adaptation later on.

In robust decision making (RDM), planners and stakeholders identify (a) a wide range of options (such as specific transport design strategies), (b) multiple scenarios of future developments that might affect the effectiveness of the options, and (c) methods, often computer models, to evaluate the effectiveness of each option under each of the scenarios. However, the evaluations do not need to be precise enough to seek optimal solutions because they are seeking those that are "robust," that is, they produce acceptable results under most scenarios and only rarely produce unacceptable results. Once an option considered robust by all stakeholders has been identified, it can be implemented and monitored with the RDM process repeated regularly, leading to adaptive adjustments of the policy. The goal of RDM is to identify a strategy that "would have been chosen with perfect predictive foresight" (Popper, Lempert, and Banks 2005). RDM reframes the question "What will the long-term future bring?" into "How can we choose actions today that will be consistent with our long-term interests?" (Lempert, Popper, and Bankes 2003).

RDM is being applied in transport planning in some countries, such as the PTOLEMY (Planning, Transport, and Land Use for the Middle East Economy) project (Dewar and Wachs 2008; Marchau, Walker, and van Wee 2010). Given RDM's ability to alleviate the need for decision makers to correctly predict the future before making a major investment, planners in the Arab countries should consider employing RDM tools to help guide the formulation of transportation and infrastructure strategy.

Sources: Authors' compilation based on Dewar and Wachs 2008; Lempert, Popper, and Bankes 2003; Marchau, Walker, and van Wee 2010; and Popper, Lempert, and Bankes 2005.

Leadership Is Central for Successful Adaptation

Effective climate change adaptation will not occur without strong leadership. International experience shows that the lead needs to be taken at the national level by a prominent ministry or senior government champion, such as the prime minister, minister of planning or economy, or state planning commission. This champion will also require the support of a strong team composed of representatives of relevant ministries, governorates, local authorities and institutions, the private sector, academia, civil society organizations, and ideally opposition parties to ensure continuity as a government changes. Clearly, this leadership approach should be adapted to the context of individual Arab countries and their circumstances. Leaders are needed at other levels of government and within civil society and private sector organizations. Leaders from all sectors need support through access to information and educational opportunities and must be treated as legitimate agents in decision-making processes. For example, in 2009, the Republic of Yemen created an interministerial panel for climate change adaptation chaired by the deputy prime minister, with ministers from 13 key ministries and other relevant actors included. Finally, the leadership must interact with other states with regard to intergovernmental issues (for example, riparian states on water flow of the Nile, Euphrates, Khabur, and other rivers).

Policy Options Are Available to Support Climate Change Adaptation

This report focuses primarily on assessing climate risk impacts and opportunities and establishes a framework for adaptation decision making. The remaining sections of this chapter focus on the range of policy interventions that are needed to increase climate resilience. The policy options addressed aim to

- Facilitate the development of publicly accessible and reliable information and analyses related to adaptation
- Support the development of human, technical, and other resources and services to support adaptation
- Provide social protection and other measures to ensure that the poor and the most vulnerable are climate resilient
- Develop a supporting policy and institutional framework for adaptation

- Build capacity to generate and manage revenue and to analyze financial needs and opportunities associated with adaptation

The next sections provide further guidance on each of these policy approaches to support adaptation decision making for policy makers. Each section emphasizes pertinent aspects for the Arab region in particular and draws from the analysis of impacts, risks, and opportunities in earlier chapters to the next level. Table 9.4 provides additional details for the implementation of these policy options in key economic sectors and areas addressed in this report.

1. Facilitate the Development of Publicly Accessible and Reliable Information and Analyses Related to Adaptation

Improve Access to Climatological Data

Access to quality weather and climate data is essential for policy makers. Without reliable data on temperature and precipitation levels, it is difficult to assess the current climate and make reliable weather forecasts and climate predictions that allow for the design of effective policies, the implementation of early warning systems, and the adaptation of key sectors on which the local and national economy depend. As indicated in chapter 2, climate station data across the Arab region is very limited compared to most other parts of the world and what data exists are often neither digitized nor publicly available. For historical reasons, a reasonable number of stations exist along the Nile and the coast of the Mediterranean Sea, but further inland coverage is very sparse. Conflict in parts of the region disrupts both the collection and sharing of data. In many areas, additional data are being gathered by various agencies but are not entered into more widely available meteorological databases. Furthermore the Arab region has not been addressed as a discrete region in climate change research assessments, such as in the Intergovernmental Panel on Climate Change reports. Typically, information must be inferred from analyses carried out in other regions.

In the short and medium term, the collection and monitoring of climate data could be improved by expanding the number of weather stations and by collaborating with other countries in the region to improve the coverage and comparability of data. This effort should be combined with a push to link climate data to impact analysis by making climate data available to policy makers and researchers. Some efforts in this direction have already started. For example, Algeria, the Arab Republic of Egypt, Lebanon, Libya, Morocco, Saudi Arabia, the Syrian Arab Republic, and Tunisia are part of the European Climate Assessment and Dataset

TABLE 9.1

Data Needs for Effective Adaptation Decision Making

Sector	Types of data needed	Key challenges	Policy options
Climatological	• Temperature • Precipitation • Air pressure • Humidity • Wind • Radiation	• Data often under governance of the Ministry of Defense or other ministries, limiting data • Insufficient data • Not linked to impact analysis	• Charge the civil authority with making climate data available • Expand data rescue and the number of weather stations • Ensure that data are readily available to policy makers and researchers for analysis
Food security	• Production levels and yields for indicator crops • Models of how food supply chains operate and how they will be affected by climate change • Imports and exports of key crops and food storage • Operation of safety net	• Data to advise on reducing vulnerability • A cross-sectoral issue, not the sole mandate of ministries of agriculture	• Identify national and regional partners for data collection and dissemination • Link findings to early warning systems
Gender	• Data disaggregated by sex, age, and location • Local knowledge and practices, for example, on local water management systems	• Data is not always disaggregated by sex and age • When it is, it is not always analyzed from a gender perspective or publicly available	• Adjust data collection systems and sets to include information on time use and division of labor • Complement with qualitative surveys • Invest in analysis of existing data
Health	• Occurrence and magnitude of climate change–related health outcomes, linking those to environmental and meteorological indicators	• Insufficient data • Insufficient linking of climate and health data • Lack of data information systems for monitoring climate and health trends • Inadequate tracking of vulnerable groups	• Develop climate-sensitive surveillance systems and evaluation techniques for health • Strengthen health management information systems • Collect and analyze information on groups vulnerable to climate change
Urban livelihoods	• Geographic location exposure (river, coast) • Population: absolute and trends • Risk zones: unstable slopes, low-lying areas, areas of high density • Governance structure • Building codes and enforcement • Economic activities and built environments	• Data is sparse and often not compiled and analyzed in a holistic way	• Systematically collect urban data and link with climate change • Make information available to all, including to local-level authorities
Water	• Water availability, salinity, and quality • River runoff • Groundwater levels • Current and future water consumption • Impacts of various policy measures on water supply and demand	• Limited capacity to monitor long-term trends in hydrometeorological data and regional climate change modeling capacity • Limited understanding of the impacts of policy responses on human behaviors	• Promote regional cooperation and sharing of data and good practices in data collection and dissemination, long-term monitoring, regional water modeling, and economic and policy analysis

Source: Authors' compilation.

(ECA&D). This project, which aims to combine collation of a daily series of observations at meteorological stations, quality control, analysis of extremes, and dissemination of both the daily data and the analysis results, is gradually being extended across the Middle East and North Africa.

Several actions can enhance the accessibility of data including digitizing data collected in the past and stored in formats that can be damaged or difficult to access, and having civil authority take responsibility for sharing data with users when meteorological services are under the governance of, for example, the ministry of defense. Many countries have websites with such data for public use. In some countries, access to the most current meteorological data may need to be restricted, but it is important that older data (for example, one month or one year) at daily or subdaily temporal resolution should eventually be made publicly available. Ideally, compilation of the information on the availability, conditions for use, and procedures to access data should be provided and regularly updated.

Improve Information on Trends in Water Availability

As highlighted in chapter 3, water is scarce in the majority of the Arab countries. All but six Arab countries (the Comoros, Iraq, Lebanon, Somalia, Sudan, and Syria) suffer from water scarcity, which is defined as less than 1,000 cubic meters of water per person per year. Information on current and future water availability and quality is therefore critical for designing adaptation responses, and requires information on river runoff, groundwater levels, and water quality, including salinity. In many parts of the Arab region, coverage of these data is poor and will need to be upgraded. Capacity is also required to monitor long-term trends in hydrometeorological data, link them with the climate data, and develop regional climate change models. Box 9.3 provides examples from Arab countries of effective water data collection, monitoring, and modeling.

Link Climate Data with Socioeconomic and Typological Data Sets

Data on climate variability and change and water availability will need to be complemented with socioeconomic data such as population growth. The data types needed for effective policy making are household data, census data, and other economic data including labor market and production data. In national, sectoral, and local data collection, it is important that social and economic information is collected in a disaggregated format to reflect location, gender, age, and socioeconomic status as these factors greatly affect exposure and ability to cope with climate risks. Ideally, microdata series would continue to be compiled so that development

BOX 9.3

Water Data Collection, Modeling, and Monitoring in Arab Countries

The Nile Forecast Center (NFC) provides hydrometeorological forecasting, monitoring, and simulation services to planning units in the government. The NFC is made of several units responsible for compiling remotely sensed and in situ data and has capacity in hydrological and water resources management modeling. One of the NFC's central units is the Nile Basin Hydrometeorological Information System, which supports management of the Nile River (http://emwis.mwri.gov.eg/recersh%20%20and%20devolpment%20-%20other-1%20-Nile%20Forecast%20Center.htm).

Morocco's IMPETUS climate monitoring network is an initiative by the German government to support management of water resources in semiarid areas in West Africa. A component of IMPETUS is a real-time network of meteorological stations in the Drâa Valley in Morocco. The network monitors several meteorological variables in the valley including precipitation, temperature, humidity, wind speed and direction, and soil moisture and humidity. The data is primarily used for climate modeling but is also made accessible to regional and local authorities (http://www.impetus.uni-koeln.de/en/morocco/).

Source: Authors' compilation.

over time can be tracked closely. Other data needs identified in this report are presented in table 9.1.

In some economic sectors, data are currently not linked with climate or other relevant data. For example, most national health surveillance systems engage in data collection related to communicable and noncommunicable diseases without linking to environmental or other indicators. By linking this data, as has been done in Kuwait, Oman, and the West Bank and Gaza (WHO EMRO 2009), policy makers can obtain a more comprehensive picture of climate risks.

The location, such as urban and rural, is also important in a changing climate. It is anticipated that more than 75 percent of the total population will live in urban areas by 2050. Most of these urban areas and conglomerations are on the coasts of Arab countries. Climate-related impacts and disasters can be extremely damaging in urban areas. For effective adaptation planning, data should be linked using geographic information system (GIS) technology to track indicators such as geographical location and exposure (for example, proximity to rivers or coasts) as well as information on current and anticipated future population size, distribution, and physical expansion of urban areas. Certain zones within urban areas such as unstable slopes, low-lying areas, or areas of particularly high density will be particularly vulnerable and should be carefully mapped. It will also be important to have information on the potential ability of an urban area to respond and cope with extreme weather events, which will depend on governance structures, building codes and their enforcement, wealth, economic activities, and built environments.

In rural areas, it is important to collect data on changes in agricultural production levels and yields for indicator crops, that is, examples from forage, vegetables, grains, and livestock categories. These data will differ among Arab countries, reflecting the varying agricultural production system in, for example, the Gulf States (alfalfa, tomatoes, goats, and soft wheat); Egypt (berseem, tomatoes, goats, and soft wheat); and Jordan (kochia, tomatoes, sheep, and soft wheat). It will also be important to understand the main food supply chains, how they operate, and how they will be affected by climate change.

Raise Awareness

Finally, data collection and analysis are an important step, but equally important are the mechanisms for disseminating this data so as to raise awareness, stimulate behavioral shifts in the population, and act or adapt as a result of the information. For example, awareness campaigns about climate-induced water scarcity may reduce household water consumption, an adaptive outcome. There are many aspects to raising awareness, including the language and terms that are used to facilitate understanding. For example, cities, local governments, and universities can employ GIS technology to develop visualizations (often maps) of vulnerabilities and risks.

Government, in particular, can play an important role in raising awareness and facilitating better understanding of the risks of climate change and its impacts. It can also encourage other stakeholder groups (such as the private sector and community leaders) to play a more active role in this regard. In particular, local organizations, opinion leaders, and educators can play critical roles in campaigns to raise awareness as well as in mainstreaming climate change into national education programs. Businesses will also have access to networks important in raising awareness. These awareness campaigns need to consider factors constraining women's access to information and should target campaigns accordingly. In this regard, female community leaders have a critical role to play in raising awareness.

2. Provide Human and Technical Resources and Services to Support Adaptation

Specialized human and technical resources are required to analyze, identify, and implement adaptive responses. Such resources can be developed through education and training, research and development, and technical improvements.

Enhance Education and Training

The need to develop new skills, knowledge, and expertise related to climate change adaptation can be met through building and expanding training, including graduate programs related to climate change and key sectors such as agriculture, health, and water. Moreover, as highlighted in chapter 2, specific training on climatology is important because a wealth of satellite information is available, but it is often not being used effectively by local institutions.

Training can also be targeted to particular areas that are subject to high climate risk. For example, the Amman Green Growth Program provided training on climate change and climate change adaptation options at the urban level in Amman, Jordan. In particularly high-risk urban and rural areas, it may be useful to initially train local officials and planning and emergency management teams on hydrometeorological disaster management, such as how to respond to floods, landslides, drought, and heat waves. These individuals can then carry out additional training of the local population to strengthen responses to these high-risk situations.

Adaptation responses can also include specialized training programs for professionals engaged in particular sectors. For example, training for water utility employees to enhance water demand management through market-based instruments such as water pricing and metering of water usage would be a useful adaptation response but will also result in broader development benefits since the region is already water scarce. In many Arab countries, rural areas are currently experiencing an out-migration of men because of climate change. For women to cope with the impacts of out-migration and be involved in decisions affecting their lives, including decisions related to climate change adaptation, special training for women in community and political participation skills, business development, general literacy, and education and extension services is required.

Encourage Research and Development

Research and development are critical to enhance understanding of current and future climate change impacts and develop new and appropriate technological responses. Key research often excludes climate-related factors. For example, studies between 1990 and 2010 found a lack of evidence-based research directly exploring the relationship between climate change and health (Habib, El Zein, and Ghanawi 2010). The available studies were geographically limited to Egypt, Lebanon, Sudan, and countries in the Gulf (Habib, El Zein, and Ghanawi 2010). There is a need to strengthen research in areas where climate change and key issues such as agriculture, gender, health, and urban and rural livelihoods intersect; such work could be carried out with existing institutions as well as new institutes or centers of excellence. For example, the United Arab Emirates' Masdar Institute of Science and Technology and King Abdullah University of Science and Technology in Saudi Arabia undertake research in local meteorology, climate projections, agricultural production, drought-resistant crops, and local methods in water reuse including aquifer recharge and desalination.

Enhance Technological Resources

Shifting to new technologies and using existing technologies more effectively could also be a mechanism to build resilience to climate change impacts. Governments have an important role to play in facilitating promotion of and access to technologies that help people to adapt to climate risks.

In water-stressed Arab countries, governments play a critical role in promoting technologies that can enhance water supply and decrease water demand. Promoting technology development is best accomplished through some combination of policy reforms that change incentives for private investment in new technology for greater climate resilience and address key market failures, combined with public financial interventions and investments (table 9.2). In terms of water supply, new and improved technology could be used for reducing water network leakage from pipe systems and to improve storage and conveyance capacity. Equally important, water demand could be reduced through, for example, drip irrigation and better metering. Desalination, an important current technology

TABLE 9.2

Technology Transfer Prospects in Agriculture and Water Sectors in the Arab Countries

Integrated policy objectives	Main benefits	Available technology and R&D	Identified barriers
• Increase agricultural productivity and efficient water use through technological approaches to adaptation in the agricultural sector	• Reduce water stress, desalination needs, and groundwater overexploitation • Increase productivity and available water • Reduce vulnerability to rainfall variability • Increase food security • Decrease land abandonment • Reduce food costs • Improve livestock management and positive externalities in health	• Reduce water losses and improve productivity through drip, sprinkler, and bubbler irrigation systems • Provide affordable micro-irrigation technology systems • Implement fog harvesting like that in the Republic of Yemen (26 small standard fog collectors) • Improve weather information, covered agriculture, alternative seeds, and saline water research	• Low technological input and low grid expansion in remote areas • Investment needs for small projects • Cross-subsidized tariff favoring irrigation water • Need for demand management programs to drive technological change and promote intersector water transfers • Need for water storage • No consideration of water footprint in trade and water added-value
• Promote investment and encourage research networks to increase wastewater treatment in urban areas for agricultural water use and natural wastewater systems for remote and isolated human settlements	• Increase available water in remote areas and reduce water stress in urban and rural areas where climate change is already reducing water availability • Reduce land and water pollution • Develop decentralized systems to allow for multiple secondary benefits (artificial wetlands, lagoons, and land recovery)	• Natural wastewater treatment systems for rural and isolated areas such as Gaza, Jordan, and Tunisia (national program since 1980s) have extended experience and technology for mechanical wastewater and reuse in agricultural systems	• Water supply–driven policy and no integrated management of hydrological cycle • Middle Eastern and Northern African countries avoid intersector water transfers (farmer lobbies) • Low sewerage networks in remote (and some urban) areas • Religious beliefs against water reuse • High quality standards • No storage facilities
• Promote technology transfer of rural renewable energy electrification (RREE) systems, integrating regional markets and the private sector especially for large desalination plants and off-grid electrification	• Increase the energy supply in remote areas • Increase baseload stability • Reduce carbon footprint and slow the rate of energy intensity • Allow for agricultural irrigation and food conservation technologies	• The Susiya and Community, Energy, and Technology (COMET) project (West Bank and Gaza) • Wind and solar projects in the Middle East and North Africa • Concentrating solar power and desalination of Ain Beni Mathar (Morocco) • Several solar- and wind-powered plants for seawater and brackish water in Algeria, Egypt, Qatar, Saudi Arabia, and Tunisia	• Subsidized oil consumption and direct subsidies in energy-intensive sectors • Low private sector involvement in RREE projects • Low participation in clean development mechanism projects • No integration of regional markets • No demand management tariffs for end consumers
• Promote autonomous renewable-energy desalination systems to improve water shortages in rural and isolated communities	• Potable water from brackish water and polluted sources • Less investment in centralized grids • Simple construction and maintenance • Abundant solar and wind sources • Increase energy supply	• Autonomous RREE desalination • Solar stills (Yemen, Rep.), RSD (Egypt), MHE (Saudi Arabia), membrane and multiple-effect distillation (pilots) • PV-RO in Ksar Ghiléne (Tunisia) and Morocco (Al Haouz, Essaouira, and Tiznit)	• Low production and expensive components (0.1–20 m^3/day and US$5–10/$m^3$) • No local industries and need to provide training • Scarce cooperation in research and development, local components, and patent transfer • International cooperation projects with scarce continuity and positive externalities

Source: Authors' compilation based on Padrón Fumero 2011.

for enhancing water supply, can result in significant local environmental impacts, including emissions to both air and water that affect human health, marine environments, and potentially economic activities such as local fisheries. Developing new desalination technologies, which reduce both air emissions and brine discharge, could help minimize these impacts and facilitate adaptation responses. New types of food storage and food transport systems to ensure food security and improve transport of agricultural goods to market are also important. These technologies can be locally derived through processes of research and development or made accessible through technology transfer. Often, however, barriers to appropriate technology development, transfer, and use must be overcome. These barriers relate to inadequate information and decision-support tools used to quantify and qualify the merits of various low-carbon or climate-resilient technologies and related investments, as well as limited local human and technical skills and the need for appropriate training. These can be overcome with approaches to enhance coordination and information sharing among governments, the private sector, and local people and through revision and clarification of laws related to environmental technology and local training and incentive programs

3. Build Climate Resilience of the Poor and Vulnerable through Social Protection and Other Measures

As highlighted in chapter 1, vulnerability to climate change largely depends on two things: (a) the scale of climate change impacts and (b) human resilience, which is determined by factors such as an individual's age, gender, or health status; a household's asset base; and one's degree of integration with the market economy. Investment in social safety nets, public services such as water supply and wastewater treatment, and housing and infrastructure in at-risk areas make poor people more resilient to a changing climate. Because these same instruments facilitate economic and social inclusion, there are also clear development benefits in investing in these measures.

Measures to ensure social protection can include insurance schemes, pensions, access to credit, cash transfer programs, and relocation programs. Additionally it is important to ensure that the poor can meet their basic needs and that there are measures in place to guarantee access to affordable health care and education.

In rural areas currently suffering from declining agricultural yields and the out-migration of men, social protection is particularly critical for women, the elderly, and children left behind. Such protections can take the form of rural pension schemes or conditional cash transfer programs similar to Bolsa Família in Brazil. Other assistance to enhance productivity can include providing access to credit or markets for agricultural and other rural produce to enhance income despite declining agricultural productivity because of climate change.

In urban areas, social services can include the provision of affordable housing away from zones at risk of climate impacts such as floods or drought. Additionally, the poor and other groups most vulnerable to the impacts of climate change will require additional measures to increase resilience such as the affordable provision of basic services such as energy (essential for heating and cooling in a variable climate), water, and public transport.

The poor and most vulnerable are particularly in need of assistance when an extreme weather event or other climate change–related crisis hits. Reducing vulnerability to climate impacts for the poorest should therefore be integrated into emergency planning and programs. This planning could include the provision of basic needs, including adequate shelter and access to food, water, and clothing.

At the local level, the livelihoods and assets of individuals and institutions can be protected through various forms of insurance such as life insurance, infrastructure insurance, or weather-based index insurance (box 9.4).

BOX 9.4

Weather-Based Index Insurance

Index insurance represents an attractive alternative for managing weather and climate risk because it uses a weather index, such as rainfall, to determine payouts. With index insurance contracts, an insurance company does not need to visit the policyholder to determine premiums or assess damages. Instead, if the rainfall recorded by gauges is below an earlier, agreed-on threshold, the insurance pays out. Such a system significantly lowers transaction costs. Having insurance allows these policyholders to apply for bank loans and other types of credit previously unavailable to them. However, if index insurance is to contribute to development at meaningful scales, a number of challenges must be overcome, including enhanced capacity, establishment of enabling institutional, legal, and regulatory frameworks, and availability of data. Droughts, floods, and other extreme events often strip whole communities of their resources and belongings. Index insurance could enable poor people to access financial tools for development and properly prepare for and recover from climate disasters.

Source: IRI 2007.

4. Develop a Supportive Policy and Institutional Framework for Adaptation

A supportive policy and institutional framework at national, sectoral, and local levels is essential for effective climate change adaptation decision making. Basic conditions for effective development such as the rule of law, transparency and accountability, participatory decision-making structures, and reliable public service delivery that meets international quality standards are conducive to effective development and adaptation action. In addition, climate change adaptation requires new policies and structures and changes to existing policies. These requirements include climate change adaptation strategies and policies at all levels, the mainstreaming of climate change considerations into existing policies, and the creation of systems for cooperative and coordinated decision making between government departments and different levels of government and in cooperation with the private sector, civil society, and other states.

A clear but coordinated governance structure is also essential to implement climate change adaptation measures. This structure must promote strong degrees of national, regional, and international collaboration among different levels of government, different sectors, and the public, private, and not-for-profit sectors.

Develop a National Adaptation Strategy

These adaptation strategies should consider a multitude of factors, which could include food security, employment, health, livelihoods, and vulnerability. Adaptation strategies should also consider the need for accessible and reliable data, human and technical resources, social protection, a supportive policy and institutional framework, and financing. Egypt's national adaptation strategy provides a good practice example (box 9.5).

Low-income countries such as Djibouti, Sudan, and the Republic of Yemen have produced NAPAs, which provide a process for least developed countries (LDCs) to identify priority adaptation activities—those that respond to their urgent and immediate needs to adapt to climate change. See table 9.3 for sample projects.

One aspect that does need strengthening in NAPAs is the gender dimension. Indeed, although globally many NAPAs acknowledge that women are among the most vulnerable to immediate and longer-term climate change impacts, few link this information to broader social, economic, and political mechanisms of gender inequality or emphasize the importance of empowering women as critical stakeholders in adaptation (UNFPA and WEDO 2009).

BOX 9.5

Egypt's National Adaptation Strategy

Egypt's national adaptation strategy primarily focuses on agriculture, water resources, and coastal areas. Agricultural recommendations include changing crop varieties and cropping schedules, skipping irrigation at different growth stages, implementing changes in farm systems and fertilization practices, developing simple and low-cost technologies suitable for the local context, establishing a special adaptation fund for agriculture, improving scientific capacity, and increasing public awareness. It also addresses the adaptive capacity of rural communities through forms of social assistance and economic diversification.

In terms of water, the national adaptation strategy recommends public awareness campaigns on water shortages or surpluses caused by climate change, the development of local area circulation models capable of assessing the impact of climate change on local and regional (Nile basin) water resources, increased capacity of researchers in all fields of climate change and its impact on water systems, and the exchange of data and information among Nile basin countries. The strategy also promotes integrated coastal management and recommends the creation of wetlands in low-lying lands, supportive protection structures (including dams), and natural sand duning systems; the management of coastal lakes; public and policy maker awareness; and use of aerial photographs and satellite images.

Source: Agaiby 2011.

Highlight the Importance of Agriculture and Water Considerations

It is useful to have separate sections of national strategies or stand-alone policies related to agriculture and water, given their importance for well-being and income generation in the Arab countries, as discussed in earlier chapters of this report.

From a process standpoint, an important mechanism to encourage a dialogue and coordinate a government response to reduce climate-induced risks could be a coordinated interministerial national policy or working group backed by specific expertise. The subject of such policies or working groups will clearly be state specific. In the highly urbanized

TABLE 9.3

Arab Least-Developed Countries Offer Top Two Priority Projects

Country	Project priority	Project title	Project sector	Sector components	Project costs (US$ thousands)
Djibouti	1	Mitigating climate change–related risks for the production system of coastal areas through an integrated, adapted, and participatory management involving grassroots organizations	Cross-sector	Coastal ecosystem, water resources, agriculture and livelihood diversification	1,000
Djibouti	2	Promoting the fencing of forest areas in Day and Mabla coupled with the introduction of improved stoves	Terrestrial ecosystems	Forest protection	294
Sudan	1	Enhancing resilience to increasing rainfall variability through rangeland rehabilitation and water harvesting in the Butana area of Gedarif State	Cross-sector	Livestock, water harvesting, and disaster management	2,800
Sudan	2	Reducing the vulnerability of communities in drought-prone areas of South Darfur State through improved water-harvesting practices	Cross sector	Vulnerability mitigation, water harvesting, and reforestation	2,500
Yemen, Rep.	1	Developing and implementing integrated coastal zone management	Coastal and marine ecosystems	Marine ecosystems	3,200
Yemen, Rep.	2	Water conservation through reuse of treated wastewater and gray water from mosques, and irrigation-saving techniques	Water resources	Water resources, agriculture	3,200

Source: Based on UNFCCC 2012.

countries of the Gulf and Mashreq, the focus is primarily on food security. In other countries, concerns for employment, livelihoods, and water scarcity appear to be highest on the political agenda.

Integrate Adaptation Consideration into Existing Policies, Plans, and Programs

It is important to integrate or mainstream climate change into all major policies, plans, and programs. Climate, economic development, and social development are interdependent—the way countries manage the economy and political and social institutions have critical impacts on climate risks. The climate and the level of adaptation of current institutions and individuals to it, in turn, are vital for the performance of the economy and social well-being. Therefore mainstreaming climate change is critical to development planning and policy formulation.

Climate mainstreaming is the processes by which climate considerations are brought to the attention of organizations and individuals involved in decision making on the economic, social, and physical development of a country (at the national, subnational, or local levels) and the processes by which climate is considered in making those decisions.

The process is based on an analysis of how climate change may affect a policy, plan, or program. This includes an analysis of the extent to which the activity under consideration could be vulnerable to risks arising from climate variability and change; the extent to which climate change risks have been taken into consideration in the course of formulating the existing policy plan or program; the extent to which the activity could lead to increased vulnerability, leading to maladaptation, or, conversely, miss important opportunities arising from climate change; and what amendments might be warranted to address climate risks and opportunities.

It is particularly important to mainstream climate change adaptation considerations into vulnerable sectors in Arab countries, such as agriculture, health, trade, tourism, and water, at all levels. For example, climate change consideration will need to be incorporated into infrastructure development in both rural and urban settings, particularly in areas prone to flash floods, flood risks, or extreme weather events.

In rural contexts, climate change considerations need to be incorporated into existing land property rights and practices. In many Arab countries, rural areas are experiencing climate-related migration particularly of men who have traditionally held inheritance and land rights. More equitable rights for women will help reduce both the vulnerability of women remaining in rural areas and their entire household.

Governments Play a Key Role in Promoting Collaboration and Cooperation

Within national governments, interministerial coordination is critical because adaptation responses often require activities involving multiple ministries and sectors. Interministerial coordination can be achieved through interministerial committees that, for example, have climate change focal points in all relevant ministries. The private sector, academia, and research institutes can also be integrated into these committees as technical advisers. For example, Egypt established a national climate change steering committee by prime ministerial decree in 2007.

Coordination among different levels of government is also essential as climate change adaptation policies will ultimately be implemented by sectoral authorities, local officials, and citizens themselves. For example, to create connections between the agriculture ministry and local farmers on climate risks and adaptation options, the government may be able to use existing farmers' associations that link directly to ministries of agriculture and agricultural research and extension services. This process would en-

BOX 9.6

Coordination of Water Management in Morocco

In Morocco, the reform of the water sector has led to significant changes since the introduction of the Water Code in 1995. Nine River Basin Organizations (and six delegations) have been created as nodal agencies for water administration at the regional level. These River Basin Organizations are legally and financially independent. They are financed through users' fees and can lend money for different local investment programs in water. The Code also created the High Council for Water and Climate, an interministerial committee to reinforce horizontal and vertical coordination among the different actors in the water sector. Gathering different representatives from the public sector, as well as nongovernment stakeholders, this council is in charge of assessing the national strategy on climate change and its impact on water resources, the national hydrological plan, and integrated water resources planning.

Source: OECD 2010.

sure clear flows of knowledge to all areas from the top down and the bottom up, relying on existing institutional mechanisms. Box 9.6 provides an example.

Regional and International Collaboration Is Also Essential for Climate Resilience

The heterogeneity of the Arab countries provides multiple opportunities for beneficial climate-related regional collaboration. Arab countries will be best equipped to address climate change if they have strong collaboration on issues such as climate-related data sharing, crisis responses, and the management of, for example, disease outbreaks, migration, shared water resources, and strong trade relationships to address food security.

Where knowledge, skills, or technology are lacking in one country, they often exist in other countries. Therefore collaboration with other Arab countries and regions, for example in health or management of shared water resources, could be particularly valuable. Arab countries could also consider establishing foundations or centers of excellence in climatology (for example, King Abdullah University of Science and Technology in Saudi Arabia) or in climate change and public health–

related disciplines. Knowledge sharing can be promoted on a regional scale through staff exchanges and through enhanced regional and international cooperation such as through the creation of an Arab knowledge network on climate change adaptation.

Opportunities to engage with existing or new initiatives within international bodies are key for improved policy making in Arab states. The World Meteorological Organization (WMO) is promoting a new large-scale initiative on climate services that could benefit from enhanced participation from Arab states.

5. Build Capacity to Generate and Manage Revenue and to Analyze Financial Needs and Opportunities

Financial resources are essential for development and to effectively adapt to climate change. Arab countries will need to invest in capacity to generate and manage climate change–related resources and to analyze their financial needs related to climate change.

Ministries, particularly those concerned with agriculture, energy, tourism, transport, and water, will need to mainstream climate change into national budgets. Moreover, current and future climate change impacts need to be taken into account in planning and costing investments—particularly long-term investment. Financial resources for climate change will need to come from domestic and international sources. For governments, national public expenditure reviews could be one tool to highlight current expenditures and hence better understand how these relate to budget estimates for climate proofing infrastructure. This information, in turn, will help governments understand what levels of additional revenues are needed to make up the shortfall and identify new revenue opportunities, such as from payments for ecosystem services, removal of subsidies, or innovative tax mechanisms.

At the local level, access to financial services can play a critical role in helping poor people widen their economic opportunities, increase their asset base, and diminish their vulnerability to external shocks such as climate change. In rural areas, simple financial services—credit and saving—can directly affect small producers' productivity, asset formation, income, and food security. Payment for ecosystem services (PES) has significant potential to enhance rural livelihoods and agricultural yields, maintain and enhance ecosystem services (such as watersheds and biodiversity), and develop long-term partnerships with the private sector. PES can contribute to disaster risk reduction, with the revenues generated serving as financial buffers for communities to climate-induced shocks.

Provided with financing opportunities and incentives, smallholders and rural communities can invest in preventing natural disasters by maintaining sand dunes, conserving wetlands, and foresting slopes as cost-effective measures, while at the same time protecting their own assets and livelihoods. Dependable revenue streams would allow them to invest in their crops and land, thus strengthening their businesses.

Funding is now more accessible from an increasing list of international sources. Particularly important for adaptation are the UNFCCC Adaptation Fund, the UNFCCC and GEF-administered Least Developed Countries Fund and Special Climate Change Fund, the Pilot Program for Climate Resilience (under the Climate Investment Funds managed by the multilateral development banks), and the many bilateral funding arrangements. The OECD (Organisation for Economic Co-operation and Development) has estimated that in 2010 about US$3.5 billion was provided by OECD members to support adaptation activities with another US$6 billion for "adaptation-related" activities. For a summary of options available, see http://www.climatefinanceoptions.org. The World Bank is already providing funding to countries through ongoing technical assistance and lending operations and the Climate Investment Funds (specifically though the Strategic Climate Fund's Pilot Program for Climate Resilience).

Applying Policy Recommendations to Address Key Climate Change Issues

In conclusion, this report is intended as a resource for Arab policy makers to begin to assess climate risks, opportunities, and actions at a regional level. This final chapter aims to provide guidance to policy makers in Arab countries on how best to move forward on this agenda at a country level. It has done so in two ways: by (a) providing a Framework for Action on Climate Change Adaptation and (b) putting forward a typology of policy approaches that are relevant to the region, to help decision makers formulate effective policy responses. Finally, a policy matrix (see table 9.4) outlines key policy recommendations covered in each of the chapters for ease of reference.

This chapter has also attempted to demonstrate that most of the actions aimed at increasing climate resilience will also have broader local development benefits by, for example, contributing to improved environmental governance and facilitating social inclusion and sustainable growth. So even in the absence of extreme events and high climate variability, they are likely to present double-wins for Arab leaders.

TABLE 9.4

A Policy Matrix for Arab Adaptation to Climate Change

Sector	Collect information on climate change adaptation and make it available	Provide human and technical resources and services to support adaptation
Climatology	• Make climate data available at daily or sub-daily temporal resolution • Compile information on availability, conditions for use, and procedures to access data • Rescue and digitize manually archived meteorological data • Extend the coverage of the observational network to ensure a minimal station density to reflect spatial variability (also beneficial to weather forecasting and early warning systems)	• Build capacity to use regional climate data information • Enhance national and regional capacity to make better use of existing international ground and satellite observation data • Promote skills in using and developing climate impacts and risks analyses • Establish regional/international centers of excellence with staff exchange programs to better share skills
Disaster risk management	• Develop consistent approaches to risk assessments at national and local levels • Perform more comprehensive, national multi-risk assessment rather than single hazard, sector, and territory specific assessments • Develop more policy-oriented scientific studies and research related to disaster risk management (DRM), environments, and ecosystems • Adopt common data standards and methodologies	• Training on climate change, natural hazards, and the adaptation and risk mitigation options at the national and local levels • Develop national strategies on integrating DRM in school curricula and public awareness activities
Water	• Ensure regular, reliable data collection on river flows, groundwater levels, water quality (particularly salinity), climate-related impacts on water, and adaptation options • Develop capacity to monitor and model long-term trends in hydrometeorological data • Develop regional climate impacts modeling capacity	• Encourage human and technical investments to promote supply and demand side management in the water sector • Support climate change and water scarcity awareness-raising programs and campaigns (to achieve both increased water efficiency and to disseminate knowledge of climate risks to water availability) • Support assessment and development of water storage and conveyance capacity • Invest in research and development of appropriate local methods in water reuse • Develop institutes for water quality protection

Provide assistance such as social protection for the poor and most vulnerable	Ensure a supportive policy and institutional framework	Build capacity to generate and manage finance and analyze financial needs and opportunities
• Empower civil authority with the responsibility of making meteorological data available for public use at minimal cost (currently services are often under the governance of the Ministry of Defense, etc.) • Link climate data with socio-economic data (including health data) to obtain the information needed to build resilience, particularly as it relates to poor and vulnerable communities	• Enhance regional collaboration on early warming systems, including use and dissemination of existing extended forecasts (available through WMO, etc.) • Engage with the WMO's new large-scale initiative on climate services. (This initiative depends on the active participation of the member states and it can assist in developing capacities critically needed in the Arab countries)	• Include hydrometeorological data collection in the government's budget, including costs related to: • Data rescue • Extending the number of weather stations • Establishing centers of excellence • International and regional cooperation • Build capacity and training
• Develop an emergency preparedness response plan to deal with emergent crises, assist the poor and vulnerable to relocate away from high risk areas, and secure basic needs such as adequate shelter, access to food, clothing, and drugs for the poor and most vulnerable	• Systematic integration of DRM policies and legislation into public investments at national and local levels, ensuring permanent emergency and response funds	• Capacity to estimate the costs, taking climate and hazard risk (including cost and benefit analysis) into account • Capacity to mobilize the needed resources from the international financial instruments
• Give priority to assistance for the poor and vulnerable in water sensitive sectors and regions • Help the poor resettle away from areas at risk of flooding (e.g., *wadis*) and/or severe drought • Help the poor acquire skills and livelihood options in economic sectors less sensitive to water deficits	• Ensure institutions and policies at national, sectoral, and regional levels consider the impacts of climate change on water resources and infrastructure • Encourage cooperation among different ministries (for example, agriculture, tourism, trade) on water conservation efforts • Support efforts to place the water scarcity and climate change government portfolios at the highest levels (prime minister's office) • Pursue cooperation with other Arab countries on water management and research and development • Make accountability and transparency high priorities in water management and service institutes	• Improve pricing of water as both a means of demand management and a source of revenue while ensuring affordable access for the poor • Incorporate the costs of climate change expenditures related to water and integrate water and climate considerations in public financial management • Develop insurance programs for high risk areas and vulnerable communities to encourage investment in climate change risk. (insuring infrastructure and lives, especially related to flood risks)

(continued on next page)

TABLE 9.4

A Policy Matrix for Arab Adaptation to Climate Change (*continued*)

Sector	Collect information on climate change adaptation and make it available	Provide human and technical resources and services to support adaptation
Rural	• Assess changes in agricultural production levels/yields for indicator crops • Model the food supply chains, model how they operate, and how they will be impacted by climate change • Monitor state of water (groundwater and salinity levels), soil conditions (depth and carbon content), and agricultural activities in "most at risk" agricultural zones (using indicator areas of marginal lands, and rainfed areas from the four regions)	• Develop knowledge and skills related to climate-resilient agricultural practices, such as growing salt-tolerant, heat-tolerant, and pest-resistant crop and livestock species, conservation agriculture; increasing irrigation efficiency; and using nonconventional water resources • Develop human and technical resources to optimize food chain systems, particularly in transport and marketing; improve value-added developments; and establish cooperatives
Urban	• Improve linkages between meteorological data and information on urban conditions including: • Exposure (river, coast), population, growth in size, physical expansion (direction and amount) • Risk zones (unstable slopes, low-lying areas, areas of high density) • Spatial information on income levels, economic activities, and built environments • Governance structure • Building codes and enforcement	• Improve training on climate change and adaptation options in urban settings • For example: Amman Green Growth Program Training on hydrometeorological disasters management (floods, landslides, drought, heat wave, etc.)

Provide assistance such as social protection for the poor and most vulnerable	Ensure a supportive policy and institutional framework	Build capacity to generate and manage finance and analyze financial needs and opportunities
• Target food price controls/subsidies during price spikes and crop failures to support the most vulnerable • Support access to markets for agricultural and other rural produce • Support development of schools and training facilities to nurture both basic academic and vocational skills and provide necessary incentives to ensure attendance	• Create a coordinated governance structure to implement climate change adaptation measures at central and local levels across ministries responsible for agriculture, water, and the economy • Develop a coordinated national policy, likely to be across ministries, supporting food security and rural livelihood developments, balancing risks with possibilities and mindful of water and energy security vulnerabilities • Create farmers' associations that link directly to ministries of agriculture and agricultural research /extension services to ensure clear flow of knowledge to all areas from the top-down and bottom-up	• Develop capacity to estimate the financial risks for not applying climate change adaptation and how to maximize risk management through available financial instruments • Enhance capacity to assess all possibilities for meeting food demand while balancing economics with geopolitical risks
• Develop affordable housing away from risk zones (flood zone, drought areas) • Provide basic services for those in affordable housing • Promote urban upgrading of self-built areas • Improve rural areas service delivery and give people choice • Ensure clear and equitable land tenure policy and regularize informal settlements in areas of low climate exposure	• Create a clear governance structure to implement climate change adaptation measures at central and local levels	• Enhance capacity to estimate the costs of taking climate risk into account in planning decisions, including the projected costs of potential damages from taking no action • Enhance capacity to mobilize the needed resources from the international financial instruments

(continued on next page)

TABLE 9.4

A Policy Matrix for Arab Adaptation to Climate Change (*continued*)

Sector	Collect information on climate change adaptation and make it available	Provide human and technical resources and services to support adaptation
Tourism	• Enhance collection, accessibility, and analysis of climate- and tourism-related data including: • Contribution of tourism to economy, employment, and trade balance • Types of tourism and their relative importance and main season; number of tourists by category and country • Vulnerability and risk assessment, and monitoring of ecosystems important to tourism: water resources, coastal and marine zones, ecosystems, biodiversity, coral reefs, and archeological and cultural sites • Tourist-oriented weather forecasts (easy access to information and knowledge-related to heat exposures and risk reduction) • Monitor the degradation of tourist sites (ecosystems, heritage sites)	• Promote technical improvements to enhance the resilience of sectors on which tourism depends, including water management, protection of natural environments, zoning of land use for minimal exposure, disaster risk management, and evacuation plans • Increase efforts to make new and existing buildings able to withstand stronger weather impact • Environmentally sensitive, low impact, coastal erosion control • Public health management and easy access to information • Increased energy reliability through diversification and shift to renewable ("green") energy • Enhance capacities through education and awareness-raising, particularly among NGOs, government, business, and local communities • Develop alternatives to traditional tourism
Ecosystem services	• Collect systematic data on the state and changes in biodiversity and ecosystems • Improve data on the distribution and status of ancestors of the crops, fruit trees, and livestock that are endemic to the Arab countries • Improve valuation of services and goods provided by ecosystems and the changes or trends for national accounting and decision-making purposes • Enhance data collection on the links between functioning ecosystems and livelihood needs of poor and marginalised communities in different countries and ecosystems • Information on the need of ecosystem services (for example, water flow) for societies to help with allocation and trade-off decisions	• Enhance human, financial, and technical capacity to research, understand and incorporate role and values of biodiversity and ecosystems in production systems • Enhance skills and knowledge to conduct economic evaluation of ecosystem services to incorporate them in development decision making • Bring knowledge and good practices from other parts of the world into the Arab countries and vice-versa, given the urgency of the need to adapt

Provide assistance such as social protection for the poor and most vulnerable	Ensure a supportive policy and institutional framework	Build capacity to generate and manage finance and analyze financial needs and opportunities
• Diversification of income by developing alternative tourism (for example, health-related tourism) • Develop social protection mechanisms and instruments to enhance the resilience of the local actors in the tourism sector to face climate change risks, such as conditional cash transfers, temporary employment programs, micro insurance schemes	• Develop an adaptation strategy for the tourism sector and integrate it into national and local strategies and policies • Adapt tourist activities to changing climate and modify the tourist season accordingly • Promote best practices in development and management of tourist facilities (energy, water, waste, etc.) • The private actors: Renovate tourist facilities, create new services, improve marketing, develop new destinations, contribute to the national efforts to rationalize water and energy use • The public actors: Coordinate efforts; issue laws and regulations on hotel renovation and construction, urban and land planning, sustainability requirements on alternative tourism; and mobilize international resources	• Tap into unique sources of revenue related to tourism and climate resilience such as payment for ecosystem services
• Improve access and rights to biodiversity and ecosystems, especially in times of droughts, fires, and as part of nomadic life • Improve sustainable management of ecosystems to minimize degradation and thus the assets of the poorest and most vulnerable • Promote alternative and diversification of livelihoods that are not ecosystem- and biodiversity-dependent, (for example, cultural tourism options, employment in small service/home industries)	• Ensure ecosystem and biodiversity management are part of the national development plans and sectoral strategies, especially for water and water-based pollution management in inland and coastal areas • Assist in internalizing the costs of biodiversity and/or ecosystem service loss/degradation • Establish gene banks and "garden" areas where ancestors of crops and livestock can be grown and managed as potential adaptation options	• Develop carbon and climate finance as part of an integrated sustainable land and water management system that helps in maintaining biodiversity and ecosystem services and as an adaptation option • Develop adaptation funding and payments for environmental services that explicitly incorporate conservation and sustainable use of biodiversity at the genetic, species, and ecosystem levels

(continued on next page)

TABLE 9.4

A Policy Matrix for Arab Adaptation to Climate Change (*continued*)

Sector	Collect information on climate change adaptation and make it available	Provide human and technical resources and services to support adaptation
Gender	• Ensure that current and future data collection is disaggregated by gender and age and that presentation and analyses of these data use this disaggregation • Analyze factors constraining women's access to information and target campaigns accordingly	• Address constraints to women's and children's access to information • Analyze gender patterns in sources of information to improve targeting of awareness campaigns, particularly where illiteracy rates are high • Increase rural and urban women's skill-development and capacity-building opportunities • Simplify and disseminate information on land ownership laws • Promote and invest in innovative new areas of business in rural economies, particularly those that emphasize/improve opportunities for women
Health	• Establish and/or strengthen information systems linking health and climate change–related outcomes • Develop climate-sensitive surveillance systems and evaluation techniques for health; that is, occurrence and magnitude of climate change–related health outcomes linked to environmental and meteorological indicators • Strengthen health-environment management information systems to enable evidence-based decision making for planning, designing, financing, and implementing adaptation program to address climate change–related burden of disease • Collect and analyze information on groups vulnerable to climate change. This includes identifying their specific vulnerabilities and characterizing risk exposures; describing their geographical locations and social and economic status; and evaluating their access to social protection services	• Strengthen expertise in climate change and public health and provide technical support to promote adaptation • Develop and strengthen a comprehensive framework for human resources for health and climate change—both technical and managerial • Encourage academic institutions to invest in climate change and health research and in the provision of technical assistance to ministries of health • Build and/or expand training/graduate programs in climate change and health sciences • Involve civil society organizations in raising public awareness on the health effects of climate change and on ways to adapt at the household and community levels

Provide assistance such as social protection for the poor and most vulnerable	Ensure a supportive policy and institutional framework	Build capacity to generate and manage finance and analyze financial needs and opportunities
• Targeted social protection including insurance schemes, rural pensions, access to credit, and cash transfer programs to take account of gender related vulnerability and in particular support female-headed households	• Reform property rights laws and practices related to land and property that account for out-migration of men • Create mechanisms to improve the enforcement of land ownership laws • Reduce significance of marital status for legal status and land/property ownership • Improve existing land access programs, especially by increasing emphasis on gender issues in access to land • Support women's collective schemes for securing land access rights	• Conduct gender-responsive budgeting to ensure that adequate financial resources are allocated to implement gender mainstreaming and other proposed policies
• Protect the poor and vulnerable through social services • Strengthen health care service delivery by upgrading primary health care, emergency, and ambulatory services to cope with emergent health crises; ensuring equitable access (both physical and financial); and improving the quality of care • Protect populations against catastrophic expenditures, and health shocks through social protection for health, especially for the poor and vulnerable, through improving targeting social spending using proxy means testing • Expand health insurance, social assistance, and safety net programs	• Create an institutional framework for health related decision making that considers climate change • Develop/update climate change-specific clinical practice guidelines and standard operating procedures covering different levels of health care: primary, secondary, tertiary, and specialized • Establish a national climate change and health steering committee consisting of focal points that represent concerned ministries and other stakeholders to initiate dialogue on climate change policies • Establish a regional platform for dealing with cross-boundary climate change–related issues within the health sector with the objectives of : (1) containing epidemics and infectious disease outbreaks; (2) facilitating technical /operational cooperation; and (3) supporting public health research	• Secure financial resources to fund potential opportunities to alleviate the burden of climate sensitive diseases • Health-sector specific public expenditure review to account for revenues and expenditures to inform budgetary decisions and sector-specific budget allocation with a climate change adaptation perspective • Health system analysis focusing on arrangements for governance, organization, financing, and delivery of health services, including both micro- and macroeconomic evaluation of climate change–related health interventions and services. This would inform dialogue on health sector reform

Note

1. For an overview of available tools to assist in climate risk analysis, see http://climatechange.worldbank.org/climatechange/content/note-3-using-climate-risk-screening-tools-assess-climate-risks-development-projects.

References

Agaiby, Ezzat Lewis Hannalla. 2011. "Egypt's Status towards Major Negotiable Issues in Climate Change." PowerPoint presentation and speech, Climate Change Central Department and National Ozone Unit, Egyptian Environmental Affairs Agency, Cairo.

Dewar, James A., and Martin Wachs. 2008. "Transportation Planning, Climate Change, and Decisionmaking under Uncertainty." Transportation Research Board of the National Academies, Washington, DC. http://onlinepubs.trb.org/onlinepubs/sr/sr290DewarWachs.pdf.

Habib, Rima R., Kareen El Zein, and Joly Ghanawi. 2010. "Climate Change and Health Research in the Eastern Mediterranean Region." *EcoHealth* 7 (2): 156–75.

IRI (International Research Institute for Climate and Society). 2007. "Poverty Traps and Climate Risk: Limitations and Opportunities of Index-Based Risk Financing." Technical Report 07-03: Working Paper, IRI, New York. http://iri.columbia.edu/publications/search.

Lempert, Robert J., Steven W. Popper, and Steven C. Bankes. 2003. *Shaping the Next One Hundred Years: New Methods for Quantitative, Long-Term Policy Analysis.* Santa Monica, CA: RAND Corporation.

Marchau, Vincent A. W. J., Warren E. Walker, and Bert P. van Wee. 2010. "Dynamic Adaptive Transport Policies for Handling Deep Uncertainty." *Technological Forecasting and Social Change* 77 (6): 940–50.

OECD (Organisation for Economic Co-operation and Development). 2010. *Progress in Public Management in the Middle East and North Africa: Case Studies on Policy Reform.* Paris: OECD.

Padrón Fumero, Noemi. 2011. "A Regional Mitigation and Adaptation Approach to Climate Change: Technology Transfer and Water Management in the MENA Region." Universidad de La Laguna, Tenerife, Canary Islands. http://webpages.ull.es/users/npadron/.

Popper, Steven W., Robert J. Lempert, and Steven C. Bankes. 2005. "Shaping the Future." *Scientific American* 292 (4): 66–71.

UNFCCC (United Nations Framework Convention on Climate Change). 2012. Least Developed Countries Portal, NAPA Priorities Database. UNFCCC, Bonn, Germany. http://unfccc.int/cooperation_support/least_developed_countries_portal/napa_priorities_database/items/4583.php.

UNFPA (United Nations Population Fund) and WEDO (Women's Environment and Development Organization). 2009. "Climate Change Connections:

A Resources Kit on Climate, Population and Gender." UNFPA, New York. http://egypt.unfpa.org/Images/Publication/2010_07/2c65caf3-4471-4d07-9dd8-71a573d57e61.pdf

WHO EMRO (World Health Organization Eastern Mediterranean Regional Office). 2009. "Epidemiological Situation: 2086 Data." WHO EMRO, Cairo.

Index

Boxes, figures, maps, and tables are indicated by *b*, *f*, *m*, and *t*, following page numbers.

T

ECO-AUDIT

Environmental Benefits Statement

The World Bank is committed to preserving endangered forests and natural resources. The Office of the Publisher has chosen to print ***Adaptation to a Changing Climate in the Arab Countries*** on recycled paper with 30 percent postconsumer fiber in accordance with the recommended standards for paper usage set by the Green Press Initiative, a nonprofit program supporting publishers in using fiber that is not sourced from endangered forests. For more information, visit www.greenpressinitiative.org.

Saved:

- 24 trees
- 11 million Btu of total energy
- 2,103 lb. of net greenhouse gases
- 11,407 gal. of waste water
- 764 lb. of solid waste